WAVELET APPLICATIONS IN CHEMICAL ENGINEERING

Wavelet Applications in Chemical Engineering

edited by

Rodolphe L. Motard
Babu Joseph

Washington University in St. Louis

KLUWER ACADEMIC PUBLISHERS
Boston / Dordrecht / London

Distributors for North America:
Kluwer Academic Publishers
101 Philip Drive
Assinippi Park
Norwell, Massachusetts 02061 USA

Distributors for all other countries:
Kluwer Academic Publishers Group
Distribution Centre
Post Office Box 322
3300 AH Dordrecht, THE NETHERLANDS

Library of Congress Cataloging-in-Publication Data

A C.I.P. Catalogue record for this book is available
from the Library of Congress.

Printed on acid-free paper.

Printed in the United States of America

Dedicated to:

Coreen and Valsa

CONTENTS

PREFACE

Increasing emphasis on safety, productivity and quality control has provided an impetus to research on better methodologies for fault diagnosis, modeling, identification, control and optimization of chemical process systems. One of the biggest challenges facing the research community is the processing of raw sensor data into meaningful information.

Wavelet analysis is an emerging field of mathematics that has provided new tools and algorithms suited for the type of problems encountered in process monitoring and control. The concept emerged in the geophysical field as a result of the need for time-frequency analytical techniques. It has since been picked up by mathematicians and recognized as a unifying theory for many of the methodologies employed in the past in physics and signal processing.

Meyer[1] states:

"Wavelets are without doubt an exciting and intuitive concept. The concept brings with it a new way of thinking, which is absolutely essential and was entirely missing in previously existing algorithms."

The unification of the theory from these disciplines has led to applications of wavelet transforms in many areas of science and engineering including:

- pattern recognition
- signal analysis
- time-frequency decomposition
- process signal characterization and representation
- process system modeling and identification
- control system design, analysis and implementation
- numerical solution of differential equations
- matrix manipulation

About a year ago, in talking to various colleagues and co-workers, it became clear that a number of chemical engineers were fascinated with this new concept. Papers reporting applications in chemical engineering were being presented at various forums. We thought it would be worthwhile to bring together these various results in an edited volume to provide a focal point for those who were interested in exploring the potential of this interesting new area of applied mathematics.

---

[1] Meyer,Y., *Book Reviews*, Bulletin of the Amer. Math. Soc., **28**,2,350-360 (1993)

The primary objective of this volume is to provide a brief glimpse of the many possible applications of time-frequency decomposition methods to various fields within chemical engineering. To familiarize readers who are new to this field, we have also included a couple of tutorial chapters introducing wavelets and providing computational examples. Thus, this volume should be accessible to even those who have not been exposed to wavelet theory previously.

BRIEF OVERVIEW OF BOOK

Chapter 1 provides an introduction to the field of wavelets with a brief history and a review of major literature references in the field. This chapter will be useful to those readers who are not very familiar with the subject.

The purpose of Chapter 2 is to provide a set of working computer software that can be used to explore the use of wavelets in data analysis. An outline of the computational algorithms along with examples using the software are given here. The actual routines can be obtained via anonymous ftp from the Washington University computer network. Instructions are given on how to access these routines. A list of other sources for public domain wavelet software are also provided.

Chapter 3 introduces the use of Frazier-Jawerth transforms which preceded the widespread application of wavelet transforms. These transforms formed the basis of the early work at Washington University in the field of process trend analysis and process fault detection.

Chapter 4 discusses one of the major new applications of wavelets to the field of process and sensor data fingerprinting. The idea is to characterize such data using decomposition as depicted in a phase-plane plot of frequency versus time. Shaded or colored boxes represent the relative energy content of a signal in each frequency interval over a particular time interval. Phase plane plots are useful for the development of new sensors, denoising and pattern detection.

In Chapter 5, Bhakshi, Koulouris and Stephanopoulos discuss the use of wavelets as attenuation functions in Artificial Neural Networks (*Wave-Nets*) and in the generation of process models from empirical data via induction. The use of wavelets allows one to break up the data at various resolutions (short term, medium term, long term, etc.) and then to look for relationships at each scale.

The application of wavelets to process control has been explored by a number of authors in the past. In Chapter 6, Chikkula and Lee provide an overview of wavelets from the perspective of process control. The application of wavelets to model predictive control is described in detail.

Wavelets have also found many applications in numerical methods and their use as basic functions for the solution of differential equations is explored by Nikolaou and You in Chapter 7. Here, they discuss the Wavelet-Galerkin method and the Laplace transform wavelet method with applications to transport models such as the ones arising in fluid flow in porous media.

The last chapter by Goirand, Wickerhauser and Farge examines in detail the implementation of wavelet packet transforms in parallel computing machines. Such techniques are useful in computer intensive applications such as the analysis of multi-dimensional turbulent flow fields.

ACKNOWLEDGMENTS

This book is the result of co-operative efforts of numerous authors. We are extremely grateful to all the authors for their wonderful co-operation and timely delivery of manuscripts and revisions. We also thank them for their patience during this publication process.

In November 1993, we organized an informal workshop on wavelets at Washington University at the suggestion of Dr. Karlene Kosanovich of DuPont. The conference allowed many of the authors of this volume to meet and exchange ideas. We want to thank the participants in this workshop for providing stimulating talks and comments.

Special thanks are due to Dr. Malden Wickerhauser of the Department of Mathematics at Washington University who introduced us and our students to the many facets of wavelets and their applications.

Financial support provided by the Chemical, Biological and Thermal Process program at the National Science Foundation is gratefully acknowledged. We also want to thank Dr. Maria Burka at NSF for her personal interest in the research on wavelets at Washington University from the very beginning.

Editing this volume required a significant time away from our families. We want to acknowledge the sacrifices made and encouragement provided by our families.

B. Joseph and R. L. Motard
Washington University in St. Louis
January, 1994

Chapter 1

Introduction to Wavelet Transform and Time-Frequency Analysis

Xue-dong Dai, B. Joseph and R. L. Motard
Washington University
Chemical Engineering Department
One Brookings Drive, Box 1198
St. Louis, Missouri 63130

Abstract

The wavelet transform has been developed in recent years and has attracted growing attention from mathematicians as well as engineers. In this introductory chapter, we would like to review briefly some basic concepts and methods of this new approach under the more general framework of time-frequency analysis methods. We will discuss the Fourier transform, the short-time Fourier transform and time-frequency distributions, followed by a discussion of wavelet theory and its variations.

1. General Introduction

Over the past decade, reliability, efficiency, safety, productivity and quality control in chemical industry have dramatically increased. More effort is being placed on improving the description and analysis of complex phenomena, from fundamental laboratory studies to industrial applications. More powerful and accurate analysis tools are required to fulfill this goal. Using traditional mathematical tools, we can study the properties of a phenomenon either in the time domain or in the frequency domain. Although the Fourier transform and its inverse allow a passage from one domain to the other, it does not give a simultaneous view of the phenomenon in both domains. When considering a non-stationary phenomenon, one of the biggest challenges we face in analysis is that the tools available do not encompass good time-frequency resolution, and are not able to capture local properties of the phenomenon accurately. As a result, it is hard to obtain the information we need to clearly describe these phenomena.

The search for time-frequency descriptions to solve the problem mentioned above has resulted in the so-called time-frequency distributions, which have been studied for decades but still have some theoretical difficulties (Boashash, 1991). The Short-time Fourier transform is currently the standard method for time-frequency analysis (Cohen, 1989). Recently, a new time-frequency analysis method, the wavelet transform, has been proposed (Rioul and Vetterli 1991) and has drawn widespread attention.

In principle, time-frequency representations characterize phenomena in a two dimensional time-frequency plane. This means that we can extract the local properties (the time and frequency information) of the phenomenon to be studied. This yields a more powerful analysis tool for handling non-stationary and time-varying aspects which characterize most complex physical phenomena.

Basically, there are two types of time-frequency representations. One is the quadratic method covering the time-frequency distributions. The other is the linear approach including the Short-time Fourier transform, and the newly-introduced wavelet transform (Figure 1.1).

In the following sections of this chapter, we are going to introduce some basic concepts of these time-frequency analysis methods. The wavelet transform and its development will be emphasized. We start from the time-frequency distributions in Section 2 to introduce the basic idea of the time-frequency analysis, and then switch to the discussion on Fourier transform and Short-time Fourier transform in Section 3. In Section 4, we begin with a brief history of wavelet transforms and point to the main references in this field. We proceed to discuss the concepts of "mother wavelet", scaling and translation operations, to the definitions and some fundamental properties of the wavelet transform in its different versions: 1) continuous, 2) discrete, 3) bi-orthogonal and semi-orthogonal, 4) orthonormal wavelet transform with the construction of the orthonormal bases by multiresolution analysis, and finally to the generalization of the wavelet transform, 5) the wavelet packets theory. Section 5 concludes this chapter with several observations on this rapidly developing technique, the

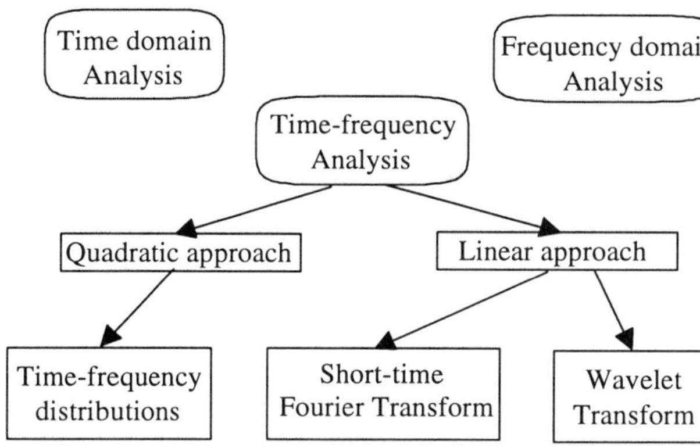

Fig. 1.1 Signal analysis tools and their relationships

wavelet theory. During the discussion, we will try to explain these fundamental mathematical concepts as well as definitions familiar to engineers. Sometimes we have to sacrifice completeness and rigor to keep the discussion clear.

2. Time-frequency distributions

Starting with the classical works of Wigner (1932) and Ville and Page (1952), there has been considerable development of the time-frequency distributions approach. The basic idea of the method is to devise a joint function of time and frequency, called a time-frequency distribution, that will describe the energy density of a signal simultaneously in both time and frequency (Cohen, 1989). In the ideal case, such a joint distribution would give the complete time-frequency information of the signal. Several different distributions have been proposed so far. A nice review on this topic has been given by Cohen (1989). Boashash (1991) also carefully reviewed the use of this approach, its recent developments and applications.

Hlawatsch and Boudreaux-Bartels (1992), on the other hand, have given a detailed comparison to linear time-frequency representations.

One of the well known distributions is the Wigner-Ville distribution. For function $f(t)$, the distribution is

$$P(t,\omega)=\frac{1}{2\pi}\int_{-\infty}^{+\infty} f^*(t-\tfrac{1}{2}\tau)e^{-j\tau\omega}f(t+\tfrac{1}{2}\tau)d\tau \tag{2.1}$$

where $f^*(t)$ is the complex conjugate of the original function $f(t)$. From the above expression, we can easily understand that the distribution $P(t, \omega)$ has a quadratic relationship with the original function $f(t)$.

Cohen (1966) developed a unified approach to represent all such distributions in a general formula,

$$P(t,\omega)=\frac{1}{4\pi^2}\iiint e^{-j\theta t-j\tau\omega+j\theta u}\phi\left(\theta,\tau\right)f^*(u-\tfrac{1}{2}\tau)f(u+\tfrac{1}{2}\tau)du\,d\tau\,d\theta \tag{2.2}$$

where $\phi(\theta,\tau)$ is an arbitrary function called the kernel. By choosing different kernels, different distributions are obtained at will. In the most recent development, Choi and Williams (1989) have presented an improvement on this topic to overcome some difficulties in the Wigner distribution.

Here is a simple example of Wigner and Choi-Williams distributions for the sum of two sine waves, $f(t)=e^{i\omega_1 t}+e^{i\omega_2 t}$. The results are plotted in the time-frequency plane as shown in Figure 2.1. From the figure, it is clear that one of the advantages of using such transforms is that the main features of the signal can be readily seen. But, we also observe that an extra peak appears in the middle of the picture. This is called the cross-terms effect, or "artifact", which is one of the problems of these distributions. So far, significant research effort has been put into trying to reduce this effect, including the contribution of Choi and Williams. There has been considerable research effort to apply these distributions in many fields where nonstationary signals arise. Forrester (1989) has applied the Wigner distribution to study vibrations of helicopter components with the aim of detecting developing failure of machine parts, in particular gear failure. By examining the vibrations of gears with and without faults he has shown that the Wigner distribution is a good discriminator and can be used to detect both the type and extent of faults. Boashash and O'Shea (1990) have proposed a methodology based on the Wigner-Ville distribution and cross

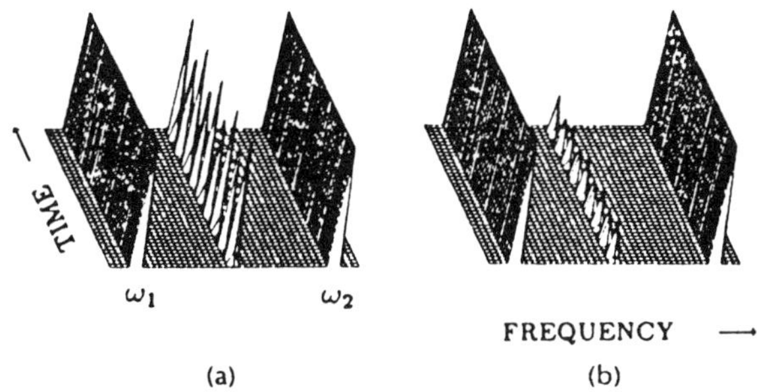

Fig. 2.1 (a) Wigner (b) Choi-Williams distribution
for the sum of two sine waves (Cohen, 1989)

Wigner-Ville distribution to detect and classify some underwater acoustic signals, i.e. the individual cylinder firing of a marine engine. In another application Choi, Williams, and Zaverri (1987) used the distribution they proposed to evaluate and classify "event-related potentials" where certain words were used to induce brain wave responses in patients. The signal obtained is represented by the distribution and used to classify the signal in terms of the types of stimuli. They found that the new distribution they proposed is effective in reducing the cross-terms effects by carefully choosing the coefficients of the distribution.

From these selected applications, we can realize the unique power of the time-frequency analysis method. By representing the distribution in a two-dimensional time-frequency plane, we can see all features of the signal in one comprehensive picture. Because of the substantial cross-terms effects together with some other theoretical difficulties, the time-frequency distributions have practical limitations. In other words, it is hard to get the true time-frequency distribution of the signal, as was the original motivation in the ideal case, so that we have difficulties in utilizing the real power of time-frequency analysis by the distribution theory. In the following section we will discuss another widely used technique, the Short-time Fourier transform, starting with the more familiar Fourier transform.

3. Fourier transform & Short-time Fourier transform

A classical analysis tool widely used is the Fourier transform. The Fourier transform and its inverse establish a one-to-one relation between the time domain, function $f(t)$, and the frequency domain, spectrum $F(\omega)$. The Fourier transform is defined by:

$$F(\omega) = \int_{-\infty}^{+\infty} f(t) e^{-i\omega t} dt \tag{3.1}$$

The inverse Fourier transform is

$$f(t) = \frac{1}{2\pi} \int_{-\infty}^{+\infty} F(\omega) e^{i\omega t} d\omega \tag{3.2}$$

Here the Fourier transform can be viewed as the decomposition of a function $f(t)$ into a sum of frequency components, the coefficients of which are given by the inner product of $f(t)$ and $e^{i\omega t}$. This transform uses sine and cosine as its bases to map a time domain function into frequency domain. Thus, the spectrum $F(\omega)$ shows the overall strength with which any frequency ω is contained in the function $f(t)$. The Fourier transform does not show how the frequencies vary with time in $f(t)$.

In order to overcome the above problem, the Short-time Fourier transform is proposed. The Short-time Fourier transform (STFT), or windowed Fourier Transform, is currently the standard method for the study of time-varying signals. The original idea came from the work done by Gabor (1946). He suggested the use of a window which slides over the signal in time, and then to compute the Fourier transform for every the portion within each window. After that, a set of extended methods based on Gabor's work have been proposed, mainly for different kinds of window functions. Here is the general definition of the Short-time Fourier transform,

$$F(t, \omega) = \int_{-\infty}^{\infty} f(\tau) g(\tau - t) e^{-j\omega\tau} d\tau \tag{3.3}$$

where $f(t)$ is the original function in time domain and $g(t)$ is the window function. Thus, STFT maps a time domain function $f(t)$ into a time-frequency domain function $F(t, \omega)$. We can see that this mapping is linear from the above definition. Actually, at time t, the STFT is the Fourier transform of the function $f(\tau)$ multiplied by a window $g(\tau\text{-}t)$. Since the

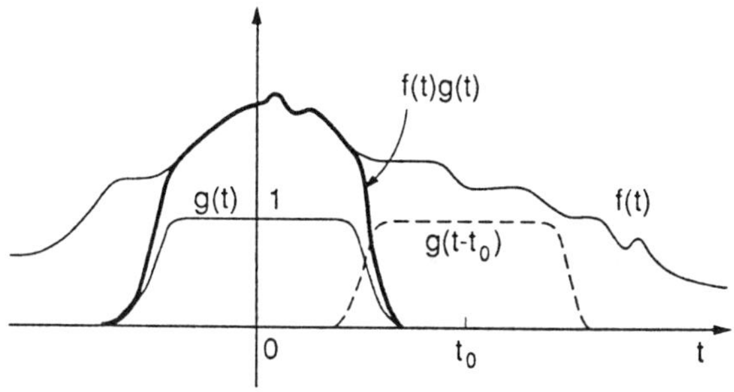

Fig. 3.1 The Short-time Fourier Transform, the window function (Daubechies, 1992)

window suppresses all function features outside a local neighborhood around time *t*, the STFT is simply a "local spectrum". Also, this window is fixed in the analysis. The above figure (Fig. 3.1) shows the effect of multiplying the function $f(t)$ with the window function $g(t)$, from which the Fourier coefficients of the product $f(t)g(t)$ are computed. The procedure is repeated for translated versions of the window function $g(t-\bullet)$.

Many researchers have found the performance of the method depends on the choice of the window function, and the choice of the window dramatically affects the appearances of, and the signal concentration in the transform (Jones and Parks, 1990).

In practical situations the spectral content of signals may change with time, and sometimes we want to see short term as well as long term variations. Then finding an appropriate window of constant-size and shape is problematic.

4. Wavelet theory

In this section, we will discuss a rapidly emerging technique -- the Wavelet transform, its extension, Wavelet packet theory, and outline their applications. This approach holds tremendous promise in the time-frequency analysis field. Hereafter the name, wavelet theory, is used to unify all the work in wavelet transform and wavelet packet theory as recommended by Rioul and Vetterli (1991).

4.1. A brief history

In the last few years, interest in the subject of wavelets has grown dramatically, both in the theoretical and applied areas. The concept of wavelets can be viewed as a synthesis of ideas which originated from several different fields, such as engineering (subband coding, quadrature mirror filter, pyramid schemes), physics (coherent states, renormalization group), and pure mathematics (Calderon-Zygmund operators). Wavelets have been recognized as a fairly simple mathematical tool with a great variety of possible applications in several different areas (Daubechies, 1992).

The wavelet transform was first proposed as a tool for signal analysis by geophysicist Morlet (1982). The numerical success of Morlet's work prompted Grossmann to make a detailed mathematical study of the continuous wavelet transform (Grossmann and Morlet, 1984). The mathematical study of the discrete case started with the introduction of "frame" by Daubechies, Grossmann and Meyer (1986, 1990). In the meantime orthonormal bases of wavelets were discovered. A first construction made by Stromberg (1981) did not receive much attention. Meyer constructed another basis in the fall of 1985, which has regarded as a theoretical breakthrough at the time. Other bases, numerically more useful because they were more concentrated, were presented by Lemarie and Battle (1987). Finally, in the fall of 1986, Mallat (1989) and Meyer realized that these orthonormal wavelet bases could be constructed systematically from a general framework named, multiresolution analysis, and ended the search for a so-called "miracle" to find an orthonormal basis (Daubechies, 1991, 1993). Furthermore, Daubechies (1988) published her famous paper for the construction of compactly supported orthonormal wavelet bases. After that, wavelet theory has attracted a lot of attention in many fields, from theoretical study to applications.

Despite of the recent development of the wavelet theory, there have been already many introductory and survey papers on this subject (Jawerth and Sweldens, 1993, Rioul and Vetterli, 1991, Strang, 1993, Flandrin, 1990), and several books. *Wavelet: Time-frequency methods and phase space* (Combes, J. M. et al eds., 1989) is the first book on this topic. The book is the proceedings of the international wavelet conference held in France. *Ten*

Lectures on Wavelets, by Daubechies (1992), contains the mathematical aspects of wavelet transform. *An Introduction to Wavelets* written by C. K. Chui (1992a) provides an introduction from the approximation theory point of view. Two edited books, *Wavelets and their applications*, edited by Ruskai, et al. (1992) and *A tutorial in theory and applications*, edited by C. K. Chui (1992b), have collected some interesting applications. Young (1993) presents his study in wavelet theory, especially in the applications of signal processing (signal and system models), in his book entitled *Wavelet theory and its application.* Also, there was a special wavelet issue of *IEEE Trans. Inform. Theory* in March of 1992 (Daubechies, et al eds., 1992). Most recently, the American Mathematical Society has released a videotape entitled *Wavelets making Waves in Mathematics and engineering*, which is an introductory lecture on wavelets given by Daubechies (1993). An electronic newsletter, *Wavelet Digest*, edited by Department of mathematics, University of South Carolina (Sweldens, ed., 1992), serves as the fastest information channel to exchange ideas and latest developments among the wavelet community.

4.2. Wavelet transform

Wavelet transform involves representing general functions in terms of simple, fixed building blocks at different scales and positions, which is named "wavelet" as first suggested by Yves Meyer and Jean Morlet (Jawerth and Sweldens, 1993). These building blocks, actually are a family of "wavelet" functions, or in short the "wavelets", are generated from a single fixed function called "mother wavelet" by translation and dilation (scaling) operations.

4.2.1. A family of wavelets

The basic building block, known as the "mother wavelet", is the function $\psi(t)$ if its Fourier transform $\Psi(\omega)$ satisfies:

$$\int_{-\infty}^{+\infty} \frac{|\Psi(\omega)|^2}{\omega} d\omega = C_\psi < +\infty \tag{4.2.1}$$

which is called the "admissibility condition". This condition implies that,

$$\int_{-\infty}^{+\infty} \psi(t)dt = 0 \tag{4.2.2}$$

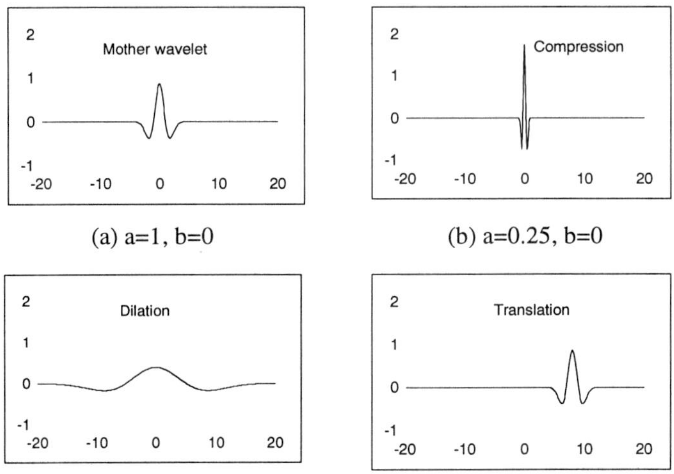

(a) a=1, b=0

(b) a=0.25, b=0

(c) a=5, b=0

(d) a=1, b=8

Fig. 4.1 The mother wavelet and its scaled and translated versions

Performing scaling and translation operations on this mother wavelet $\psi(t)$ creates a family of scaled and translated versions of the mother wavelet function,

$$\psi_{a,b}(t) = \frac{1}{\sqrt{|a|}}\psi(\tfrac{t-b}{a}) \qquad (4.2.3)$$

where a is the scaling and b is the translation parameters respectively, and the factor $\frac{1}{\sqrt{|a|}}$ is used to ensure that the energy of the scaled and translated versions are the same as the mother wavelet.

Figure 4.1(a) shows an example of a mother wavelet $\psi(t)$, the Mexican hat function (Daubechies, 1992),

$$\psi(t) = \frac{2}{\sqrt{3}}\pi^{-\frac{1}{4}}(1-t^2)\exp(-\tfrac{t^2}{2})$$

In fact, the scaling operation is nothing more than performing "stretching" and "compressing" operations on the mother wavelet, which in turn can be used to catch the different frequency information of the function to be analyzed. The "compressing" version in Figure 4.1(b) is used to fit the high frequency needs, and the "stretching" version in Figure 4.1(c) is for low frequency requirements. The translation operation, on the other hand, involves "shifting" of the mother wavelet along the time axis. Then the translated version is used to catch the time information of the function to be analyzed, shown in Figure 4.1(d). In this way, a family of scaled and translated wavelets are created according to the different scaling and translation parameters a and b, which serves as the base, the building blocks, for representing the function to be analyzed. We may notice that the shape of these functions looks like small waves, thus the name, "wavelet".

In contrast to the Short-time Fourier transform, which uses a single analysis window function, the wavelet transform can use short windows at high frequencies and long windows at low frequencies resulting from the family of "wavelet" functions mentioned above. This is called "constant - Q" or constant relative bandwidth frequency analysis. Thus, wavelet transform are capable of "zooming-in" on short-lived high frequency phenomena, and "zooming-out" on long-lived low frequency phenomena. This is one of the main advantages of wavelet transform over the Short-time Fourier transform. There are several types of wavelet transforms according to the type of input signal, time and scale parameters, namely, continuous, discrete, bi-orthogonal and semi-orthogonal, and orthonormal bases version.

4.2.2. Continuous wavelet transform

The mathematical aspects of the continuous wavelet transform are similar to those of the Short-time Fourier transform. Given a mother wavelet function $\psi(t)$, we can define the continuous wavelet transform CWT_f of a function $f(t)$ by

$$CWT_f(a,b) = \langle f, \psi_{a,b} \rangle = |a|^{-1/2} \int_{-\infty}^{\infty} f(t) \psi^* \left(\tfrac{t-b}{a} \right) dt \quad a, b \in R, a \neq 0 \qquad (4.2.4)$$

where the asterisk denotes the complex conjugate, and we use the notation $\langle . \rangle$ for the standard scalar product, as shown by

$$\langle f, g \rangle = \int_{-\infty}^{+\infty} f\ (t) g^*(t) dt$$

Here we allow time t and the scale-time parameters a, b to change continuously. Again, this is a linear transform of the same form as the Short-time Fourier transform. However, one of the differences between these two transforms is that the Short-time Fourier transform is always a complex function in the time-frequency domain, regardless of the choice of the window function, even for a real function $f(t)$, while the wavelet transform defined above is real if the function $f(t)$ is real and the mother wavelet is real too. Then, the reconstruction formula is as follows,

$$f(t) = C_{\psi}^{-1} \int_{-\infty}^{+\infty} \int_{-\infty}^{+\infty} CWT_f(a,b)\psi_{a,b}(t)\frac{dadb}{a^2} \tag{4.2.5}$$

where C_ψ is called the "admissibility constant" that depends only on $\psi(t)$ and is defined in Eq. (4.2.1). The mother wavelet $\psi(t)$ can be chosen to be either real or complex according to the requirements of the application. In some cases, a complex wavelet transform can be very useful (Daubechies, 1991).

Roughly speaking, a wavelet transform coefficient, $CWT_f(a,b)$, at a particular scale and translation represents how well the function f, and the scaled and translated wavelet $\psi_{a,b}(t)$ match. If they are similar to each other, then this coefficient will be large in magnitude.

4.2.3. Discrete wavelet transform

In the discrete case, wavelets can be redundant, bi-orthogonal and semi-orthogonal, or orthonormal, in terms of the functional dependency among wavelets.

For discrete wavelets, scale-time parameters a and b are discretized: $a = a_0^m$ and $b = nb_0a_0^m$. And this family of wavelets $\{\psi_{m,n}(t)\}$ is given by

$$\psi_{m,n}(t) = a_0^{-m/2}\psi(a_0^{-m}t - nb_0) \qquad m,n \in Z \tag{4.2.6}$$

So, we have the discrete version of wavelet transform,

$$(DWT_f)_{mn} = \langle f, \psi_{mn} \rangle = a_0^{-m/2} \int_{-\infty}^{\infty} f(t)\psi(a_0^{-m}t - nb_0)dt \tag{4.2.7}$$

As a_0 tends to 1 and b_0 tends to 0, the discrete version approaches a continuous version.

In many cases, the discrete version Eq. (4.2.6) is "redundant": the family of wavelets $\{\psi_{m,n}(t)\}$ are not linearly independent. It is then called *redundant* wavelets. This redundancy can lead to robustness, in the sense that one can

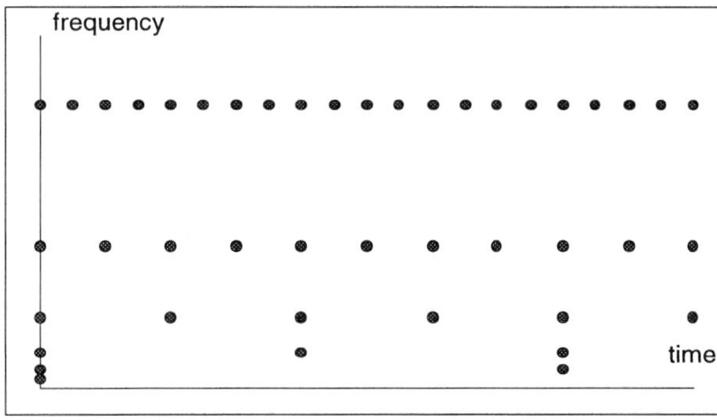

Fig. 4.2 The time-frequency space lattice of the wavelet transform

store the wavelet coefficients $<f, \psi_{m,n}>$ with low precision, and still reconstruct f with comparatively much higher precision (Daubechies, 1992). This will be an advantage for some applications.

The discrete wavelet transform represents a function $f(t)$ by a countable set of wavelet coefficients, which correspond to points on a two-dimensional grid or lattice of discrete points in the scale-time domain indexed by m and n. The narrow (high frequency) wavelets are translated by small time steps in order to cover the whole time range, while the wider (low frequency) wavelets are translated by large time steps. Figure 4.2 shows schematically the lattice of time-frequency localization centers corresponding to the $\psi_{m,n}$.

In the case of a continuous wavelet transform, the reconstruction formula Eq. (4.2.5) shows that a function $f(t)$ can be completely and easily reconstructed from $CWF_f(a, b) = \langle f, \psi_{a,b} \rangle$. However, for the redundant discrete version, there is no guarantee that $f(t)$ can be reconstructed from these discrete wavelet transforms in general. It would be expected that reconstruction is still possible if the discrete lattice has a very fine mesh. For very coarse meshes, the coefficients will not contain enough information for the reconstruction of the original function f from these coefficients. There exist threshold values for the lattice parameters (a, b). For those who satisfy the condition of these thresholds we can have a numerically stable reconstruction formula. This leads to the concept of a "frame" as follows (Daubechies, Grossmann and Meyer, 1986).

If the wavelets form a frame, i.e. if there exists constants $A > 0$ and $B < \infty$ such that

$$A\|f\|^2 \le \sum_{m,n} |\langle f, \psi_{m,n}\rangle|^2 \le B\|f\|^2 \qquad (4.2.8)$$

then we can have the reconstruction formula,

$$f = \frac{2}{A+B} \sum_{m,n} \langle f, \psi_{m,n}\rangle \psi_{m,n} + R \qquad (4.2.9)$$

R is the error in the reconstruction. For A/B close to 1, the frame is called a tight frame, $R \approx 0$, and a simple reconstruction formula is obtained. When $A = B = 1$, the family of wavelet then forms an *orthonormal* basis.

The orthogonality condition puts a strong restriction on the construction of wavelets. It is difficult to generate wavelets that are compactly supported, symmetric and orthogonal. To overcome this limitation, the generalization to *bi-orthogonal* wavelets has been introduced by Cohen (1992), Daubechies and their colleagues. Also, Chui and his colleagues (Chui, 1992a) presented another group of wavelets, named *semi-orthogonal* wavelet with different conditions. That is, if $\{\psi_{m,n}\}$ only satisfies the semi-orthogonal conditions,

$$\psi_{j,k} \perp \psi_{l,m}, \quad j \neq l,\ j, k, l, m \in Z$$

then $\{\psi_{m,n}\}$ are *semi-orthogonal* wavelets. This condition means that these scaled and translated wavelets at different scales are orthogonal among each other, but not within the same scale.

When we reduce the redundancy as much as possible, $\{\psi_{m,n}\}$ becomes linearly independent and forms an *orthonormal* basis. The orthonormal wavelet transform, i.e. wavelet transform using orthonormal wavelet basis, plays an important role in both theory and practice. Mallat (1989) and Meyer's theory of multiresolution analysis gives a systematic way to construct these orthonormal wavelet bases. This is discussed next.

4.2.4. The orthonormal wavelet transform and multiresolution analysis

A particularly interesting class of wavelets are the orthonormal wavelets. When choosing the mother wavelet function ψ, scaling and translation parameters a_0 and b_0 specifically, Daubechies (1988) has developed conditions under which $\{\psi_{m,n}\}$ forms orthonormal bases. Typically, $a_0 = 2$ and $b_0 = 1$ are chosen (however a_0 and b_0 are not restricted to these values). Thus, from Eq. (4.2.6), substituting these values, we have a family of orthonormal wavelets, which now becomes an orthonormal basis:

$$\psi_{m,n}(t) = 2^{-m/2}\psi(2^{-m}t - n) \tag{4.2.10}$$

Then, the orthonormal wavelet transform, using the orthonormal basis, for a function f is that

$$(OWT_f)_{mn} = \langle f, \psi_{mn} \rangle = 2^{-m/2} \int_{-\infty}^{\infty} f(t)\psi(2^{-m}t - n)dt \tag{4.2.11}$$

and the formula for reconstruction of f is simply given by

$$f = \sum_{m,n \in Z} \langle f, \psi_{m,n} \rangle \psi_{m,n} \tag{4.2.12}$$

In this case both the transformation as well as reconstruction are complete, because the family of wavelets form an orthonormal basis. Now let us consider the construction of the mother wavelet and the corresponding orthonormal basis.

A standard example of a function ψ that satisfies the conditions is the Haar function:

$$\psi_H(t) = \begin{cases} 1, & \text{for } 0 \le t \le 1/2 \\ -1, & \text{for } 1/2 \le t \le 1 \\ 0, & \text{otherwise} \end{cases} \tag{4.2.13}$$

The corresponding $\{\psi_{m,n}\}$ formed by Eq. (4.2.10) constitutes an orthonormal basis, the Haar basis. Another simple example of ψ, is the Shannon wavelet:

$$\psi_S(t) = \frac{\sin(2\pi t) - \sin(\pi t)}{\pi t} \tag{4.2.14}$$

However, these two are not very useful in practice, since the first one has very bad frequency localization and the second one has very bad time localization. Fortunately, there are three groups of wavelets with much nicer properties (smooth, fast decay, compactly supported, etc.). These are the Meyer wavelet, the Battle-Lemarie wavelets and the most commonly used, Daubechies wavelets (Jawerth and Sweldens 1993).

Multiresolution analysis provides a formal approach to constructing the orthonormal basis. The idea of multiresolution analysis is to write a function f as a limit of successive approximations, each of which is a smoother version of f. The successive approximations thus correspond to different resolutions, hence the name multiresolution analysis.

A multiresolution analysis of $L^2(R)$ is defined as a sequence of closed subspaces V_j of $L^2(R)$, $j \in Z$, with the following properties:

C1. $V_j \subset V_{j-1}$,

C2. $f(t) \in V_j \Leftrightarrow f(2t) \in V_{j-1}$,

C3. $f(t) \in V_0 \Leftrightarrow f(t+1) \in V_0$,

C4. $\bigcap_{j \in Z} V_j = \{0\}$, and $\overline{\bigcup_{j \in Z} V_j} = L^2(R)$,

C5. There exist $\phi \in V_0$ such that $\{\phi(t-n)\}_{n \in Z}$ is an orthonormal basis for V_0.

Here condition C1 presents the successive subspaces that are used to represent the different resolutions (scales). Within these subspaces, condition C2 indicates that when the function $f(t)$ is stretched appropriately, it will become the function in the next larger scale, while C3 means that the integer translation of $f(t)$ still keeps the function within the same scale (same subspace). Condition C4 guarantees the completeness of these subspaces. Finally, in condition C5, the function ϕ is called the scaling function of the multiresolution analysis.

By doing the scaling and translation, for each j, we have a collection of functions $\{\phi_{j,n}\}_{n \in Z}$.

$$\phi_{j,n}(t) = 2^{-j/2}\phi(2^{-j}t - n) \qquad (4.2.15)$$

These form an orthonormal basis for V_j. Multiresolution analysis allows us to approximate a given function f by a function f_j in each of the subspaces V_j. We call f_j the approximation at resolution j. Also because of condition C4, it is guaranteed that there are such approximations which converge to the original function, or $f = \lim_{j \to -\infty} f_j$.

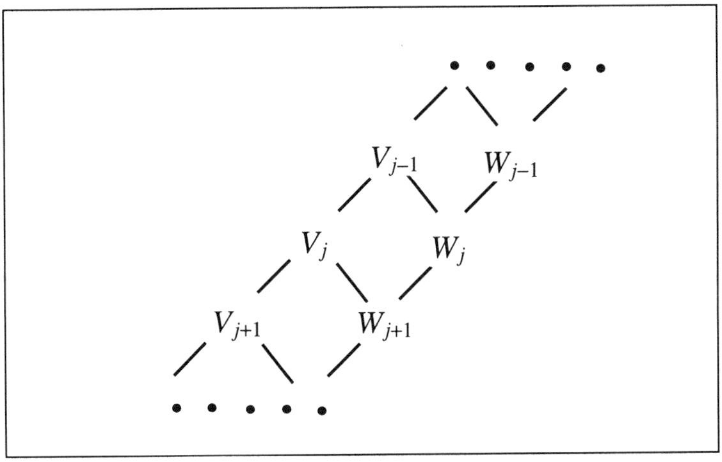

Fig. 4.3. The multiresolution decomposition of subspaces

Furthermore, we use W_j to denote a subspace as the orthogonal complement of V_j in V_{j-1}, i.e. it satisfies

$$V_{j-1} = V_j \oplus W_j \quad \text{and} \quad V_j \perp W_j \tag{4.2.16}$$

where the symbol $\oplus$ stands for direct sum. It means that the subspace W_j contains the "detail" information needed to go from an approximation at resolution j to an approximation at resolution j-1. Then applying the above equation recursively, we are able to "travel" along these successive subspaces (Fig. 4.3).

A function ψ is an *orthogonal wavelet* if the collection of functions $\{\psi(t-k)\}_{k\in Z}$ is an orthonormal basis of W_0. With similar scaling and translation, we have the collection of wavelet functions $\{\psi_{j,k}\}_{j,k\in Z}$, i.e.

$$\psi_{j,k}(t) = 2^{-j/2}\psi(2^{-j}t-k) \tag{4.2.17}$$

Again, for each j, $\{\psi_{j,k}\}_{k\in Z}$ is an orthonormal basis for W_j. Since the spaces W_j are mutually orthogonal, $\{\psi_{j,k}\}_{j,k\in Z}$ is then an orthonormal basis of $L^2(R)$, which is our desired result -- the orthonormal wavelet basis. Here ψ is called the *mother wavelet*. Actually, the ψ can be solved from the following equations. There exists c_n such that

$$\phi(t) = \sum_n c_n \phi(2t - n) \tag{4.2.18}$$

Define then

$$\psi(t) = \sum_n (-1)^n c_{n+1} \phi(2t + n) \tag{4.2.19}$$

These are the major ideas of multiresolution analysis which shows the construction of an orthonormal wavelet basis from it. Also, this construction procedure provides the theoretical basis for constructing a fast algorithm for wavelet transform. It means we do not need to calculate the transformation and reconstruction directly by Eq. (4.2.11) and Eq. (4.2.12). Instead, a recursive algorithm can be built by traveling along those successive subspaces forward (transformation) and backward (reconstruction). A more detailed discussion of these computational aspects will be given in Chapter 2.

When studying the orthonormal wavelet basis closely, we will notice that this basis is particularly well suited for analyzing signals with short-lived high frequency phenomena superposed on much lower frequencies. However, this is not always the situation in practice. This leads to the question: Do we have other choices for the orthonormal basis to meet the requirements of a specific application, which retains the nice properties of the wavelet basis? In the next section we will find an answer from the results proposed by Coifman and Wickerhauser (1992b). We will introduce the signal dependent basis in Chapter 4, based on the results of wavelet packets.

4.3. Wavelet packets

As the natural extension of the wavelet transform theory, Coifman, Meyer and Wickerhauser (1992a,b; Wickerhauser, 1992, 1993) first proposed their generalization results as Wavelet Packets to design efficient schemes for representing and compressing signals and images. The idea is to construct a library of orthonormal bases for $L^2(R)$ by generalizing the method of multiresolution decomposition, and then the "best basis" related to the given signal is obtained by searching the library, which has the lowest information cost. The library of orthonormal bases also has time-frequency localization properties which can be reasonably well controlled.

4.3.1. A library of orthonormal bases

The library construction begins with considering the quadrature mirror filter h(n) (Wickerhauser, 1991). Define a summable sequence $h = \{h_n\}$ satisfying

$$\sum_n h_{n-2k} h_{n-2l} = \delta_{k,l}, \qquad \sum_n h_n = \sqrt{2} \tag{4.3.1}$$

and let $g = \{g_k\}$ where $g_k = (-1)^k h_{1-k}$. Then introduce summing and differencing operators F_0, F_1:

$$F_0\{s_k\}(2i) = \sum_k s_k h_{k-2i} \tag{4.3.2}$$

$$F_1\{s_k\}(2i) = \sum_k s_k g_{k-2i} \tag{4.3.3}$$

and their adjoints F_0^*, F_1^* which satisfy

$$F_0^* F_0 + F_1^* F_1 = I, \quad F_0 F_0^* = F_1 F_1^* = I, \quad F_0 F_1^* = F_1 F_0^* = 0 \tag{4.3.4}$$

where the mapping F defined by $F = F_0 \oplus F_1$ is orthogonal.

Now these two summing and differencing operators are applied to splitting the functional space. We see that the classical multiresolution analysis is obtained by splitting V_{j-1} with these operators into V_j and W_j (Eq. 4.2.16) and doing the same for V_j recursively. The wavelet packets are the basis functions that we obtain if we not only apply these two operators to splitting the V_j spaces, but also W_j spaces, recursively as well. In the frequency domain, this splitting operation corresponds to dividing the frequency interval represented by the original space into a lower and upper part. So, the wavelet packets allow more flexibility in adapting the basis to the time and frequency contents of a function (a signal).

Since the library of wavelet packet bases is constructed in this way, it is naturally organized as subsets of a binary tree. This tree structure is shown in Figure 4.4. Each row is computed from the row above it by one application of either F_0 or F_1, which we think of as "summing" (s) or "differencing" (d) operations, respectively. Thus, for example, the subblock $\{ss_0, ss_1\}$ comes from the application of F_0 to $\{s_0, s_1, s_2, s_3\}$, while $\{ds_0, ds_1\}$ comes similarly from applying F_1. The two descendent s and d subblocks on row $n+1$ are determined by their mutual parent on row n. On the other hand, the parents on row n can be calculated by applying the adjoints F_0^* and F_1^*.

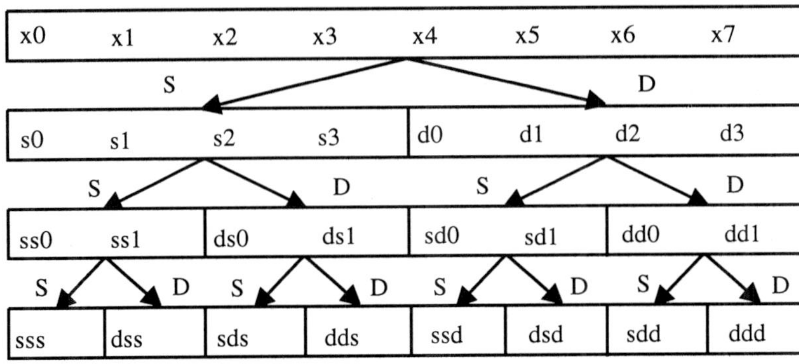

Fig. 4.4. Wavelet packets organized as a binary tree
(Coifman and Wickerhauser, 1992)

The leaves of every connected subtree give an orthonormal basis. Figure 4.5 presents the configuration of the subtrees for the wavelet basis (a), and the other unnamed basis (b). From the picture of the wavelet basis, we can understand the meaning of "generalization" of the wavelet packet theory more clearly, i.e. for the wavelet basis, the "summing" and "differencing" operations are always applied to the "s" side ($\{s_0, s_1, s_2, s_3\}$, $\{ss_0, ss_1\}$, i.e. the V_j spaces, and keep the "d" side ($\{d_0, d_1, d_2, d_3\}$, $\{ds_0, ds_1\}$), i.e. the W_j spaces, while for the constructing the whole wavelet packet library, s and d operations have been applied to every subtrees (both "s" side and "d" side) in the library until they reach the bottom.

It is easy to construct a fast wavelet packet transform algorithm by applying the same low and band pass filters to the coefficients of the functions of both V_j and W_j in a iterative manner.

4.3.2. The best basis algorithm

We now have a library of orthonormal bases and any of these basis sets can be used to represent a function (e.g. a signal segment to be analyzed) completely. This gives us more freedom in deciding which basis we will use to represent any given function. There must be a basis among them which can represent the function in the most efficient way, i.e. using the least number of coefficients. In order to find this "best basis", Wickerhauser and

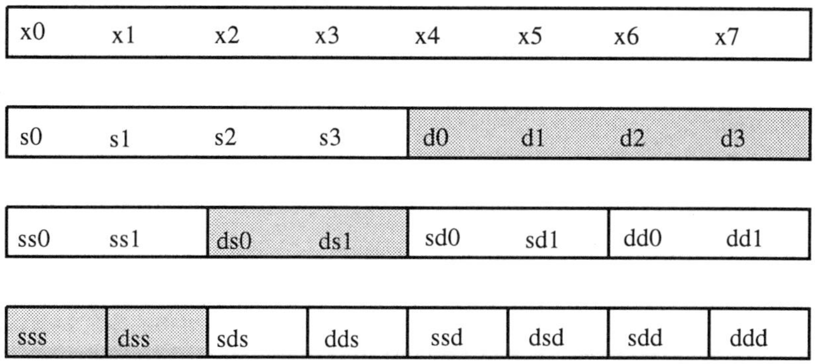

(a) The wavelet basis in the library

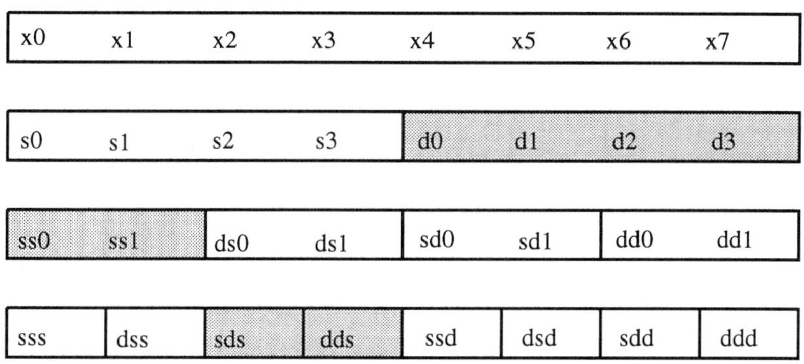

(b) An orthonormal basis in the library

Fig. 4.5. Examples of the bases in the wavelet packets library (Coifman and Wickerhauser, 1992)

his colleagues introduced the concept of the information cost function. Then, the cost function is used as a measure to search for its minimum over all bases in the wavelet packet library to obtain such a "best basis" for the function. The information cost function is defined to be real-valued. It should be large when the coefficients are roughly the same size and small when the most coefficients are negligible but only a few coefficients need to be retained. One of the cost functions suggested is the entropy cost function defined as follows (Wickerhauser, 1991).

For a vector $x = \{x_j\}$, the Shannon-Weaver entropy is defined as

$$H(x) = -\Sigma_j p_j \log p_j \tag{4.3.5}$$

where $p_j = \frac{|x_j|^2}{\|x\|^2}$, set $p \log p = 0$ if $p = 0$. A classical explanation about entropy is that $\exp H(x)$ is proportional to the number of coefficients needed to represent the vector to a fixed mean square error.

And then the best basis for the vector x is the basis which minimizes the cost function of the wavelet packets transform. A fast search for a minimum information cost function algorithm has been constructed by Wickerhauser. For an N-sample signal, we can obtain all the coefficients for all bases in the library by applying only the two operations F_0 and F_1, as defined in Eq. (4.3.2) and Eq. (4.3.3), recursively. This results in 2^N different orthonormal bases in the library and takes $O(N \log N)$ operations. The search for the best basis uses an additional $O(N)$ operations.

4.4. Brief outline of past applications

Even though wavelet theory is still a fast evolving area, its applications are growing rapidly (Combes, J. M. et al. eds., 1989; Rioul and Vetterli, 1991; Daubechies et al. eds, 1992). The applications concern many fields: biomedical signal analysis, solution of partial differential equations, fractal structure analysis (Bacry et al., 1993), turbulent flow studies, signal and image singularity analysis, neural networks, observational cosmology (Hjorth, et al., 1992, Goupil, et al., 1991, Slezak, 1989), computer vision (Mallat, 1989, Manjunath and Chellappa, 1993), seismic signal analysis, sonar and radar signal analysis, speech analysis (Yang, et al., 1992), noise reduction (Donoho and Johnstone, 1992a, 1992b, Wornell and Oppenheim, 1992), music synthesis, etc. Some of the applications in chemical engineering are covered in the subsequent chapters of this book.

Grossmann and Morlet (1984) were first to use wavelets in seismic signal analysis. Tuteur (1989) has applied the continuous wavelet transform (CWT) to detect ventricular delayed potentials (VLP) in the electrocardiogram. The VLP can be seen clearly on a specific scale of the transformed signal. The results show the ability of the wavelet transform in detecting weak signals embedded in a stronger background, without precise knowledge of their shape, location, time extent, and spectrum. Khadra and his colleagues (1991) presented another application by using CWT to analyze heart sounds, especially the high frequency murmurs. They showed that CWT has a much better time resolution than the corresponding spectrogram and Wigner distribution. This allows a diagnostic process to be developed to classify normal and abnormal heart sounds and then detect the heart diseases. Huang and Solorzano (1991) describe results on using the wavelet transform to preprocess acoustic broadband signals in a system that discriminates between different classes of acoustic bursts. H. Kikuchi and his colleagues (1991) applied the wavelet transform to detecting detonation in a vehicle engine for precise real-time control of ignition advancement. T. Kikuchi and Sato (1992) described the application of wavelet transform for a high resolution ultrasonic measurement. They have been trying to apply this method for detecting positions of point scatterers in random media, and it was found by calculations and experiments that this method can be used for characterization of random media. D. Morlet and his research group (1991) investigated the contribution to the prediction of ventricular tachycardia (VT) of wavelet transformation. Desai and Shazeer (1991) demonstrated the use of wavelet decomposition in extracting relevant information from passive acoustic signals. Crowe and his group (1992) described a preliminary investigation into its application to the study of both ECG and heart rate variability data.

Meneveau (1991) studied the turbulence problem by wavelet transform. A decomposition of turbulent velocity fields into modes that exhibit both localization in wavenumber and physical space is performed. Benzi and Vergassola (1991) applied wavelet transform to the numerical simulation of a two-dimensional decaying turbulent flow. The results show that many different scaling behaviors are present in the vorticity field. Yamada and Ohkitani (1991) applied the orthonormal wavelet expansion method to the study of the data of atmospheric turbulence. Farge and his colleagues (1992) proposed to use wavelet packet bases to compress two-dimensional turbulent flows.

Bakshi and Stephanopoulos (1993), Zhang and Benveniste (1992) in parallel proposed a class of networks call wavelet networks by combining the class of feedforward neural networks and results from the wavelet theory. The activation functions of the new network are derived from the class of wavelets. McAulay and Jian Li (1992) applied wavelet transform to do the preprocessing before classification with neural networks.

Mallat and Zhong (1990) presented an important application based on multiresolution analysis. They described an algorithm that reconstructs one-D signals and 2-D images from their sharper variation points at dyadic scales. The sharper variation points of a signal, measured at different scales, can be detected from the local maxima of its wavelet transform. Thus, they showed that the wavelet maxima representation is a new organization of the image information. It leads to the possibility of developing algorithms uniquely based on edges for solving image processing problems. Mallat and Hwang (1992) further studied the problem. They have proven that the local maxima of wavelet transform modules detect the locations of irregular structures. An application to noise removal in one- and two-dimensional signals was described.

Frich and Messer (1992) have observed that the application of the wavelet representation to detection problems is expected to improve the detector performance. Friedlander and Porat (1992) have studied the performance of detection algorithms based on linear time-frequency transforms, i.e. STFT, CWT, and Daubechies' orthonormal wavelet transform. They concluded that the transforms are similar in their ability to separate the three components of the test signal. But the wavelet transform does not appear to offer any clear advantages over STFT for detection purposes under their model assumptions. Ghosh, Deuser and Beck (1992) presented a classifier system for short-duration oceanic signals obtained from passive sonar, which exhibit variability in both temporal and spectral characteristics even in signals obtained from the same source. Wavelet-based feature extractors are shown to be superior to the more commonly used autoregressive coefficients and power spectral coefficients for describing these signals.

Wickerhauser and his colleagues (Wickerhauser, 1992; Coifman et al. 1992a) have applied their wavelet packets to design efficient schemes for compressing acoustic signals and images, and showed the efficiency of this approach. Hopper and Preston (1992) investigated a number of techniques including wavelet packets theory developed for fingerprint identification.

More recently, Pati and Krishnaprasad (1992) have developed a wavelet representation for linear, time invariant, causal, dynamic systems. Elias-Juarez and Kantor (1992) have demonstrated use of wavelets in model predictive control. The main results of their paper is the advantage of using a wavelet basis in constraints handling. They illustrate that wavelets can be used to impose constraints on local output behavior. Lee (1992), on the other hand, has reported that the use of the Haar basis has computational benefits in applications of model predictive control with infinite output horizon.

Beylkin, Coifman and Roklin (1991) have applied the wavelet approach to study a variety of integral operators. Alpert (1992) has reported the use of a wavelet like basis in the analysis of various linear operators.

From the applications shown above, we understand that wavelet theory does have some special advantages in the analysis of nonstationary signals, and provides a great opportunity to new applications.

5. Conclusion

In this chapter we have briefly introduced some basic concepts of the theory of time-frequency analysis methods including time-frequency distributions, Short-time Fourier transform, and the developments of Wavelet transform and Wavelet packets. From the above discussion, we might make the following observations

- Time-frequency analysis methods are useful for analyzing nonstationary signals (phenomena).

- Time-frequency distributions (TFD) introduce a powerful idea for using time-frequency information. Some of the applications have already shown the potential promise of recognizing signal features by using time-frequency representation.

- Short-time Fourier transform (STFT) is now considered as the standard tool for nonstationary signal analysis. Because STFT and wavelet transform are linear in the signal, they do not exhibit the "artifact" typical of TFD. But the time-frequency resolution in STFT has to be fixed since a fixed window shape and size are used in the analysis. This is the main difference between STFT and WT in applications.

- The Wavelet transform (WT) offers the distinguishing advantage of multi-resolution analysis for signals. Here we can use both long and short windows to catch the signal features.

- As the generalization of WT, wavelet packets (WP) together with the best basis algorithms can represent the signal in a most compact way, which can be seen most clearly in the time-frequency plane by comparing the transform results of WT and WP. Therefore, WP provides a more general tool for applications.

Indeed, the wavelet theory opens a new window for study of the non-stationary phenomena. Both the theoretical and practical results are emerging exponentially. We expect to see more and more theoretical developments as well as applications in our chemical engineering field and many other areas, combining with the other existing methodologies. Finally, we would like to quote the conclusion made by Daubechies (1993) in her videotaped lecture: the wavelet transform is a very useful tool, but is not a miracle. We cannot expect to use it to solve any unknown problems we face. Our studies so far on wavelet applications confirm this.

Acknowledgements

The authors would like to acknowledge the support of NSF grant # CTS - 9108649.

6. References

Alpert, B. K. "Wavelets and other bases for fast numerical linear algebra." in *Wavelets: A Tutorial in Theory and Applications* Chui, C. K. ed., Academic Press, (1992) : 181-216.

Bacry, E., J. F. Muzy and A. Arneodo " Singularity spectrum of fractal signals from wavelet analysis: exact results" *Journal of Statistical Physics*, 70, 3-4 (1993) : 635-674.

Bakshi, B. R. and G. Stephanopoulos " Wave-Net: a multiresolution, hierarchical neural network with localized learning." *AIChE J.,* 39, 1 (1993) : 57-81.

Battle, G. "A block spin construction of ondelettes. Part I: Lemarie functions." *Comm. Math. Phys.*110, (1987) : 601-615.

Benzi, R. and M. Vergassola "Optimal wavelet analysis and its application to two-dimensional turbulence." *Fluid Dynamics Research*, 8, 1-4 (1991) : 117-126.

Beylkin, G., R. Coifman and V. Rokhlin "Fast wavelet transforms and Numerical Algorithms." *Comm. Pure Appl. Math.,* 44, (1991) : 141-183.

Boashash, B. and P. O'Shea "A methodology for detection and classification of some underwater acoustic signals using time-frequency analysis techniques." *IEEE Trans. Acoust., Speech, Signal Processing,* 38, 11 (1990) : 1829-1841.

Boashash, B. "Time-frequency signal analysis." *Advances in Spectrum Analysis and Array Processing* Volume I, Haykin, Simon. Ed. Prentice Hall, Englewood Cliffs, NJ., (1991) : 418-517.

Choi, H. I. and W. J. Williams "Improved time-frequency representation of multicomponent signals using exponential kernels." *IEEE Trans. Acoust., Speech, Signal Processing,* 37, 6 (1989) : 862-871.

Choi, H. I., W. J. Williams and H. Zaveri "Analysis of event related potentials: time frequency energy distribution." *Biomedical Sciences Instrumentation* , 23, (1987) : 251-258.

Chui, C. K. *An Introduction to Wavelets.* Academic Press, Boston, 1992a.

Chui, C. K. ed. *Wavelets: A tutorial in theory and applications.* Academic Press, Boston, 1992b.

Cohen, A. "Biothogonal wavelets." *Wavelets: A tutorial in theory and applications. Chui*, C. K. ed. Academic Press, Boston, (1992) : 123-152.

Cohen, L. "Time-frequency distributions -- A review." *Proceedings of the IEEE,* 77, 7 (1989) : 941-981.

Cohen, L. "Generalized phase-space distribution functions." *J. Math. Phys.,* 7 (1966) : 781-786.

Coifman, R. R., Y. Meyer, S. Quake and M. V. Wickerhauser "Signal processing and compression with wavelet packets." *Progress in Wavelet Analysis and Applications*, Meyer and Roques eds., Editions Frontieres, Toulouse, France, (1992a) : 77-93.

Coifman, R. R. and M. V. Wickerhauser "Entropy-based algorithms for best basis selection." *IEEE Trans. Information Theory,* 38, 2 (1992b) : 713-718.

Combes, J. M. et al. eds. *Wavelets, Time-Frequency Methods and Phase Space* Springer-Verlag, Berlin, 1989.

Crowe, J. A. et al. "Wavelet transform as a potential tool for ECG analysis and compression." *Journal of Biomedical Engineering*, 14, 3 (1992) : 268-272.

Daubechies, I. "Orthonormal basis of compactly supported wavelets." *Comm. Pure & Applied Math.,* 41 (1988) : 909-996.

Daubechies, I. "The wavelet transform, time-frequency localization and signal analysis." *IEEE Trans. Information Theory,* 36, 5 (1990) : 961-1005.

Daubechies, I. "The wavelet transform: a method for time-frequency localization." *Advances in Spectrum Analysis and Array Processing* Volume I, Haykin, Simon Ed. Prentice Hall, Englewood Cliffs, NJ., (1991) : 366-417.

Daubechies, I. *Ten Lectures in Wavelets* Society for Industrial and Applied Mathematics, Philadelphia, Pennsylvania, 1992.

Daubechies, I. *Wavelets Making Waves in Mathematics and Engineering.* Videotape, American Mathematical Society, 1993.

Daubechies, I., A. Grossmann and Y. Meyer "Painless non-orthogonal expansions." *J. Math. Phys.,* 27 (1986) : 293-309.

Daubechies, I., S. Mallat and A. S. Willsky eds. *Special Issue on Wavelet Transforms and Multiresolution Signal Analysis, IEEE Trans. Information Theory,* 38, 2 (1992).

Desai, M. and D. J. Shazeer "Acoustic transient analysis using wavelet decomposition." *IEEE Conference on Neural Networks for Ocean Engineering*, (1991) : 29-40.

Donoho, David L. and Iain M. Johnstone "Ideal Spatial Adaptation by Wavelet Shrinkage." Preprint, Department of Statistics, Stanford University, (1992a).

Donoho, David L. "De-Noising by Soft-Thresholding." Preprint, Department of Statistics, Stanford University, (1992b).

Elias-Juarez, A. and J. C. Kantor "On the application of wavelets to model predictive control." *Proceedings American Control Conference,* Chicago, (1992) : 1582-1586.

Farge, M. et al. "Improved predictability of two-dimensional turbulent flows using wavelet packet compression." *Fluid Dynamics Research,* 10, 4-6 (1992) : 229-250.

Flandrin, P. "Wavelets and related time-scale transforms." *SPIE Advanced Signal-Processing Algorithms, Architectures, and Implementation,* 1348, (1990) : 2-13.

Forrester, B. D. "Use of the Wigner-Ville distribution in helicopter fault detection." *Proc. ASSPA 89,* (1989) : 78-82.

Gabor, D. "Theory of communication." *J. Inst. Elec. Eng.,* 93, (1946) : 429-441.

Ghosh, J., L. Deuser and S. D. Beck "A neural network based hybrid system for detection, characterization, and classification of short-duration oceanic signals." *IEEE Journal of Oceanic Engineering*, 17, 4 (1992) : 351-363.

Goupil, M. J., M. Auvergne and A. Baglin "Wavelet analysis of pulsating white dwarfs" *Astronomy and Astrophysics*, 250, 1 (1991) : 89-98.

Grossmann, A. and J. Morlet "Decomposition of hardy functions into square integrable wavelets of constant shape." *SIAM J. Math. Anal.,* 15, (1984) : 723-736.

Hjorth, P. G. et al. "Wavelet analysis of 'doublet quasar' flux data" *Astronomy and Astrophysics*, 255, 1-2 (1992) : 20-23.

Hlawatsch, F. and G. F. Boudreaux-Bartels "Linear and quadratic time-frequency signal representations." *IEEE Signal Processing Magazine,* (Apr. 1992) : 21-67.

Hopper, T. and F. Preston "Compression of grey-scale fingerprint images." *DCC '92. Data Compression Conference* , Storer, J. A. and Cohn, M. eds., (1992) : 309-318.

Huang, W. Y. and M. R. Solorzano "Wavelet preprocessing of acoustic signals." *Conference Record of the Twenty-Fifth Asilomar Conference on Signals, Systems and Computers,* 2, (1991) : 1114-1118.

Jawerth, B. and Wim Sweldens "An overview of wavelet based multiresolution analysis." Research report 1993:1, Department of Mathematics, University of South Carolina (1993).

Jones, D. L. and T. W. Parks "A high resolution data-adaptive time-frequency representation." *IEEE Trans. Acoust., Speech, Signal Processing,* 38, 12 (1990) : 2127-2135.

Khadra, L., M. Matalgah, B. El-Asir and S. Mawagdeh "The wavelet transform and its applications to photocardiogram signal analysis." *Med. Inform.,* 16, 3 (1991) : 271-277.

Kikuchi, T. and S. Sato "Experimental studies on ultrasonic measurements of scattering media by using wavelet transform." *Japanese Journal of Applied Physics, Supplement*, 31, 1 (1992) : 115-117.

Kikuchi, H. et al. "Fast wavelet transform and its application to detecting detonation." *IEICE Transactions on Fundamentals of Electronics, Communications and Computer Sciences*, E75-A, 8 (1992) : 980-987.

Lee, J. H., M. S. Gelormino, and M. Morari "Model predictive control of Multi-rate sampled-data systems: A state-space approach." *International J. of Control,* 55, 1 (1992) : 153-191.

Mallat, S. and S. Zhong "Signal characterization from multiscale edges." *NYU, Comput. Sci. Tech. Rep.,* (Nov. 1991) : 891-896.

Mallat, S. "Multiresolution approach to wavelets in computer vision." *Wavelets, Time-Frequency Methods and Phase Space* Combes, J. M. et al. eds. Springer-Verlag, (1989) : 313-327.

Mallat, S. "A theory for multiresolution signal decomposition: the wavelet representation." *IEEE Trans. Pattern Analysis and machine intelligence,* 11, (1989) : 674-693.

Mallat, S. and W. L. Hwang "Singularity detection and processing with wavelets." *IEEE Trans. Information Theory,* 38, 2 (1992) : 617-643.

Manjunath, B. S. and R. Chellappa "A unified approach to boundary perception: edges, textures, and illusory contours" *IEEE Transactions on Neural Networks,* 4, 1 (1993) : 96-108.

McAulay, A. D. and Jian Li "Wavelet data compression for neural network preprocessing." *Proceedings of the SPIE - The International Society for Optical Engineering,* 1699, (1992) : 356-365.

Meneveau, C. "Analysis of turbulence in the orthonormal wavelet representation." *Journal of Fluid Mechanics,* 232, (1991) : 469-520.

Morlet, D. et al., "Time-scale analysis of high-resolution signal-averaged surface ECG using wavelet transformation." *Proceedings. Computers in Cardiology,* (1991) : 393-396.

Morlet, J., G. Arens, I. Fourgeau, and D. Giard "Wave propagation and sampling theory." *Geophysics,* 47, (1982) : 203-236.

Page, C. H. " Instantaneous power spectra." *J. Appl. Phys.,* 23, (1952) : 103-106.

Pati, Y. C. and P. S. Krishnaprasad "Frames generated by subspace addition." Technical report, University of Maryland, Systems Research Center, (1992).

Rioul, O. and M. Vetterli "Wavelets and signal processing." *IEEE Signal Processing Magazine,* (Oct. 1991) : 14-38.

Ruskai, M. B. et al. eds. *Wavelets and their applications.* Jones and Barlett, Boston, 1992.

Slezak, E., A. Bijaoui and G. Mars "Identification of structures from galaxy counts: use of the wavelet transform." *Astronomy and Astrophysics,* 227, (1990) : 301-316.

Strang, G. "Wavelet transforms versus Fourier transforms." *Bulletin of the American mathematical society,* 28, 2 (1993) : 288-305.

Stromberg, J. O. "A modified franklin system and higher order spline systems on R^n as unconditional bases for hardy spaces." *Conf. in Honor of A. Zygmund,* Beckner, W et al. eds. II (1981) : 475-493.

Sweldens, Wim et al. eds. *Wavelet Digest* Department of mathematics, University of South Carolina, wavelet@math.scarolina.edu. 1992.

Titeur, F. B. "Wavelet transformations in signal detection." *Wavelets, Time-Frequency Methods and Phase Space* Combes, J. M. et al. eds. Springer-Verlag, (1989) : 133-139.

Wickerhauser, M. V. *Lectures on Wavelet Packet Algorithms* Department of Mathematics, Washington University in St. Louis, (1991).

Wickerhauser, M. V. "Acoustic signal compression with wavelet packets." *Wavelets -- A Tutorial in Theory and Applications* Chui, C. K. ed. Academic Press, Boston, (1992) : 679-700.

Wickerhauser, M. V. "Best-adapted wavelet packet bases." *Different Perspectives on Wavelets* Daubechies, I. ed. American Mathematical Society, San Antonio, Texas, (1993) : 155-171.

Wigner, E. P. "On the quantum correction for thermodynamic equilibrium." *Phys. Rev.,* 40, (1932) : 749-759.

Wornell, G. W. and A. V. Oppenheim "Estimation of fractal signal from noisy measurements using wavelets." *IEEE Trans. Signal Processing,* 40, 3 (1992) : 611-623.

Yamada, M. and K. Ohkitani "Orthonormal wavelet analysis of turbulence." *Fluid Dynamics Research,* 8, 1-4 (1991) : 101-115.

Yang, X., K. Wang and S. A. Shamma "Auditory representations of acoustic signals" *IEEE Transactions on Information Theory,* 38, 2, pt.2 (1992) : 824-839.

Young, Randy K. *Wavelet theory and its application.* Kluwer Academic Publishers, MA., 1993.

Zhang, Q. and A. Benveniste "Wavelet networks." *IEEE Trans. Neural Networks,* 3, 6 (1992) : 889-898.

Chapter 2

Computational Aspects of Wavelets and Wavelet Transforms

S. Palavajjhala, R.L. Motard, and B. Joseph
Washington University
Department of Chemical Engineering
One Brookings Drive, Campus Box 1198
St. Louis, MO 63130-4899, USA.

Abstract

This chapter is a tutorial on the computational aspects of wavelets and wavelet transforms. Algorithms for the construction of various types of wavelets are implemented on MATLAB. The implementation of the periodic discrete wavelet transform on MATLAB is explained. Various graphical representations of the wavelet coefficients/wavelet transform are illustrated. This chapter should be useful to understand some preliminary implementational aspects of wavelet analysis.

1. Introduction

The objective of this chapter is to provide an easy-to-read tutorial on the computational aspects of wavelets and wavelet transforms. This chapter is by no means a complete account of all aspects of the broad field of wavelet analysis. We try to explain various mathematical concepts using simple examples, and illustrate their implementation on MATLAB®[1] (version 4.0). Some programs included in this chapter do not use the best algorithm. We provide pointers to references that give better/efficient algorithms when appropriate.

---

[1]MATLAB is a trademark of The Math Works, Inc.

Wavelet transform of a function $f(t) \in L^2(R)$ (the space of absolutely square integrable functions) is defined as the projection of this function onto a set $\{\psi_{a,b}(t) = |a|^{-1/2}\psi((t-b)/a);\ a,\ b \in R\}$ of analyzing functions that are called the wavelets. The forward wavelet transform involves computation of the inner products $\langle f(t), \psi_{a,b}(t)\rangle$ for all $a,\ b \in R,\ a \neq 0$. The inner products are called the wavelet coefficients and are defined as:

$$\langle f(t), \psi_{a,b}(t)\rangle = |a|^{-1/2} \int_{-\infty}^{\infty} f(t)\psi\left(\frac{t-b}{a}\right) dt = (T_{a,b}^{wav} f(t))(a,b). \tag{1}$$

The function $f(t)$ is reconstructed from the wavelet coefficients as follows:

$$f(t) = c_\psi^{-1} \int_{-\infty}^{\infty}\int_{-\infty}^{\infty} \langle f(t), \psi_{a,b}(t)\rangle \psi_{a,b}(t) \frac{da\,db}{a^2}. \tag{2}$$

Where,

$$c_{\psi} = \int_{-\infty}^{\infty} \frac{|\hat{\psi}(\omega)|^2}{|\omega|} d\omega. \tag{3}$$

Section 2 describes different types of wavelets and their properties. In section 3, wavelet transforms are defined and classified. Discrete wavelet transform implementation using Mallat's fast algorithm (Mallat, 1989b) is explained in section 4. Different ways of displaying wavelet coefficients/wavelet transforms graphically are then presented. The discussion in this chapter is restricted to one-dimensional, real-valued wavelets and wavelet transforms only. Each section contains a brief outline of the theory, followed by a MATLAB program illustrating it. Figures/results obtained using these programs are also presented.

2. Wavelets - Classification and Properties

2.1 WAVELETS

Wavelets, or analyzing wavelets, are the building blocks of wavelet transforms, just as trigonometric functions of different frequencies are the building blocks used in Fourier transforms. Wavelets are generated by the dilation and the translation of a single prototype function called the *mother wavelet.* The mother wavelet is an absolutely integrable function and is denoted as $\psi(t)$. Thus, the family of functions $\{\psi_{a,b}(t)\}$, defined by

$$\psi_{a,b}(t) = |a|^{-1/2}\,\psi\left(\frac{t-b}{a}\right) \tag{4}$$

where $a, b \in R$ and $a \neq 0$, are wavelets. a is called the dilation parameter, and b the translation parameter. For the wavelets to be useful analyzing functions, the mother wavelet must possess certain desirable properties. From the inversion formula given in equation (2), the condition for a function $\psi(t)$, with Fourier transform $\hat{\psi}(\omega)$, to be a wavelet is

$$c_\psi = \int_{-\infty}^{\infty} \frac{|\hat{\psi}(\omega)|^2}{|\omega|} d\omega < \infty. \tag{5}$$

This condition is called the *"admissibility" condition* (Grossmann et al, 1985; see Daubechies (1992) for a derivation of equation (5)). If the wavelet $\psi(t)$ is an absolutely integrable function, as is usually the case, then its Fourier transform $\hat{\psi}(\omega)$ is continuous. If $\hat{\psi}(\omega)$ is continuous, c_ψ can be finite only if

$$\hat{\psi}(0) = 0 \Leftrightarrow \int_{-\infty}^{\infty} \psi(t)\,dt = 0. \tag{6}$$

A wavelet must therefore be an oscillatory function with zero mean. Equation (5) also suggests that $|\hat{\psi}(\omega)|^2$ decays at least as $1/|\omega|$. Therefore, the wavelet $\psi(t)$ has a decay of at least $1/|t|^{1-\varepsilon}$, for some $\varepsilon > 0$. In practice, much stricter decay conditions are imposed on the wavelet $\psi(t)$. In addition to the admissibility condition, there are other desirable properties for a function to be a useful wavelet.

- *Smoothness*: $\psi(t)$ should be a smooth function (n times differentiable) with continuous derivatives.
- *Good time localization*: $\psi(t)$, together with its derivatives, should decay rapidly.
- *Good frequency localization*: $\hat{\psi}(\omega)$ should decay sufficiently fast as frequency $\omega \to \infty$ and $\hat{\psi}(\omega)$ should be sufficiently flat near $\omega = 0$. The flatness at $\omega = 0$ is related to the number of vanishing moments of $\psi(t)$. The lth moment of a wavelet is defined as

$$\int_{-\infty}^{\infty} t^l \psi(t)\,dt. \tag{7a}$$

And, a wavelet is said to have M vanishing moments if

$$\int_{-\infty}^{\infty} t^l \psi(t)\, dt = 0 \tag{7b}$$

for l=0,1,...,M. Equation (7b) is equivalent to

$$\frac{d^l \hat{\psi}(\omega)}{d\omega^l}\Big|_{\omega=0} = 0 \qquad l = 0,1,\ldots,M. \tag{8}$$

Wavelets with larger number of vanishing moments results in more flatness when frequency ω is small.

These properties of wavelets, along with the admissibility condition, suggest that: (i) wavelets are band-pass filters (the frequency response decays sufficiently fast at large values of frequency ω, and is zero at ω=0), (ii) $\psi(t)$ is the impulse response of this filter (which again decays sufficiently fast as t increases), and it is a zero-mean oscillatory function. Usually, wavelets are assumed to be absolutely square integrable functions.

2.2 TIME-FREQUENCY LOCALIZATION

Associated with each wavelet there is a time localization and a frequency localization defined, respectively, by

$$\sigma_t^2 = \int_{-\infty}^{\infty} (t-\bar{t})^2 |\psi(t)|^2 dt \text{ and } \sigma_\omega^2 = \int_{-\infty}^{\infty} (\omega-\overline{\omega})^2 |\hat{\psi}(\omega)|^2 d\omega. \tag{9}$$

In equation (9)

$$\bar{t} = \int_{-\infty}^{\infty} t\, |\psi(t)|^2 dt \text{ and } \overline{\omega} = \int_{0}^{\infty} \omega\, |\hat{\psi}(\omega)|^2 d\omega. \tag{10}$$

The variance and the mean defined in equations (9) and (10) compute moments of $|\psi(t)|^2$ and $|\hat{\psi}(\omega)|^2$ (and not of $\psi(t)$ and $\hat{\psi}(\omega)$). This is because these quantities are like probability density functions in Quantum Mechanics. The wavelet $\psi(t)$ is centered around $(\bar{t},\overline{\omega})$ in the time-frequency plane and contains information from the time interval $(\bar{t}-\sigma_t, \bar{t}+\sigma_t)$ and the frequency interval $(\overline{\omega}-\sigma_\omega, \overline{\omega}+\sigma_\omega)$. The two intervals define a time-frequency cell (also called the Heisenberg's cell) whose area is restricted by the uncertainty principle, i.e., $\sigma_t \times \sigma_\omega \geq 1/2$. The time localization and the frequency localization remain unchanged with translation of the wavelet. However, with dilation, the time and the frequency localization change as follows:

$$\sigma_{a,t}^2 = a^2\sigma_t^2 \text{ and } \sigma_{a,\omega}^2 = \sigma_\omega^2 / a^2. \tag{11}$$

Thus, the wavelets have good time localization for small values of a. Small values of a corresponds to high frequency, or small scales. Scale can be defined as the inverse of frequency. For large values of a, the wavelets have good frequency localization.

2.3 CLASSIFICATION

Wavelets can broadly be classified into: *Continuous wavelets* and *Discrete wavelets.* The dilation parameter a and the translation parameter b vary continuously in the case of continuous wavelets (see equation (4)), but are restricted to a discrete lattice in the case of discrete wavelets. The parameters a and b are discretized as

$$a = a_0^m, \quad b = nb_0a_0^m \tag{12}$$

$m,n \in Z$. Typically, $a_0 = 2$ and $b_0 = 1$ is used, though a_0 and b_0 need not be restricted to these values. Equation (12) suggests that the translation parameter is proportional to the dilation parameter. For small (large) values of dilation parameter a or m, the wavelet $\psi(t)$ is concentrated and small (large) values of translation parameter b or n are necessary to cover an interval. Substituting a=2^m and b=$n2^m$ in equation (4)

$$\psi_{m,n}(t) = 2^{-m/2}\psi\left(2^{-m}t - n\right). \tag{13}$$

Discrete wavelets can further be classified into: *non-orthogonal, semi-orthogonal,* or *orthogonal wavelets.* Discrete wavelet that satisfy

$$A\|f(t)\|^2 \le \sum_{m,n} \left|\langle f(t), \psi_{m,n}(t)\rangle\right|^2 \le B\|f(t)\|^2 \quad (A > 0, B < \infty, f(t) \in L^2(R)) \tag{14}$$

are called *frames* (Daubechies, 1990). A and B are called the *frame bounds.* The ratio of the frame bounds B/A is a measure of redundancy present in the wavelet representation. Frames with $B/A \approx 1$ have little redundancy and they offer good reconstruction of $f(t)$ (defined in equation (2)) from the wavelet coefficients $\langle f(t),\psi_{m,n}(t)\rangle$. Such frames are called *tight frames.* Frames with B/A>>1 are linearly dependent and redundant. They are called *loose frames.* These frames offer reconstruction from wavelet coefficients that are robust to noise.

Non-orthogonal wavelets are linearly dependent and redundant frames. *Semi-orthogonal wavelets* (Chui, 1992) are frames that are complete in $L^2(R)$, and are orthogonal at different scales, i.e.,

$$\psi_{j,k} \perp \psi_{l,m} \tag{15}$$

for $j \neq l, j, k, l, m \in Z$. These wavelets are also orthogonal at the same scale ($j=l$) when their supports do not intersect (large $|k\text{-}m|$). *Orthogonal wavelets* are linearly independent, complete in $L^2(R)$, and orthogonal. These wavelets seize to be a frame if a single element is removed. Such frames are called *exact frames*. The frame bounds A and B for orthonormal wavelets are equal to 1, and the reconstruction of $f(t)$ from $\langle f(t), \psi_{m,n}(t) \rangle$ is exact.

Continuous and discrete wavelet can also be classified into *Real-Valued* and *Complex-Valued* wavelets. The real part and the complex part of a complex-valued wavelet satisfy the admissibility condition given in equation (6). These wavelets have been used to study complex-valued functions (Grossmann et al , 1985, Meyer, 1993). Frequency response of a dynamic system is an example of such a function. In this chapter, we restrict the discussion to real-valued wavelets and functions only.

2.4 PROPERTIES

Some important properties associated with wavelets are: smoothness, number of vanishing moments, symmetry, and time and frequency localization. *Smoothness* of a wavelet is determined by the number of existing derivatives of the wavelet. When approximating/analyzing a function, it is recommended that the smoothness of the wavelet and the function should match reasonably well. This is done so that a more compact representation is obtained. *Number of vanishing moments* is defined as the number of moments of the wavelet that are zeros (see equation (7b)). Larger number of vanishing moments is equivalent to the vanishing of the derivatives of the Fourier transform of the wavelet at the origin to a much higher order (see equation (8)). Wavelets with a larger number of vanishing moments tend to enhance the sparsity of certain matrix operators to a greater extent (Beylkin et al, 1991). Beylkin et al exploit this property of wavelets to develop a class of fast computing algorithms. *Symmetric* wavelets do not distort the phase information of a signal transform and are important in image analysis (Mallat, 1989b). *Compactly supported* wavelets are well localized (are non-zero over a finite interval only) in the time domain or in the frequency domain. This property of wavelet is particularly useful in estimating local Hölder regularity, and detecting local singularity in signals and images (Mallat and Hwang, 1992). This property is also responsible for the development of fast wavelet transform and reconstruction algorithms. Wavelets with different properties are available in the literature. However, in practice it is not possible to construct wavelets with an arbitrary combination of the above properties. For instance, a C^{∞} wavelet with compact support in both time and in frequency cannot be constructed. Similarly, it is not possible to construct a compactly supported, symmetric, and orthogonal wavelet (see Jawerth and Sweldens (1993) for more examples). The choice of wavelet in an application is usually made by matching the properties of the wavelets with those required by the application.

2.5 EXAMPLES

There are a number of wavelets with different properties available in the literature. The Haar wavelet (Haar, 1910), Daubechies' orthonormal wavelets of compact support (Daubechies, 1989), Daubechies' Coiflets (Beylkin et al, 1991; Daubechies, 1992), Chui-Wang's block-spline semi-orthogonal wavelets (Chui and Wang, 1992), Battle-Lemarie's orthonormal wavelets (Battle, 1987), Cohen et al's biorthogonal wavelets (Cohen et al, 1993), Shannon's wavelet, Meyer's wavelet, Mexican hat wavelet, Alpert's wavelets (Alpert, 1992), and Morlet's wavelet (Morlet et al, 1982) are some examples. We will not be able to illustrate properties of all these wavelets. We pick illustrative examples from the broad classification of wavelets described before. In particular, we illustrate the Mexican Hat wavelet (an example of non-orthogonal wavelet); Haar wavelet, Daubechies' compactly supported orthonormal wavelets, Battle-Lemarie's wavelet (examples of orthogonal wavelets); and Chui-Wang's semi-orthogonal wavelet.

2.6 SCALING FUNCTION

One way of constructing wavelets is by using a *scaling function* and *multiresolution analysis* (Mallat, 1989b). In multiresolution analysis, the space of square integrable functions $L^2(R)$ is decomposed into closed subspaces V_j with the following properties:

- $V_j \subset V_{j+1}$ - The coarser subspace is contained in a finer subspace.

- $\bigcap_{m\in Z} V_m = \{0\}$ - Separation Condition and

  $\bigcup_{m\in Z} V_m = L^2(R)$ - Condition for Completeness.
- $f(t)\in V_m \Leftrightarrow f(2t)\in V_{m-1}$ - Scaling property.
- There exists a scaling function $\phi(t)\in V_0$ such that

$$\phi_{m,n}(t) = 2^{-m/2}\phi\left(2^{-m}t-n\right) \tag{16}$$

  for $m,n\in Z$ satisfy equation (14) and are linearly independent. Such functions are also known as *Riesz basis*.

Since $\phi(t)\in V_0\subset V_1$, and $\phi(2t)$ is a basis for the subspace V_1, there exists a finite sequence $h(k)$ such that the scaling function $\phi(t)$ can be written as a linear combination of $\phi(2t)$. The scaling function therefore satisfies the two-scale difference equation

$$\phi(t) = \sum_k h(k)\phi(2t-k). \tag{17}$$

The scaling function $\phi(t)$ has the following property when $\int_{-\infty}^{\infty}\phi(t)\,dt = 1$ and $\sum_k h(k) = 1$:

  If $H(\omega) = \sum_k h(k)e^{-ik\omega}$, then $H(0)=1$ and $H(\pi)=0$, i.e., the frequency response of the scaling function is that of a low-pass filter and the sequence $\{h(k)\}$ is therefore known as the low-pass filter coefficients.

Let W_{m-1} be the orthogonal complement of V_{m-1} in V_m. Then V_m is the direct sum of V_{m-1} and W_{m-1}. Thus,

$$V_m = V_{m-1}\oplus W_{m-1}, \text{ and } V_{m-1}\perp W_{m-1}. \tag{18}$$

Substituting this recursive relationship in the completeness condition suggests that the direct sum of the infinite subspaces W_j is the whole of $L^2(R)$, i.e.,

$$\bigoplus_j W_j = L^2(R). \tag{19}$$

If $\psi(t)$ is the Riesz basis for the space W_0, which is a subspace of V_1, then the wavelet can be written as a linear combination of $\phi(2t)$. Therefore, the wavelet satisfies the equation

$$\psi(t) = \sum_k g(k)\phi(2t-k). \tag{20}$$

The finite sequence $\{g(k)\}$ is known as the band-pass filter coefficients, as the frequency response of the wavelet is like that of a band-pass filter. Equation (20) suggests that the wavelet can be constructed once the scaling function is constructed using equation (17). This is the idea used in the construction of wavelets via multiresolution analysis. We will see examples of such a construction later in this chapter.

2.7 DISCRETE NON-ORTHOGONAL WAVELETS AND FRAMES

In this section, we will illustrate the properties of the Mexican hat wavelet. Computation of frame bounds and their values for the Mexican hat wavelet can be found in Daubechies (1992, 1990).

The Mexican Hat Wavelet

The Mexican hat wavelet, defined as

$$\psi(t) = \frac{2}{\sqrt{3}}\pi^{-1/4}\left(1-t^2\right)e^{-t^2/2}, \tag{21}$$

is the second derivative of a Gaussian. This wavelet has two vanishing moments. Its Fourier transform is

$$\hat{\psi}(\omega) = \frac{2}{\sqrt{3}}\pi^{-1/4}\,\omega^2\, e^{-\omega^2/2}. \tag{22}$$

The Fourier transform of the translation of $\psi(t)$ by n is

$$e^{-in\omega}\hat{\psi}(\omega) \tag{23}$$

and that of the dilation by $2^{-m/2}$ is

$$2^{m/2}\hat{\psi}\left(2^m\omega\right). \tag{24}$$

Figure 1 shows the translated Mexican hat wavelet at two values of dilation parameter, m=0 and m=-1. The translated wavelets have a non-zero inner-product or overlapping area. Therefore, these wavelets are non-orthogonal. The peak frequency of the Mexican hat wavelet shifts to higher frequencies when the dilation parameter is decreased (see Figure 2). If a range of values for the dilation and translation parameters are used, and if the

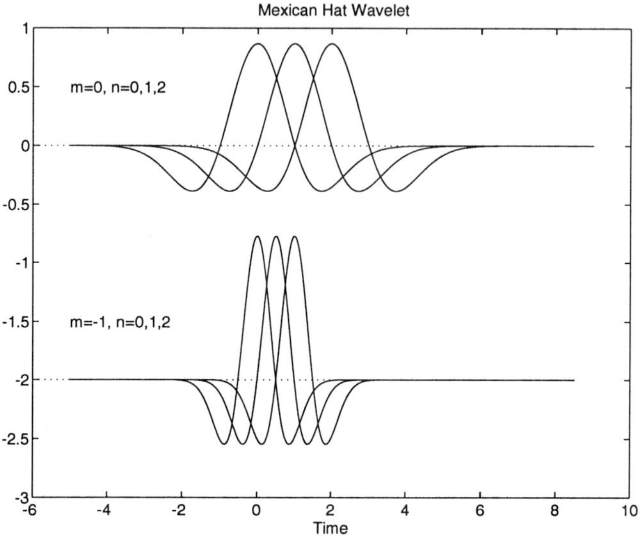

Figure 1 Mexican hat wavelet at different dilations and translations.

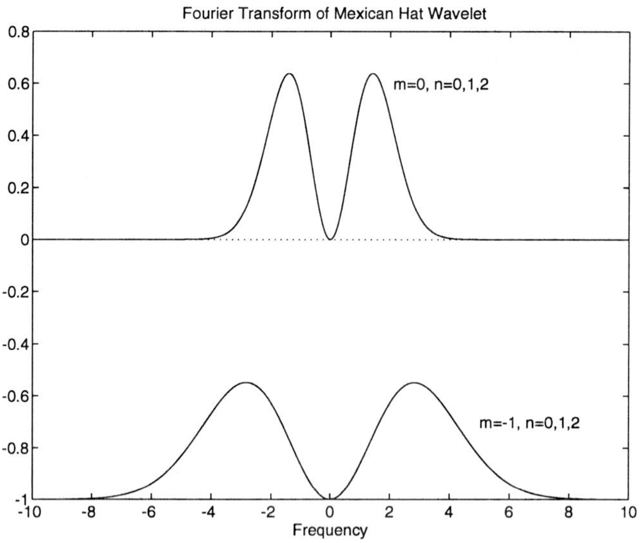

Figure 2 Fourier transform of the Mexican hat wavelet at different dilations.

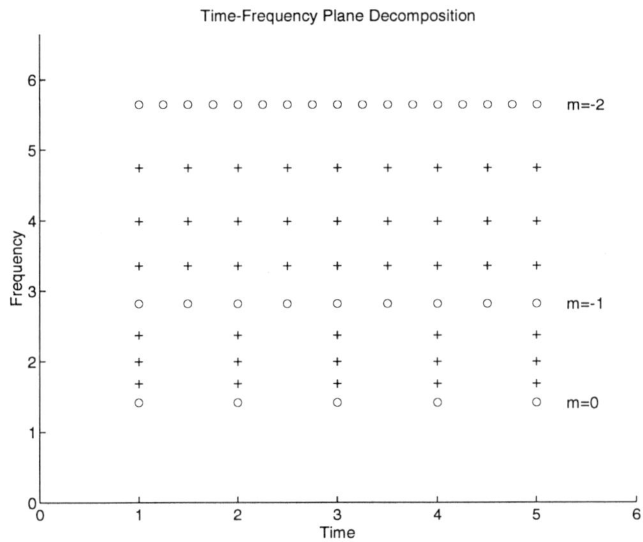

Figure 3 Time-Frequency plane decomposition obtained by dilation and translation of the Mexican hat wavelet.

location of the peak values of the wavelets in the time and the frequency domains are marked, then we obtain a time-frequency lattice. Figure 3 shows such a time-frequency lattice generated using MATLAB function *fxt*. For integer values of dilation, the frequency localization (also called an octave - denoted by 'o' in Figure 3) varies logarithmically. Each such octave can be further decomposed into *multiple-voices* according to the following equation:

$$\psi^{v}(t) = 2^{-(v-1)/N}\psi\left(2^{-(v-1)/N}t\right). \tag{25}$$

In Figure 3, N=4, i.e. 4 voices per octave, is used. Voices are denoted by '+' in Figure 3. Each octave (along the frequency axis) is partitioned into 4 logarithmically spaced intervals, while the intervals along the time axis remain unchanged. The number of voices per octave should be selected depending on the redundancy required in a representation.

2.8 DISCRETE ORTHOGONAL WAVELETS

Haar Wavelet

The Haar wavelets, defined by:

$$\psi(t)=\begin{cases}1 & 0\le t<1/2\\ -1 & 1/2\le t<1\\ 0 & otherwise\end{cases} \tag{26}$$

and equation (13), is an example of orthonormal wavelets. The Haar scaling function is defined by

$$\phi(t)=\begin{cases}1 & 0\le t<1\\ \\ 0 & otherwise\end{cases}. \tag{27}$$

This wavelet has compact support in time, however, it has just 1 vanishing moment and is piecewise constant. Therefore, the Fourier transform of this wavelet does not have compact support. The Haar wavelet is a simple example that can be used to understand many aspects of wavelet analysis. Since these functions are not continuous, they have little practical use. The influence of varying the dilation and the translation parameters is illustrated in Figure 4. Decreasing the dilation parameter m results in contraction of the wavelet (scaling function), and increasing this parameter results in dilating the wavelet (scaling function). Increasing the translation parameter n shifts the wavelet (scaling function) along the time axis. The projection of the wavelets at different translation and dilation is zero. Hence, these wavelets are orthogonal. Furthermore, the energy of the wavelet (defined as the integral of the wavelet squared) is equal to unity; hence, the wavelets are normal.

Daubechies Orthonormal Wavelets of Compact Support

Construction of Daubechies' Wavelets

One of the most widely used wavelets are those constructed by Daubechies (1989). These wavelets are orthonormal, compactly supported, they have maximum number of vanishing moments for the support, and are reasonably smooth. Daubechies (1989) develops the algorithm for the construction of these wavelets using Mallat's multiresolution analysis, and

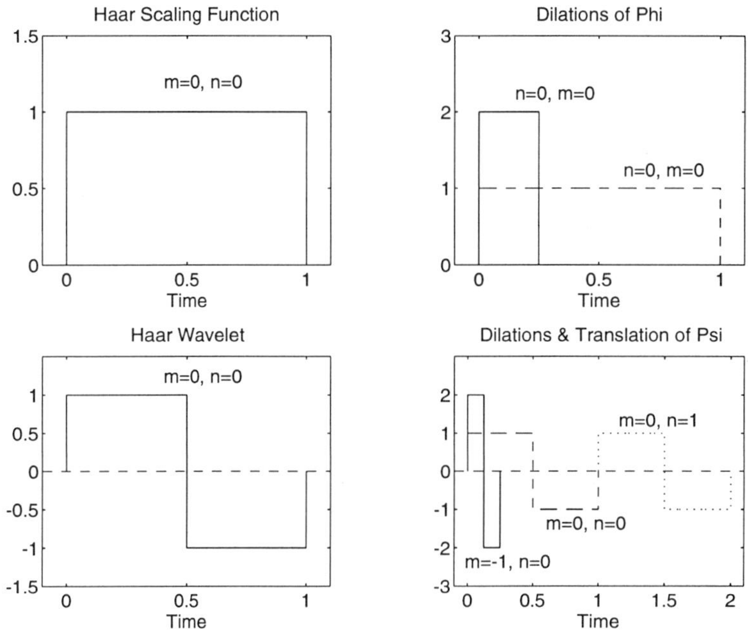

Figure 4 Dilation and translation of the Haar scaling function and the Haar wavelet.

estimates their support and smoothness. Strichartz (1992) gives a comprehensive summary of the construction of Daubechies' orthonormal wavelet bases. Daubechies' orthonormal wavelets are presented in the form of low-pass filter coefficients. There is no other know compact representation for these wavelets. The low-pass and the band-pass filter coefficients (defined in equations (17) and (20)) satisfy the following conditions:

$$\textit{Orthogonality}: \sum_n h(n-2k)h(n-2l) = \delta_{kl}. \tag{28}$$

$$\textit{Normality}: \sum_n h(n) = 2^{1/2}. \tag{29}$$

Regularity : Suppose $m_0(\omega)=2^{-1/2}\sum_n h(n)e^{in\omega}$ then

$$m_0(\omega) = \left[\frac{1}{2}(1+e^{i\omega})\right]^N \left[\sum_n f(n)e^{in\omega}\right]. \tag{30}$$

In equation (30), $\sum_n |f(n)||n|^{\varepsilon} < \infty$ for some ε>0, and
$\sup_{\omega \in R}\left|\sum_n f(n)e^{in\omega}\right| < 2^{N-1}$.

The integer N>1 defines the support as well as the Hölder regularity of the scaling function and the wavelet. A function is said to be Hölder regular and belongs to C^r if there exists a constant C such that for all t and all $h>0$

$$|f(t+h) - f(t)| \le Ch^r. \tag{31}$$

r is called the Hölder exponent. Daubechies scaling function and wavelet belong to C^r, where $r=r(N)$ and $\lim_{N\to\infty} N^{-1}r(N) = \gamma > 0$. The value of γ for these wavelets is approximately 1/5, i.e., if the wavelet is to have m continuous derivatives, then the length of its support must be 5*m (Strichartz, 1992; Meyer, 1993).

The scaling function and the wavelet satisfy equations (17) and (20). Furthermore,

$$g_N(n) = (-1)^n h_N(1-n). \tag{32}$$

We now describe the steps involved in the construction of Daubechies' wavelets. Details of the algorithm can be found in Daubechies (1989, 1992) and Strichartz (1992).

Step 1 : Construct a polynomial of order N-1

$$P_N(y) = \sum_{n=0}^{N-1} p_n y^n \tag{33}$$

where, $p_0 = 1$, and $p_k = \binom{N+k-1}{k}$, k=0, 1,2 ,..., N-1.

Step 2 : Construct a polynomial Q with minimal order so that

$$\left|Q(e^{i\omega})\right|^2 = P_N(\sin^2(\omega/2)). \tag{34}$$

Step 3 : The polynomial Q is extracted from $|Q(e^{i\omega})|^2$ by retaining all the zeros inside the unit circle. This corresponds to a minimal phase filter design.

Thus,

$$Q_N(e^{i\omega}) = \sum_{n=0}^{N-1} q_N(n) e^{in\omega} . \tag{35}$$

Step 4 : m_0 is obtained using

$$_N m_0 = \left[\frac{1}{2}(1+e^{i\omega})\right]^N \sum_{n=0}^{N-1} q_N(n) e^{in\omega} , \tag{36}$$

and the low-pass filter coefficients are obtained by equating the above equation to

$$_N m_0 = 2^{-1/2} \sum_{n=0}^{2N-1} h_N(n) e^{in\omega} . \tag{37}$$

Therefore,

$$h_N(n) = 2^{1/2}\left[\frac{1}{2}(1+e^{i\omega})\right]^N * q_N , \quad n = 1,2,\cdots,2*N-1 \tag{38}$$

where, $*$ denotes convolution.

Step 5 : $g_N(n)$ is computed using equation (32).

The MATLAB function *daubonw* implements the above procedure. The input argument to the function is N. The output arguments are the filter coefficients h and g.

Daubechies' Cascade Algorithm

The cascade algorithm is a recipe for constructing the scaling function from the filter coefficients. The cascade algorithm assuming piecewise constant interpolation is described here.

Step 1 : We start with the sequence η_0=[1 0 0 0 0 ... 0]. Set j=1.

Step 2 : η_j is computed by the convolution of the sequence η_{j-1} with the low-pass filter coefficients $h(n)$:

$$\eta_j(i) = \sum_l h(i-2*l+2)\,\eta_{j-1}(l) \quad l = 1,2,\cdots 2*N*2^{j-1} . \tag{39}$$

Or, the even points can be computed by the convolution of η_{j-1} with the even filter coefficients:

$$\eta_j(2^{-j}2k) = \sum_l h(2(k-l))\,\eta_{j-1}(2^{-j}l), \tag{40}$$

and the odd points are computed by the convolution of

η_{j-1} with the odd filter coefficients:

$$\eta_j\left(2^{-j}(2k+1)\right)=\sum_l h(2(k-l)+1)\eta_{j-1}\left(2^{-j}l\right). \tag{41}$$

If the size of the low-pass filter coefficients $h(n)$ is $2*N$, and if

$$N_{j,-}=\frac{1}{2}\left[N_{j-1,-}+N_-\right] \qquad N_{j,+}=\frac{1}{2}\left[N_{j-1,+}+N_+\right], \tag{42}$$

then the support of $\eta_j \subset [N_{j,-}, N_{j,+}]$, where, $h(n)=0$ for $n<N_-$ and $n>N_+$.

Step 3 : j=j+1. Repeat Step 2.

Usually, eight iterations yield a reasonably smooth function η_8. This function is a good approximation to the scaling function $\phi(t)$. Once $\phi(t)$ is known, the wavelet $\psi(t)$ can be constructed using equation (20). The support of the wavelet

$$supp\ \psi(t) \subset \left[\frac{1}{2}(1-N_+-N_-),\frac{1}{2}(1+N_+-N_-)\right]. \tag{43}$$

The MATLAB function *cascade* implements this algorithm. The input argument to this function is the low-pass filter coefficients. The output arguments are the scaling function and the wavelet. When Daubechies low-pass filter coefficients with N=1 are used as the input argument, the Haar scaling function and wavelet graphed in Figure 4 is obtained. The scaling function and the wavelet obtained using Daubechies low-pass filter coefficients with N=2 are shown in Figures 5 and 6. Both the scaling function and the wavelet are smoother than those in the Haar case. However, smoothness is obtained at the cost of the support of these functions. As the value of N is increased, smoother scaling functions and wavelets are obtained, and their support increases.

Fourier Transform of Daubechies Scaling Functions and Wavelets

The algorithm described in Daubechies (1989, 1992) is used to compute the Fourier transform of the scaling function from the low-pass filter coefficients. This, along with the band-pass filter coefficients, are then used to compute the Fourier transform of the wavelets. The algorithm is summarized next.

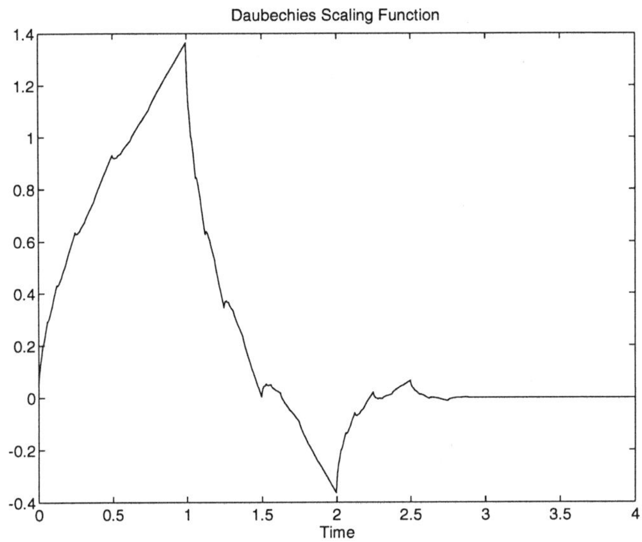

Figure 5 Daubechies' scaling function (length 4) with extremal phase and highest number of vanishing moments compatible with their support width.

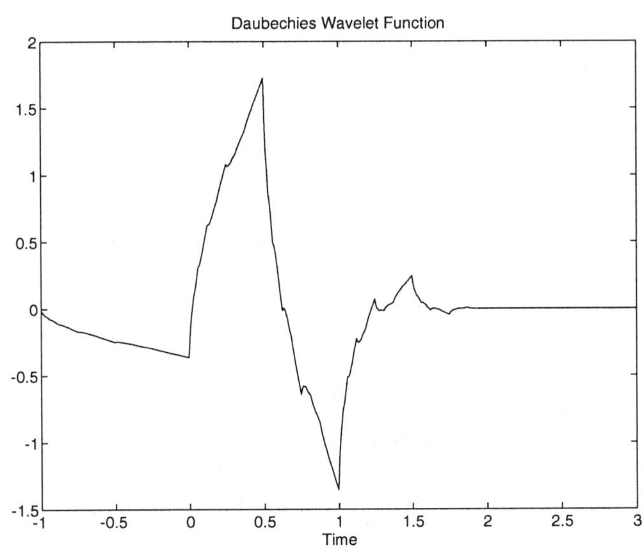

Figure 6 Daubechies' wavelet (length 4) with extremal phase and highest number of vanishing moments compatible with their support width.

Step 1 : The Fourier transform of the scaling function is computed using the recursion formula

$$\hat{\phi}(\omega) = 2^{-1/2} \sum_{n} h(n) e^{in\omega/2} \phi(\omega/2). \tag{44}$$

Since,

$$m_0(\omega) = 2^{-1/2} \sum_{n} h(n) e^{in\omega} \tag{45}$$

we have

$$\hat{\phi}(\omega) = \left[\prod_{j=1}^{\infty} m_0\left(2^{-j}\omega\right) \right] \hat{\phi}(0). \tag{46}$$

Step 2 : The Fourier transform of the wavelet is evaluated using

$$\left|\hat{\psi}(\omega)\right| = 2^{-1/2} \left| \sum_{n} g(n) e^{in\omega/2} \right| \left|\hat{\phi}(\omega/2)\right|. \tag{47}$$

The MATLAB function *ftscale* uses the procedure described to compute the Fourier transform of Daubechies' scaling function and wavelet. Figures 7 and 8 graph the Fourier transform of Daubechies' scaling function and wavelet using N=2. The scaling function and the wavelet have a frequency response that is not compactly supported, but has fast decay. However, the corresponding functions in the time domain have compact support (see Figures 5 and 6). As N is increased, the size of the filter increases, the support in the time domain increases, and the decay in the frequency domain is faster.

Battle-Lemarie Wavelets

Battle (1987) constructs orthonormal wavelet bases with exponential decay using block-splines (B-splines). The wavelet obtained is a polynomial of degree N-2, and has N vanishing moments. Here we illustrate the construction of this wavelet using linear B-splines. The scaling function is given by

$$\phi(t) = \begin{cases} 1-|t|, & 0 \le |t| \le 1 \\ 0 & otherwise \end{cases} \tag{48}$$

and has compact support. Fourier transform of the normalized scaling function (see Daubechies, 1992; Chui, 1992a) is

$$\hat{\phi}_1(\omega) = \sqrt{3}(2\pi)^{-1/2} \frac{4\sin^2\omega/2}{\omega^2[1+2\cos^2\omega/2]^{1/2}}. \tag{49}$$

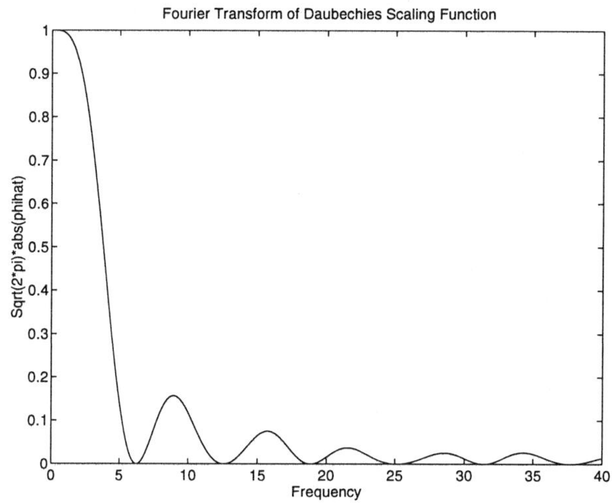

Figure 7 Fourier transform of Daubechies' scaling function with N=2 (filter size 4).

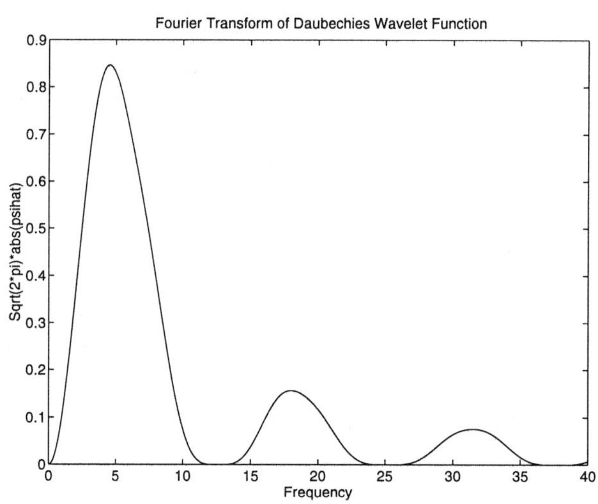

Figure 8 Fourier transform of Daubechies' wavelet with N=2 (filter size 4).

Equation (49) can be considered as the Fourier transform of the convolution of the sequence $\{c_n\}$ with the scaling function $\phi(t)$ (where, $\{c_n\}$ is the inverse Fourier transform of $[1+2\cos^2\omega/2]^{-1/2}$). Thus,

$$\phi_1(t) = \sqrt{3}\sum_n c_n\, \phi(t-n). \tag{50}$$

The Fourier transform of the wavelet is (Daubechies, 1992)

$$\hat{\psi}(\omega) = \sqrt{3}\, e^{i\omega/2} \sin^2 \omega/4 \left[\frac{1+2\sin^2\omega/4}{(1+2\cos^2\omega/2)(1+2\cos^2\omega/4)} \right]^{1/2} \hat{\phi}(\omega/2). \tag{51}$$

Then,

$$\psi(t) = \frac{\sqrt{3}}{2} \sum_n (d_{n+1} - 2d_n + d_{n-1})\phi(2t-n) \tag{52}$$

and d_n is the inverse Fourier transform of

$$\left[(1-\sin^2\omega/4)(1+\cos^2\omega/2)^{-1}(1+\cos^2\omega/4)^{-1}\right]^{1/2}. \tag{53}$$

The wavelet thus obtained (see Figure 10) does not have compact support in time, though it decays sufficiently fast. It will be shown later in this section that Chui-Wang construct compactly supported wavelets using B-splines but sacrifice orthogonality. The script files *batlem.m* and *batwave.m* implement the algorithm described to construct Battle-Lemarie wavelet using linear splines. Figures 9 and 10 graph Battle-Lemarie's scaling function and wavelet respectively.

2.9 DISCRETE SEMI-ORTHOGONAL WAVELETS

Semi-orthogonal wavelets, or pre-ondelettes, are a family of wavelets which are orthogonal at different scales and for wavelets with non-overlapping support at the same scale. Chui and Wang (1991) and Micchelli (1991) independently introduced construction of semi-orthogonal wavelets. Here we shall illustrate Chui-Wang's semi-orthogonal wavelets.

Chui and Wang use B-splines to construct wavelets. The difference between Battle's and Chui-Wang's approaches lies in orthogonalization of the scaling function. Battle orthogonalizes the B-spline and uses these scaling functions to construct orthogonal wavelets. While, Chui and Wang use the B-spline wavelets without orthogonalization to construct the semi-

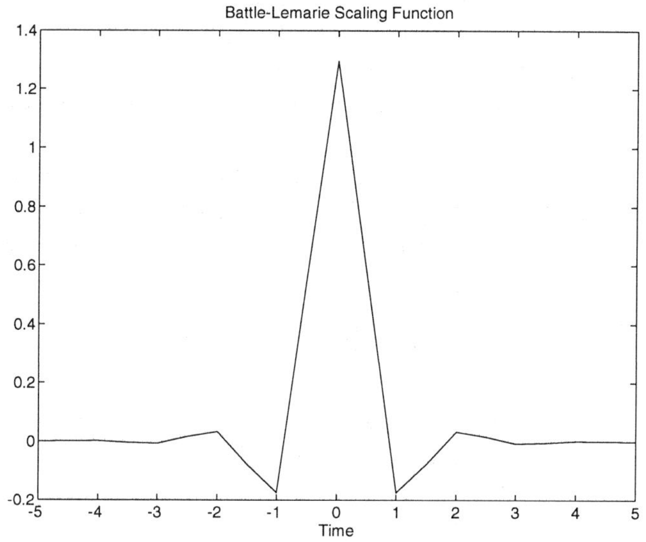

Figure 9 Battle's orthogonal scaling function constructed using linear splines.

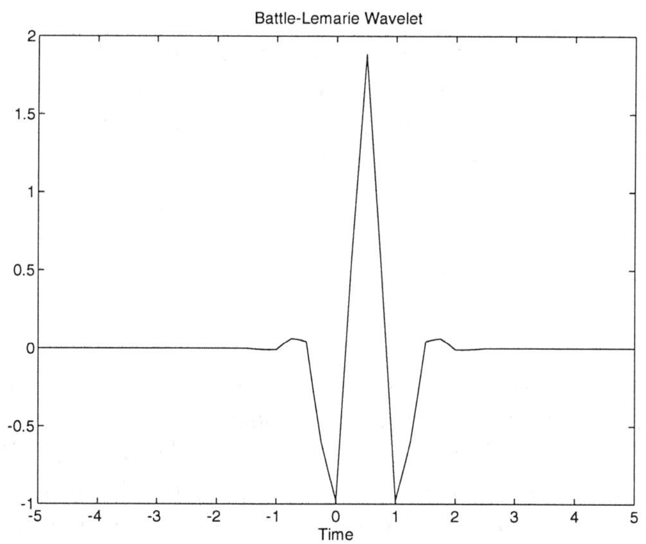

Figure 10 Battle's orthogonal wavelet constructed using linear splines.

orthogonal B-spline wavelets. These wavelets have great computational value since the wavelets constructed have more convenient properties (like compact representation) as well as explicit position-scale representation. Construction of Chui-Wang's wavelet using linear splines is illustrated next.

Step 1 : The scaling function used is the cardinal B-spline of order 2 (or linear splines). This function is denoted by N_2, and can be constructed by two fold convolution of $N_1=\chi_{[0,1)}$. The scaling function satisfies

$$N_2(t) = \sum_{j=0}^{2} 2^{-1} \binom{2}{j} N_2(2t-j). \tag{54}$$

Step 2 : The B-spline wavelet is constructed from the scaling function using the relationship

$$\psi_2(t) = \sum_{j=0}^{4} q_j N_2(2t-j). \tag{55}$$

Where, $q_j = (-1)^j 2^{-1} \sum_{l=0}^{2} \binom{2}{l} N_4(j-l+1)$.

The support of the wavelet so constructed is [0,3].

The script file *bspline.m* implements Chui-Wang's wavelet. Figures 11 and 12 graph Chui-Wang's linear spline scaling function and wavelet respectively. The MATLAB script file *bspline.m* can be extended to construct wavelets using higher order splines.

3. Wavelet Transformation

Wavelet Transformation (WT) is a tool that cuts data, functions, or operators, into different frequency components, and then studies each component with a resolution matched to its scale (Daubechies, 1992). To perform this transformation a set of functions $\{\psi_{a,b}(t)=|a|^{-1/2}\psi((t-b)/a);\ a, b\in R\}$ called *wavelets* are used. Wavelets are generated by the dilation and translation of a single prototype function. The prototype function is called the *mother wavelet*, and is denoted by $\psi(t)$. *Continuous Wavelet Transform* (CWT) of a function $f(t)\in L^2(R)$ involves computation of the inner product

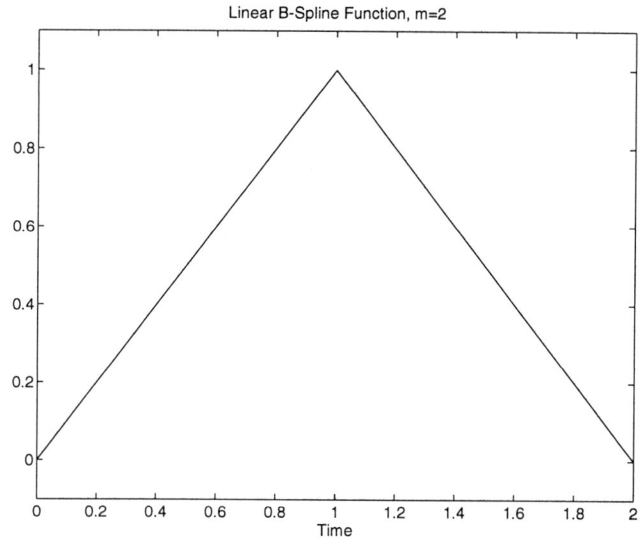

Figure 11 Chui-Wang's B-Spline scaling function constructed using linear splines.

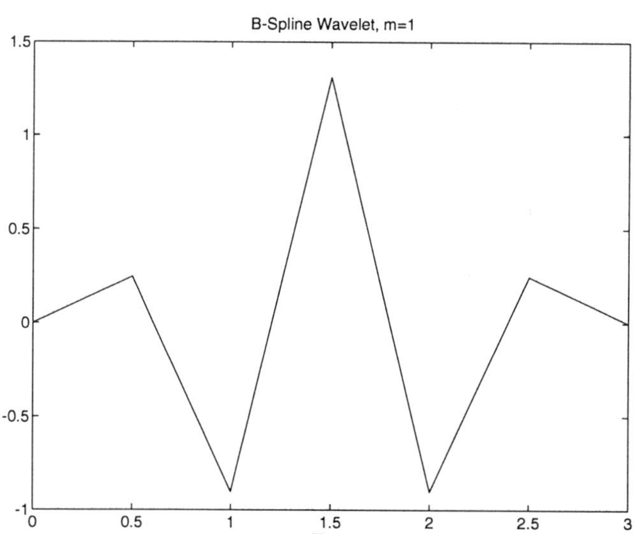

Figure 12 Chui-Wang's B-Spline Wavelet constructed using linear splines.

$\langle f(t),\psi_{a,b}(t)\rangle$ for $a,b\in R$, $a\neq 0$. The inner products for various values of parameters a and b are called the *Continuous Wavelet Coefficients* and are defined as:

$$\left\langle f(t),\psi_{a,b}(t)\right\rangle = (\mathbf{T}f(t))(a,b) = |a|^{-1/2}\int_{-\infty}^{\infty} f(t)\psi((t-b)/a)\,dt. \tag{56}$$

Discrete Wavelet Transformation (DWT) of the function $f(t)$ involves computation of the above inner product at discrete values of dilation and translation parameters. These inner products are called the *Discrete Wavelet Coefficients* and are defined as:

$$\left\langle f(t),\psi_{m,n}(t)\right\rangle = \mathbf{T}_{m,n}(f(t)) = 2^{-m/2}\int_{-\infty}^{\infty} f(t)\psi\left(2^{-m}t-n\right)dt. \tag{57}$$

Equations (56) and (57) suggest that like Fourier transforms, where a function is projected onto trigonometric functions with different frequencies, the WT projects a function onto analyzing functions $\psi_{a,b}(t)$, or $\psi_{m,n}(t)$, with different dilation and translation parameters. The WT maps a time domain function onto the two-dimensional time-scale domain.

DWT can be classified, based on whether the function $f(t)$ is continuous or discrete in time, into: the *Wavelet Series Transform (WST)* and the *Discrete Time Wavelet Transform (DTWT)*. Another classification of DWT is based on the properties of wavelets used for the transformation. We use orthogonal wavelet bases here and we discuss only this class of DWT. For other DWT algorithms we refer to the papers by Shensa (1992) and Rioul and Duhamel (1992). These papers discuss and compare various discrete and continuous wavelet transform algorithms. For the orthogonal case, the reconstruction formula for the function $f(t)\in L^2(R)$ from the discrete wavelet coefficients is

$$f(t) = \sum_{m,n\in\mathbf{Z}} \psi_{m,n}(t)\left\langle f(t),\psi_{m,n}(t)\right\rangle. \tag{58}$$

The parameters m and n vary between $[-\infty,\infty]$. However, in practice signals are measured at a finite resolution, and therefore, the value of m should be restricted between $[0,L]$. $m=0$ corresponds to the finest scale and $m=L$ corresponds to the coarsest scale. Substituting this in equation (58)

$$f(t) = \sum_{m\in[1,L],n\in Z} \psi_{m,n}(t)\left\langle f(t),\psi_{m,n}(t)\right\rangle + \phi_{L,n}(t)\left\langle f(t),\phi_{L,n}(t)\right\rangle. \tag{59}$$

For scale m and translation n, the wavelet $\psi_{m,n}(t)$ and the scaling function $\phi_{m,n}(t)$ are defined in equations (13) and (16). Equation (59) contains two right hand side terms. The first term is the projection of $f(t)$ on wavelets at different scales or resolutions. The second term is the projection of the function on the coarsest scaling function. Thus, DWT can be interpreted as a set of signals resulting from smoothing of a function $f(t)$ from scale m=1 to scale m=L. The term *level* is sometimes used to refer to a scale in this decomposition.

3.1 MALLAT'S FAST WAVELET ALGORITHM

Consider discrete samples of a function $f(t) \in L^2(R)$. Let us set the scale m of this sampled function equal to 0. A smoother version (scale m=1) of this function is obtained by projecting it on the scaling functions $\phi_{1,n}(t)$ for all $n \in Z$ (see equation (59)). The information lost in obtaining the smooth version is captured by the projections on the wavelets $\psi_{1,n}(t)$ for all $n \in Z$. This concept of writing a sampled function as a limit of successive approximations, each of which is a smoothed version, leads to the development of a fast wavelet algorithm (Mallat, 1989). This algorithm is applicable to both orthogonal and non-orthogonal wavelet transforms. Here we explain the algorithm for the orthogonal case.

Let the data sequence c_0 denote the projection of a continuous function $f(t)$ on the scaling functions $\phi_{0,n}(t)$, $n \in Z$. Here we assume the sequence c_0 to be the equally spaced samples of the function $f(t)$. More accurate methods for estimating the sequence c_0 are given in Rioul and Duhamel (1992). Define the projection $\mathbf{P_m}$ on the scaling functions at scale m as:

$$\mathbf{P_m} f(t) = \sum_n c_{m,n} \phi_{m,n}(t) \tag{60}$$

where, $c_{m,n} = \langle f(t), \phi_{m,n}(t) \rangle = \sum_k h(k-2n) c_{m-1,k}$ and

$$h(n) = 2^{-1/2} \int_{-\infty}^{\infty} \phi(t/2) \phi(t-n)\, dt.$$

Thus, the sequence c_m, which is the projection of the function $f(t)$ on the scaling functions $\phi_{m,n}(t)$, can be computed using

$$c_m = \mathbf{H} c_{m-1}. \tag{61}$$

Operator **H** can be interpreted as convolution of the sequence c_{m-1} with the filter coefficients $h(n)$ followed by decimation (dropping even terms). Further, define the projection $\mathbf{Q_m}$ on the wavelets as:

$$\mathbf{Q_m} f(t) = \sum_n d_{m,n} \psi_{m,n}(t) \tag{62}$$

where, $d_{m,n} = \langle f(t), \psi_{m,n}(t) \rangle = \sum_k g(k-2n) c_{m-1,k}$ and

$$g(n) = 2^{-1/2} \int_{-\infty}^{\infty} \psi(t/2) \phi(t-n)\, dt.$$

The sequence d_m, which is the projection of the function $f(t)$ on the wavelets $\psi_{m,n}(t)$, can be computed using

$$d_m = \mathbf{G} c_{m-1}. \tag{63}$$

Operator **G** can be interpreted as convolution of the sequence c_{m-1} with the filter coefficients $g(n)$ followed by decimation. The wavelets and the scaling functions at a given scale m form orthonormal basis for the subspaces V_m and W_m respectively. Using equation (18), the following recursive decomposition algorithm is obtained:

$$\mathbf{P_{j-1}} f = \mathbf{P_j} f + \mathbf{Q_j} f \tag{64}$$

$$c_j = \mathbf{H} c_{j-1} \text{ and } d_j = \mathbf{G} c_{j-1}. \tag{65}$$

Figure 13 (a) depicts two stages in the decomposition algorithm. The projections on the scaling function are recursively decomposed to obtained coarser approximations. The projections on the wavelets at each scale contain the finer details.

The reconstruction algorithm is similar to the decomposition algorithm, except the steps are now reversed:

$$\mathbf{P_{j-1}} f = \mathbf{P_j} f + \mathbf{Q_j} f \tag{66}$$

$$c_{j-1} = \mathbf{H}^* c_j + \mathbf{G}^* d_j. \tag{67}$$

Figure 13 (b) depicts the reconstruction from the two stage decomposition shown in Figure 13 (a). Finer details captured by the wavelets are added to the coarser approximations to reconstruct the function. $\mathbf{H}^*$ and $\mathbf{G}^*$ are the adjoint of operators **H** and **G** respectively.

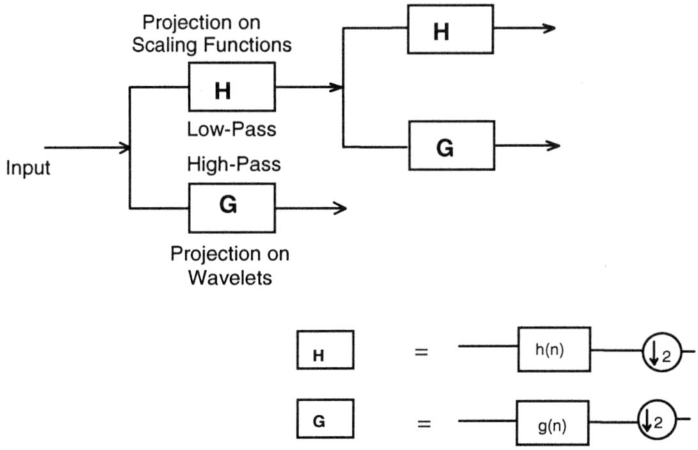

(↓ 2 mean downsample by 2 (drop even terms))

Figure 13 (a) Block diagram of discrete wavelet decomposition.

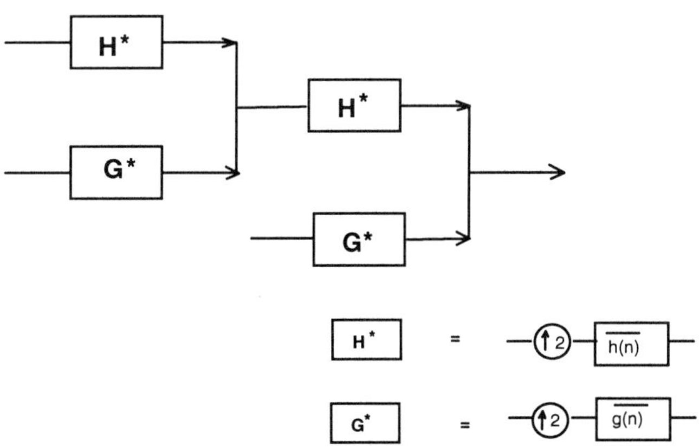

($\overline{h(n)}$ is the complex conjugate of h(n); ↑ 2 means upsample by 2 (pad even terms with 0).)

Figure 13(b) Block diagram of signal reconstruction.

Figure 13 Discrete wavelet decomposition and reconstruction.

Operators **H** and **G** used in the wavelet decomposition are low-pass and band-pass filters respectively. Filter coefficients for these filters, $h(n)$ and $g(n)$ respectively, with different properties -- like orthogonality, symmetry, smoothness, support in time domain, support in frequency domain, number of vanishing moments -- were discussed in section 2 of this chapter.

3.2 WAVELET PACKET ALGORITHM

Coifman and Wickerhauser (1992) introduce a general class of orthonormal bases by constructing a *library* of modulated waveforms out of which various bases can be extracted. Further, they also show that this library of orthonormal bases is a binary tree, and they develop a *fast* algorithm to search for the *best* basis (best in the sense of minimum information cost) from this library. The library of bases is called the *Wavelet Packet* and the decomposition of a signal onto these bases is called the *Wavelet Packet Decomposition.*

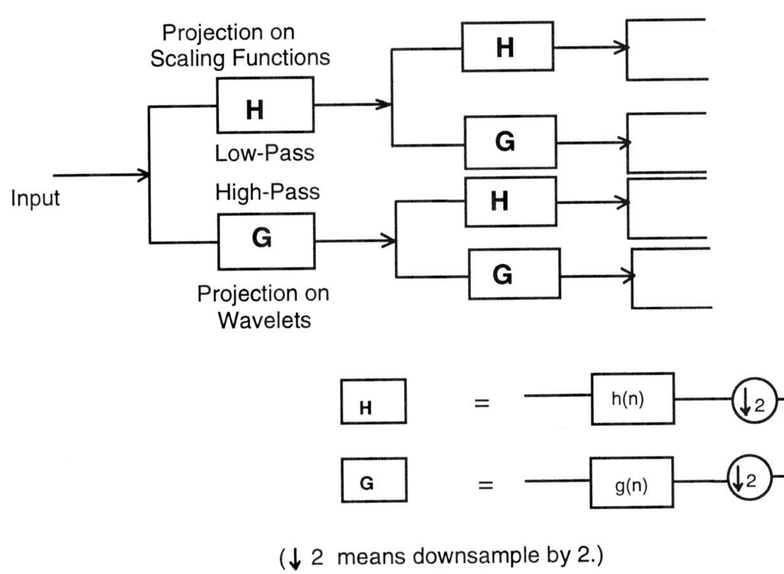

Figure 14 Wavelet Packet Decomposition.

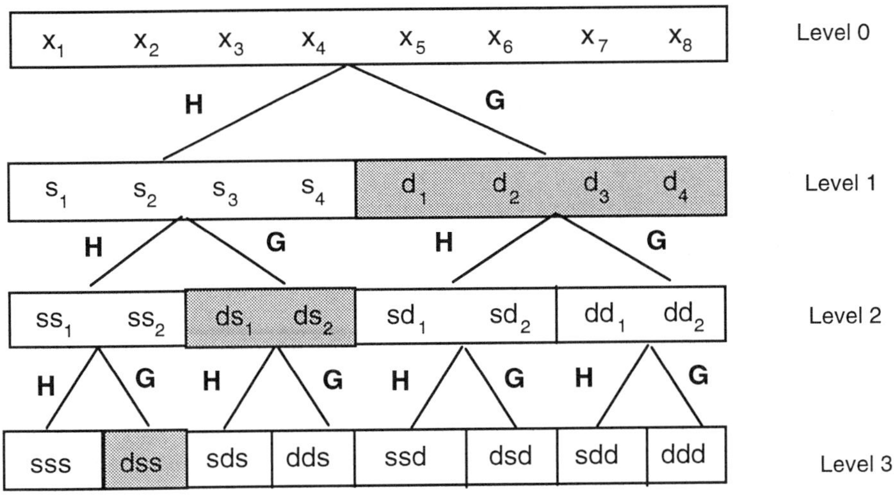

Figure 15 Wavelet packet decomposition of a sequence. (Wickerhauser, 1991)

The idea behind wavelet packet decomposition is illustrated in Figure 14. The projections of a function on the scaling functions and the wavelets are recursively decomposed to obtain a binary tree.

Consider the wavelet packet decomposition of a function sampled at 8 equally spaced points. Let the sampled function be represented by the sequence $\mathbf{x}=\{x_1, x_2, \ldots, x_8\}$. The wavelet packet decomposition of this sequence is obtained by applying filters **H** and **G** recursively to form a binary tree shown in Figure 15. The root of the tree is a cell consisting of the original sequence **x**. The filters **H** and **G** are applied recursively on the parent cell to yield two child cells.

In Figure 15, the left child cell $\mathbf{s}=\{s_1, s_2, s_3, s_4\}$ and the right child cell $\mathbf{d}=\{d_1, d_2, d_3, d_4\}$ at Level 1 are obtained by applying the filters **H** and **G** respectively to the sequence **x**. The sequence **x** is assumed to be periodic. Its period is assumed to be equal to the length of the signal. The sequences **s** and **d** for real valued filters $h=\{h_1\ h_2\ h_3\ h_4\}$ and $g=\{g_1\ g_2\ g_3\ g_4\}$ are computed as follows:

	x_1	x_2	x_3	x_4	x_5	x_6	x_7	x_8
s_1	h_1	h_2	h_3	h_4	0	0	0	0
s_2	0	0	h_1	h_2	h_3	h_4	0	0
s_3	0	0	0	0	h_1	h_2	h_3	h_4
s_4	h_3	h_4	0	0	0	0	h_1	h_2
d_1	g_1	g_2	g_3	g_4	0	0	0	0
d_2	0	0	g_1	g_2	g_3	g_4	0	0
d_3	0	0	0	0	g_1	g_2	g_3	g_4
d_4	g_3	g_4	0	0	0	0	g_1	g_2

(Read as : $s_1 = h_1\ x_1 + h_2\ x_2 + h_3\ x_3 + h_4\ x_4$)

The matrix formed by the filter coefficients in the first four rows is the operator **H**. The matrix formed by the last four rows is the operator **G**. Thus, **s=Hx** and **d=Gx**.

The wavelet packet decomposition shown in Figure 15 can be interpreted in terms of approximating sequences. The sequence **s** can be considered to be a coarser approximation of the original sequence **x**. Let the filters **H** and **G** be constructed using orthonormal basis. Let the normalization be done so that the total energy before and after applying these filters to a sequence remains unchanged, i.e., the transformation is loss-less. The energy of a sequence is defined as the sum of the elements squared. Therefore, the error in approximating sequence **x** by sequence **s** can be measured as the energy of sequence **d.** Furthermore, the coarser approximation of elements $\{x_1, x_2\}$ is element s_1 and the error in approximation is energy of element d_1. Carrying this argument further to other levels in the tree, the first element at Level 3 is the coarsest approximation to the original sequence, and the error (energy loss) associated with this approximation is given by the energy of the shaded area in Figure 15. The first element at Level 3 along with the shaded elements in Figure 15 constitute the wavelet representation of the original sequence **x**.

The reconstruction of the original sequence **x** from sequences **s** and **d** is done as follows :

	s_1	s_2	s_3	s_4	d_1	d_2	d_3	d_4
x_1	h_1	0	0	h_3	g_1	0	0	g_3
x_2	h_2	0	0	h_4	g_2	0	0	g_4
x_3	h_3	h_1	0	0	g_3	g_1	0	0
x_4	h_4	h_2	0	0	g_4	g_2	0	0
x_5	0	h_3	h_1	0	0	g_3	g_1	0
x_6	0	h_4	h_2	0	0	g_4	g_2	0
x_7	0	0	h_3	h_1	0	0	g_3	g_1
x_8	0	0	h_4	h_2	0	0	g_4	g_2

(Read as : $x_1 = h_1 s_1 + h_3 s_4 + g_1 d_1 + g_3 d_4$)

The matrix formed by the first four columns is the operator $\mathbf{H}^*$, and that formed by the last four columns is the operator $\mathbf{G}^*$. Since the filter coefficients are real the adjoint operators of **H** and **G** are simply their transpose. Thus, $\mathbf{x}=\mathbf{H}^T\mathbf{s}+\mathbf{G}^T\mathbf{d}$.

The following are some MATLAB functions useful to perform wavelet decomposition and reconstruction: wavelet transform (*wtrans*), wavelet packet decomposition (*wttree*), reconstruction of the low-frequency component (*reconpl*), reconstruction of the high-frequency component (*reconph*), and the reconstruction of the high- or low-frequency (*reconphl*) component.

3.3 ILLUSTRATIVE EXAMPLE

Example 1 : Wavelet Packet Decomposition of a sequence.

Consider the wavelet packet decomposition of the following sequence: {0, 0, 0.0945, 0.5443, 0.9582, 1.3387, 1.6892, 2.0115} Let **H** and **G** be the Haar filters, i.e., $h=\{1/\sqrt{2},1/\sqrt{2}\}$ and $g=\{1/\sqrt{2},-1/\sqrt{2}\}$. Though the above sequence is non-periodic, the filter size used here is such that no problem is posed by assuming the sequence to be periodic. Table 1 shows the wavelet packet decomposition obtained by executing the following commands:

```
» inp_sig=[0 0 0.0945 0.5443 0.9582 1.3387 1.6892 2.0115];
» levels=3;
» trans_sig = wttree(inp_sig,levels,'h',1);
```

Table 1
Wavelet Packet Decomposition of a Sequence (trans_sig).

x_1 0.0	x_2 0.0	x_3 0.094	x_4 0.544	x_5 0.958	x_6 1.339	x_7 1.689	x_8 2.012	**Level 0**
s_1 0.0	s_2 0.452	s_3 1.624	s_4 2.617	d_1 0.0	d_2 -0.318	d_3 -0.269	d_4 -0.228	**Level 1**
ss_1 0.319	ss_2 2.999	ds_1 -0.319	ds_2 -0.702	sd_1 -0.225	sd_2 -0.352	dd_1 0.225	dd_2 -0.029	**Level 2**
sss 2.346	dss -1.895	sds -0.722	dds 0.270	ssd -0.408	dsd 0.090	sdd 0.138	ddd 0.180	**Level 3**

Since, in this example x_1=0 and x_2=0, we obtain s_1=0 and d_1=0, i.e., in this case there is no error (energy loss) involved in moving from a finer to a coarser scale. Also, observe that by approximating x_3=0.0945 and x_4=0.5443 by $s_2 = 0.4518$ we introduces an error given by the energy of d_2=-0.3180 (i.e., $x_3^2+x_4^2 = s_2^2+d_2^2$, and d_2^2 is the energy loss in the approximation). The other elements in the tree can be interpreted in a similar manner.

Assuming a non-periodic signals to be periodic in the wavelet transform computation described above, especially for the cases when there are a large number of filter coefficients, leads to *end effects*. The filter coefficients at the end of the signal *wrap-around* and lead to distortion of the time and frequency localization of the signal. Methods used to overcome this problem include:

- Constructing wavelets that avoid the wrap-around problem (Alpert, 1992).
- Using folding functions that make the signal periodic (Wickerhauser, 1991).
- Using different filter coefficients at the ends (Cohen et al, 1993).

4. Graphical Display of the Wavelet Transformation

There are three commonly used methods to present the results of wavelet transform:

- Multiresolution plots.
- Phase-Plane display of wavelet coefficients.
- Time-Frequency-Amplitude plots.

We illustrate these methods for an example signal called LADY. This signal is borrowed from Mallat et al's software *Wave1*.

4.1 MULTIRESOLUTION PLOT

The wavelet coefficients at each scale (or level) are reconstructed back to the time domain. These time domain signals at different resolutions constitute the multiresolution plot. Given all the reconstructed signals the MATLAB function *mra* graphs such a plot. For example:

```
» load lady.asc
» [trans_sig,recon_sig_l,recon_sig_h]=wtrans(lady,8,'d10',2);
» mra(lady,recon_sig_l,recon_sig_h,-1)
```

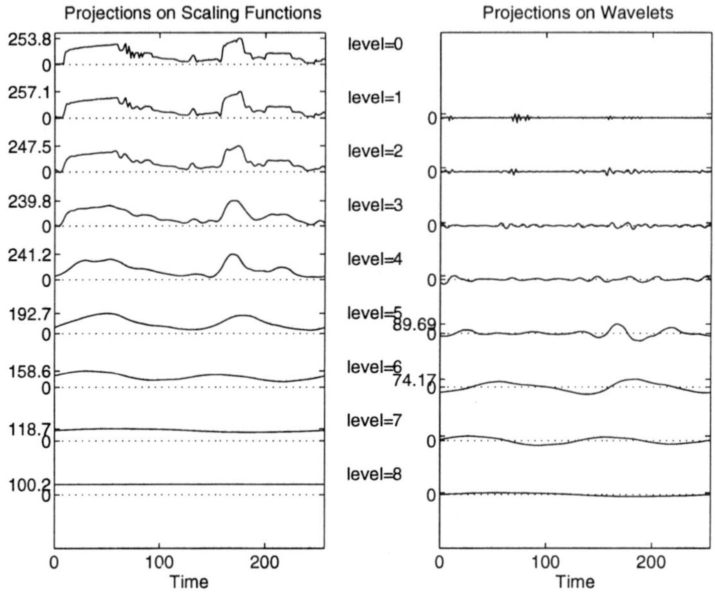

Figure 16 Multiresolution plot generated using the signal LADY.

displays the multiresolution plot shown in Figure 16. In Figure 16, level 0 corresponds to the original signal. The projections on the scaling function at various levels are displayed on the left hand side in Figure 16, and on the right hand side are displayed the projections on the wavelet. The projections on the wavelet (say, at level 1) capture the features lost in moving from one level (level 0) to the next coarser level (level 1). The projections on the scaling functions are smoother versions of the original signal. The signal at larger values of level are smoother than those at smaller values.

4.2 PHASE-PLANE PLOT

In the phase-plane plot the wavelet coefficients are displayed on a two-dimensional time-scale plane. A color determined by wavelet coefficients fill rectangular cells in this plot. The width of each of these cells is twice the standard deviation in the time domain of the wavelet (in a given scale). The height of each cell is twice the standard deviation in the frequency domain of the wavelet (in that given scale). Each cell is centered (over) on point $(\bar{t},\overline{\omega})$ as determined by the scale for each wavelet. For orthonormal basis these cells yield a disjoint cover of the time-scale plane (Coifman and Wickerhauser, 1992). The function *phaseplane* makes such a plot. The wavelet transform of a signal is required as the input argument. The ordinate for this plot is the level and not the true frequency. However, one could compute the mean and variance of the wavelet in the frequency domain and make a similar plot on the time-frequency plane. The following commands

```
» load lady.asc
» time=[0:255];
» [trans_sig]=wtrans(lady,8,'d10',1);
» phaseplane(trans_sig,time);
```

create a phase-plane plot (shown in Figure 17) for the signal LADY. The function *phaseplane* does not account for the distortion in time localization due to end-effects.

4.3 3-D TIME-LEVEL-AMPLITUDE PLOT

The reconstructed low-frequency signals or the wavelet coefficients can be displayed on the time-frequency-amplitude plot. Such a graph is an useful visual tool to observe the decay of the wavelet coefficients, their localization, and the multiresolution picture of the original signal. Here we illustrate the multiresolution plot on the time-level-amplitude plane. Executing the following command create the graph shown in Figure 18.

```
» load lady.asc
» [trans_sig,recon_sig_l,recon_sig_h]=wtrans(lady,8,'d10',2);
» z=[lady';recon_sig_l];
» mesh(z')
» view(-60,30)
» title('Plot of Reconstructed Signal at Different Levels')
» xlabel('Levels'), ylabel('Time')
```

Figure 18 shows the decay of the signal features from the finest scale to the coarsest scale. Again illustrating the filtering of high-frequency components of the signal while moving to the higher levels.

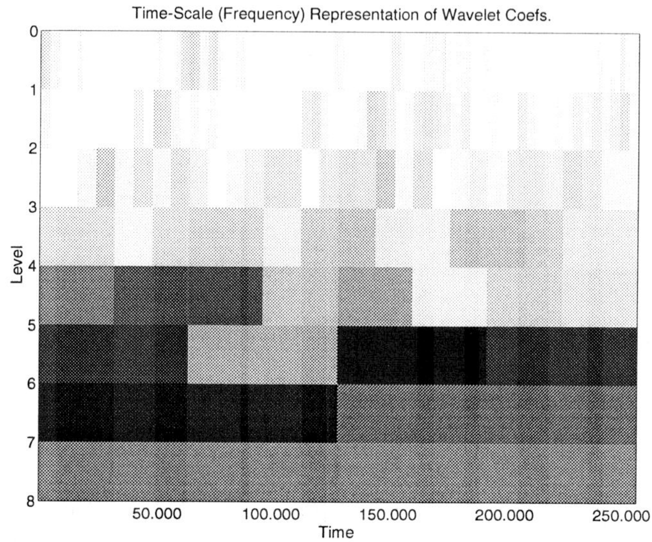

Figure 17 Phase-plane (level-time) plot of the wavelet coefficients of the signal LADY.

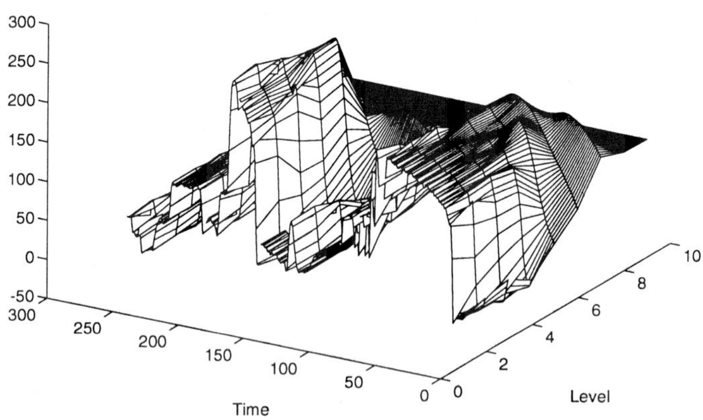

Figure 18 Plot of reconstructed signals of LADY at different levels on the time-level-amplitude plane.

5. Other Public Domain Wavelet Software

A number of public domain wavelet transform software are available for non-commercial, research, and educational purposes. Commercial software have been developed by the group at Yale and Aware Inc., Cambridge, MA, USA. Aware has also developed a wavelet compression technique on a chip. Some wavelet transform public domain software available are:

- *Wavelet Packet Laboratory (WPLab)* : Can be obtained from ceres.math.yale.edu. This software implements wavelet packet decomposition and reconstruction, and the best basis selection.

- *Wave1 and Wave2* : These two packages, written in C, process 1-dimensional and 2-dimensional signals using wavelet transformation. This software is available via anonymous ftp from cs.nyu.edu. The software is a good supplement to the works of Mallat and Zhong (1992), and Mallat and Hwang (1992).

- *Wavelet Transforms for S* : S software to compute 1-dimensional and 2-dimensional wavelet transforms using Mallat's pyramid algorithm and Daubechies compactly supported wavelets is developed by Guy P Mason. These programs are available via anonymous ftp from gdr.bath.ac.uk.

- *Wavelet Toolbox* : There are a number of MATLAB Toolboxes -- a collection of MATLAB programs -- that implement wavelet transforms:

    - Jeffrey C Kantor, Univ. of Notre Dame (anonymous ftp from control.cheg.nd.edu).
    - Neil Getz , Univ. of California, Berkeley (getz@robotics.Berkeley.edu).
    - Carl Taswell, Stanford Univ., Stanford (anonymous ftp from simplicity.stanford.edu).
    - Ramesh Gopinath, Rice University, Houston (anonymous ftp from cml.rice.edu).
    - David Donoho and Iain Jonhstone, Stanford Univ., Stanford (anonymous ftp from playfair.stanford.edu).

Useful information (including: papers, programs, source codes, archives of Wavelet Digest, and messages) on wavelets and wavelet transforms is available through Gopher server of the University of South Carolina, Department of Mathematics. Wavelet Digest is a electronic newsletter on wavelets edited by Wim Swelden and Bjorn Jawerth, Department of Mathematics, Univ. of South Carolina. To subscribe to this newsletter send electronic mail to wavelet@math.scarolina.edu with "subscribe" as the subject.

6. Concluding Remarks

To end this chapter we quote Y. Meyer (1993) :

"Wavelets are without any doubt an exciting and intuitive concept. This concept brings with it a new way of thinking, which is absolutely essential and was entirely missing in the previously existing algorithms. For that reason, wavelets will be more appealing than other algorithms to engineers who are still using a conventional Fast Fourier Transform (FFT) or Joint Photographic Experts Group (JPEG) in a routine way and do not even know subband coding"

Wavelets are useful tools, they have great potential in the future, but will not perform miracles.

Acknowledgments:

Financial support from the National Science Foundation, grant number CTS-91-08649, is gratefully acknowledged. The first author would like to thank Prof. M.V. Wickerhauser, Dept. of Mathematics, Washington Univ., St. Louis, and Dr. Arun Kumar, Southwestern Bell, Technology Resources, St. Louis, for helpful discussions.

7. Notation

Abbreviations

CWT	Continuous Wavelet Transformation.
DWT	Discrete Wavelet Transformation.
DTWT	Discrete Time Wavelet Transformation.
WST	Wavelet Series Transformation.

Symbols

A, B	Frame bounds.
$a\ (m)$	Dilation parameter.
$b\ (n)$	Translation parameter.
C^r	Space of functions with regularity *r*.
C^∞	Space of continuous infinitely differentiable functions.
$c_{m,n}$	Projection on a scaling function at scale *m* and translated by *n*.
c_ψ	A constant that is a function of the wavelet.
d_m	Wavelet coefficients at scale *m*.
$f(t)$	Function or a signal.
g	High-pass filter coefficients.
$\mathbf{G}$	High-pass filtering followed by decimation.
h	Low-pass filter coefficients.
$\mathbf{H}$	Low-pass filtering followed by decimation.
L^2	Space of square integrable functions.
l^2	Space of square summable sequences.
M	Number of vanishing moments.
m_0	Fourier transform of the low-pass filter coefficients.
N_m	Block-spline of order m.
$\mathbf{P_m}$	Projection on scaling functions at scale *m*.
P_N	Polynomial of order *N-1*.
$\mathbf{Q_m}$	Projection on wavelets at scale *m*.
Q_N	Trigonometric polynomial with N coefficients.
R	Field of Real Numbers.
R^n	Field of Real vectors of size *1* x *n*.
r	Local Hölder regularity.
t	Time.

$\bar{t}$	Center of Heisenberg's cell along the time-axis.
V_j	Closed subspace of $L^2(R)$ at scale j.
W_j	Orthogonal complement of V_j in V_{j+1}.
Z	Field of Integers.

Greek Characters

$\chi_{[0,1)}$	Characteristic function (or a unit pulse) on the interval [0,1).
$\phi(t)$	Scaling function.
$\phi_{m,n}(t)$	Scaling function at scale m and translation n.
$\hat{\phi}(\omega)$	Fourier transform of the scaling function.
η_l	Piecewise constant function obtained after l recursions.
σ_t (σ_ω)	Standard deviation in the time domain (frequency domain).
$\sigma_{a,t}$ $(\sigma_{a,\omega})$	Standard deviation in the time domain (frequency domain) at scale a.
ω	Frequency.
$\bar{\omega}$	Center of Heisenberg's cell along the frequency-axis.
$\psi(t)$	Wavelet.
$\psi_{m,n}(t)$	Wavelet at scale m and translation n.
$\psi^v(t)$	Wavelet with multiple-voices.
$\hat{\psi}(\omega)$	Fourier transform of the wavelet.

Special Notation

$\|\ldots\|$	L^2-norm (l^2-norm) of a continuous function (vector).
$\langle \ldots, \ldots \rangle$	Inner product.
$*$	Convolution.
$\in$	Belongs to.
$\oplus$	Direct sum.
$\cap$	Intersection of.
$\cup$	Union of.
$\subset$	Subset of.
$\perp$	Orthogonal to.
»	MATLAB prompt.

8. References

1. Akansu, A.N., and R.A. Haddad, *Multiresolution Signal Decomposition - Transforms, Subbands, and Wavelets*, Academic Press, Inc., San Diego, CA (1993).
2. Alpert, B.K., "Wavelets and Other Bases for Fast Numerical Linear Algebra", in *Wavelets: A Tutorial in Theory and Applications*, Chui, C.K. (ed.), Academic Press, 181-216 (1992).
3. Auscher, P., G. Weiss, and M.V. Wickerhauser, "Local Sine and Cosine Bases of Coifman and Meyer and the Construction of Smooth Wavelets", in *Wavelets : A Tutorial in Theory and Application*, Chui C K (ed.), Academic Press, New York (1992).
4. Battle, G., "A Block Spin Construction of Ondelettes. Part I: Lemarie Functions", *Comm. Math. Phys.*, **110**, 601-615 (1987).
5. Beylkin, G., R. Coifman, and V. Rokhlin, "Fast Wavelet Transforms and Numerical Algorithms", *Comm. Pure Appl. Math.*, **44**, 141-183 (1991).
6. Chui, C.K., *An Introduction to Wavelets*, Academic Press, New York (1992a).
7. Chui, C.K., "Wavelets - with Emphasis on Spline-Wavelets and Applications to Signal Analysis", in *Approximation Theory, Spline Functions and Applications*, Singh, S.P., (ed.), Kluwer Academic Publishers, Netherlands, 19-39 (1992b).
8. Chui, C.K. (ed.), *Wavelets : A Tutorial in Theory and Applications*, Academic Press, New York (1992c).
9. Chui, C.K., and J.Z. Wang, "On Compactly Supported Spline Wavelets and a Duality Principle", *Trans. Amer. Math. Soc.*, **330**(2), 903-915 (1992).
10. Cohen, A., "Biorthogonal Wavelets", in Chui, C.K. (ed.), *Wavelets: A Tutorial in Theory and Applications*, Academic Press, 123-152 (1992).
11. Cohen. A., I Daubechies, and J. Feauveau, "Biorthogonal Bases of Compactly Supported Wavelets", *Comm. Pure and Appl. Math.*, **45**(5), 485-560 (1992).
12. Cohen, A., I Daubechies, B Jawerth, and P. Vial, "Multiresolution Analysis, Wavelets and Fast Algorithms on an Interval", *C.R. Acad. Sci Paris*, t. 316, Serie I, 417-421 (1993).
13. Coifman, R., Y. Meyer, and M.V. Wickerhauser, "Wavelet Analysis and Signal Processing", in *Wavelets and their Applications*, Ruskai et al, Jones and Bartlett, Boston, 153-178 (1992).

14. Coifman. R., and M.V. Wickerhauser, "Entropy-Based Algorithms for Best Basis Selection", *IEEE Trans. on Inf. Theory*, **38**(2), 713-718 (1992).
15. Combes, J.M., A. Grossmann and Ph. Tchamitchian (eds.), *Wavelets-Time-Frequency Methods and Phase Space*, Proceedings of the Int. Conf., Marseille, Dec. 1988, Springer-Verlag, Berlin (1989).
16. Daubechies, I., "Orthonormal Bases of Compactly Supported Wavelets", *Comm. Pure Appl. Math.*, **41**, 909-996 (1988).
17. Daubechies, I., "The Wavelet Transform, Time-Frequency Localization and Signal Analysis", *IEEE Trans. Inform. Theory*, **36**(5), 961-1005 (1990).
18. Daubechies, I., *Ten Lectures on Wavelets*, CBMS-NSF regional conference series in applied mathematics, SIAM, Philadelphia (1992).
19. Daubechies, I., A. Grossmann, and Y. Meyer, "Painless Non orthogonal Expansions", *J. Math. Phys.*, **27**(5), 1271-1283 (1986).
20. DeVore, R.A., and B.J. Lucier, "Wavelets", *Acta Numerica*, **1**(1), 1-56 (1991).
21. Gabor, D., "Theory of Communication", *J. Inst. Electr. Engrg.*, London, **93**(III), 429-457 (1946).
22. Grossmann, A., and J. Morlet, "Decomposition of Hardy functions into Square Integrable Wavelets of Constant Shape", *SIAM J. Math.*, **15**, 723-736 (1984).
23. Grossmann, A., J. Morlet and T. Paul, "Transforms Associated to Square Integrable Group Representation-I : General Results", *Journal of Mathematical Physics*, **26**(10), 2473-2479 (1985).
24. Haar, A., "Zur Theorie der orthogonalen Funktionen-Systeme", **69**, Math. Ann., 331-371 (1910).
25. Heil, C.E., and D.F. Walnut, "Continuous and Discrete Wavelet Transforms", *SIAM Review*, **31**(4), 628-666 (1989).
26. Jawerth, B., and W. Sweldens, "An Overview of Wavelet Based Multiresolution Analyses", preprint, Univ. of South Carolina, Columbia (1993).
27. Mallat, S., "Multiresolution Approximation and Wavelets", *Trans. Amer. Math. Soc.*, **315**(1), 69-88 (1989a).
28. Mallat, S., "A Theory for Multiresolution Signal Decomposition : The Wavelet Representation", *IEEE Trans. on Pattern Analysis and Machine Intelligence*, **11**(7), 674-693 (1989b).
29. Mallat, S., "Zero-Crossings of a Wavelet Transform", *IEEE Trans. Inform. Theory*, **37**(4), 1019-1033 (1991).

30. Mallat, S., and W.L. Hwang, "Singularity Detection and Processing with Wavelets", *IEEE Trans. Inform. Theory*, **38**(2), 617-643 (1992).
31. Mallat, S., and S. Zhong, "Wavelet Transform Maxima and Multiscale Edges", in Ruskai, M.B., et al. (eds.), *Wavelets and Their Applications*, Jones and Bartlett Publishers, 67-104 (1992).
32. Meyer, Y., *Book Reviews*, *Bulletin of the Amer. Math. Soc.*, **28**(2), 350-360 (1993).
33. Meyer, Y., *Wavelets - Algorithms and Applications*, Translated and Revised by Ryan, R.D., SIAM, Philadelphia (1993).
34. Micchelli, C., "Using the Refinement Equation for the Construction of Pre-Wavelets", *Numer. Alg.*, **1**, 75-116 (1991).
35. Morlet, J., G. Arens, I. Fourgeau, and D. Giard, "Wave Propagation and Sampling Theory", *Geophysics*, **47**, 203-236 (1982).
36. Rioul, O., and M. Vetterli, "Wavelets and Signal Processing", *IEEE Signal Proc. Mag.*, **8**(4), 14-38 (1991).
37. Rioul, O., and P. Duhamel, "Fast Algorithms for Discrete and Continuous Wavelet Transforms", *IEEE Trans. on Inform. Theory*, **38**(2), 569-586 (1992).
38. Ruskai, M.B., G. Beylkin, R. Coifman, I. Daubechies, S. Mallat, Y. Meyer, and L. Raphael (eds.), *Wavelets and their Applications*, Jones and Bartlett, Boston (1992).
39. Shensa, M.J., "Wedding the à trous and Mallat Algorithms", *IEEE Trans. on Signal Process.*, **40**(10), 2464-2482 (1992).
40. Strang, G., "Wavelets and Dilation Equations: A Brief Introduction", *SIAM Review*, **31**(4), 614-627 (1989).
41. Strang, G., "Wavelet Transforms Versus Fourier Transforms", *Bulletin of the Amer. Math. Soc.*, **28**(2), 288-305 (1993).
42. Strichartz, R.S., "How to Make Wavelets", *Amer. Math. Monthly*, to appear (1992).
43. Strichartz, R.S., "Construction of Orthonormal Wavelets", preprint, Cornell University, Itaca (1992).
44. Wickerhauser, M.V., "Picture Compression by Best Basis Sub-Band Coding", preprint, Yale University (1990).
45. Wickerhauser, M.V., "Lectures on Wavelet Packet Algorithms", preprint, Washington University (1991).
46. Wickerhauser, M.V. , "Acoustic Signal Processing with Wavelet Packets", in *Wavelets : A Tutorial in Theory and Application*, Chui C K (ed.), Academic Press, New York, 679-700 (1992b).

9. Appendix I : MATLAB Routines for Discrete Wavelet Transformation

The following is the list of MATLAB script files that are used in this chapter.

	Wavelet Tools
diltrans	Dilate and/or translate a function.
Daubechies' Wavelet	
daubonw	Constructs Daubechies orthonormal wavelet of compact support.
cascade	Generates Daubechies scaling function and wavelet from filter coeffs.
ftscale	Computes Fourier transform of Daubechies scaling function and wavelet from filter coeffs.
Battle's Wavelet	
batlem.m	Constructs Battle linear B-spline scaling function.
batwave.m	Constructs Battle linear B-spline wavelet.
Chui-Wang's Wavelet	
bspline.m	Constructs Chui-Wang's semi-orthogonal wavelet using linear B-spline.

	Wavelet Transform Tools
wtrans	Computes wavelet transform and/or reconstruction of a signal.
wttree	Computes wavelet packet decomposition.
recon_tre	Reconstructs signal for coefficients of a wavelet packet basis.
reconphl	Reconstructs signal from high or low frequency component.
reconpl	Reconstructs signal from the low-frequency component.
reconph	Reconstructs signals from the high-frequency component.
cdps	Convolution followed by decimation of a periodic signal.

	Graphic Display Tools
phaseplane	Plots the wavelet coefficients on the time-scale plane.
mra	Plots the multiresolution decomposition of a signal.

9.1 Commands and Their Function

In this section, we describe the function of each MATLAB script file, its input and output arguments, and we give some examples. An on-line help for each file is also available (type *help <command>*).

Wavelet Tools

Daubechies' Wavelet

---

daubonw

Function

Constructs Daubechies' orthonormal wavelet of compact support.

Input and Output Arguments

[h,g]=daubonw(N)

N - Twice the size of the filter coefficients.
h - Low-pass filter coefficients.
g - High-pass filter coefficients.

Example

```
» [h,g]=daubonw(2);
» format long;
» h'

ans =

   0.48296291314453
   0.83651630373781
   0.22414386804201
  -0.12940952255126

»g'

ans =

  -0.12940952255126
  -0.22414386804201
```

0.83651630373781
-0.48296291314453

---

cascade

Function
Computes and plots Daubechies' scaling function and the wavelet.

Input and Output Arguments
[phi,psi]=cascade(h)

phi - Scaling function.
psi - Wavelet.
h - Low-pass filter coefficients.

Example
» [h,g]=daubonw(2);
» [phi,psi]=cascade(h);
Result shown in Figures 5 and 6.

---

ftscale

Function
Computes and plots the Fourier transform of Daubechies' scaling function and wavelet.

Input and Output Arguments
[phihat,psihat]=ftscale(h)

h - Low-pass filter coefficients.
phihat - Fourier transform of the scaling function.
psihat - Fourier transform of the wavelet.

Example
» [h,g]=daubonw(2);
» [phihat,psihat]=ftscale(h);
Result shown in Figures 7 and 8.

Wavelet Transform Tools

---

wtrans

Function
Computes wavelet transform and/or reconstruction of a signal.

Input and Output Arguments

```
trans_sig=wtrans(inp_sig,levels,'filter_name')
[trans_sig,recon_sig_l,recon_sig_h]=wtrans(inp_sig,levels,'filter_name'
                                                   ,mra)
```

inp_sig	-	Signal to be transformed. The size of this signal should be 2^N.
levels	-	Number of levels till which to decompose (<N).
'filter_name'	-	Filter to be used in decomposition. Available filters include: Daubechies' filters: d4, d6, d8, d10, d12, d14, d16, d18, d20. Coiflets: c6, c8, c24, c30. Haar filter: haar.
mra	-	Use 1 for decomposition only. Use 2 for decomposition and reconstruction.
trans_sig	-	Contains the wavelet coefficients. trans_sig is a vector of size 2^N when mra is not used. trans_sig is a matrix of size levels x 2^N when mra is used.
recon_sig_l	-	Reconstruction of the low frequency components. Size of matrix is levels x 2^N when mra=2.
recon_sig_h	-	Reconstruction of the high-frequency components. Size of the matrix is levels x 2^N when mra=2.

When mra=1, only one output argument is required.

Example

```
» load lady.asc
» [trans_sig,recon_l,recon_h]=wtrans(lady,8,'d10',2);
```

Result plotted in Figures 16, 17 and 18.

wttree

Function

Computes the wavelet packet decomposition of a signal.

Input and Output Arguments

[trans_sig]=wttree(inp_sig,levels,'filter_name')

inp_sig - Signal to be transformed. The size of this signal should be 2^N.

levels - Number of levels till which to decompose (<N).

'filter_name' - Filter to be used in decomposition. Available filters include:

Daubechies' filters: d4, d6, d8, d10, d12, d14, d16, d18, d20.

Coiflets: c6, c8, c24, c30.

Haar wavelet: haar.

trans_sig - Contains the wavelet packet coefficients. trans_sig is a matrix of size levels x 2^N.

Example

```
» load lady.asc
» trans_sig=wttree(lady,8,'d10');
```

recon_tre

Function

Reconstructs the signal from coefficients of a wavelet packet basis.

Input and Output Arguments

out_sig=recon_tre(best_coef,'filter_name')

best_coef - Contains coefficients of a wavelet packet basis. It should be a matrix of size N x 2^N.

filter_name - Filter to be used for reconstruction. Filter available include:

Daubechies' filters: d4, d6, d8, d10, d12, d14, d16, d18, d20.

Coiflets: c6, c8, c24, c30.

Haar filter: haar.

out_sig - Reconstructed signal. Returned as a vector of size 2^N.

Example

```
» load lady.asc
» trans_sig=wtrans(lady,8,'d10',1);
» new_lady=recon_tre(trans_sig,'d10');
```

---

reconpl, reconph, reconphl

Function

reconpl reconstructs the low frequency component.
reconph reconstructs the high frequency component.
reconphl reconstructs a signal from high or low frequency component.

Input and Output Arguments

out_sig=reconpl(inp_sig,level,h,g,N)
out_sig=reconph(inp_sig,level,h,g,N)
out_sig=reconphl(inp_sig,level,position,h,g,N)

inp_sig - Low-pass/high-pass/low and high pass components of a signal.
level - Level till which the inp_sig has been decomposed.
h - Low-pass filter coefficients.
g - High-pass filter coefficients.
N - Half the size of the filter.

---

cdps

Function

Convolution of a periodic signal with a low-pass and a high-pass filter coefficients followed by decimation (dropping even terms).

Input and Output Arguments

out_sig=cdps(inp_sig,h,g,N)

inp_sig - A periodic signal.
h - Low-pass filter coefficients.
g - High-pass filter coefficients.
N - Half the size of the filters.

Graphic Display Tools

---

phaseplane

Function
Plots the wavelet coefficients on a time-level (scale) plane.

Input and Output Arguments
handle=phaseplane(trans_sig,time)

trans_sig	-	Matrix of dimensions levels x 2^N containing the wavelet coefficients.
time	-	The time axis (default time axis has sample time 1).
handle	-	Graphics handle that can be used to change properties of the plot.

Example
» load lady.asc
» [trans_sig]=wtrans(lady,8,'d10,1);
» phaseplane(trans_sig);
Plot shown in Figure 17.

---

mra

Function
Makes a multiresolution plot of a signal.

Input and Output Arguments
mra(signal,recon_sig_l,recon_sig_h,label_out)

signal	-	Original signal.
recon_sig_l	-	Reconstruction of the projections on the scaling function. Matrix of size levels x 2^N.
recon_sig_h	-	Reconstruction of the projections on the wavelet. Matrix of size levels x 2^N.
label_out	-	Option to label inside or outside the figure. label_out=1 labels inside the figure, and label_out=-1 labels outside the figure.

Example

```
» load lady.asc
» [trans_sig,recon_sig_l,recon_sig_h]=wtrans(lady,8,'d10',2);
» mra(lady,recon_sig_l,recon_sig_h,-1);
```

Plot shown in Figure 18.

9.2 Accessing the Programs

The MATLAB script files used in this chapter, along with programs for best basis selection, two-dimensional periodic wavelet transforms, and discrete wavelet transform on an interval, may be obtained by anonymous FTP (Internet file transfer protocol) from wuarchive.wustl.edu [128.252.135.4]. This archive is located within Washington University in St. Louis, Missouri, USA. The programs are stored in compressed form. To access the file execute the following UNIX commands.

ftp *wuarchive.wustl.edu*
login: *anonymous*
Password: ⟨*your e-mail address*⟩
cd *doc/techreports/wustl.edu/math/software*
get *toolbox.readme*
binary
get *toolbox.tar.Z*
bye

The file '*toolbox.readme*' contains instructions to retrieve the MATLAB script files from the file '*toolbox.tar.Z*'

Chapter 3
Trend Analysis Using the Frazier-Jawerth Transform

A. Joshi, A. Kumar[*] ,and R. L. Motard
Chemical Engineering Department
Washington University
One Brookings Drive, Box 1198
St. Louis, MO 63130-4899

Abstract

This chapter discusses the Frazier-Jawerth Transform (FJT), a new transform for time-frequency analysis (Frazier and Jawerth, 1985) and a new method for its on-line implementation. Algorithms for FJ decomposition and reconstruction of 1-D signals are also included. FJT bears close resemblance to the wavelet transform technique which is enjoying much attention lately (Science, August 1990). The theory of frames has been shown to subsume the FJ and the wavelet transforms (Heil and Walnut, 1989). It is now widely believed that these new time-frequency techniques may replace the Fourier transform and older time-frequency techniques in some applications in future. Potential applications of interest to chemical engineers include analysis of dynamic systems, process control, solution of partial differential equations, geoexploration data processing, vibration/acoustic testing methods, artificial neural networks etc. FJT has been employed as a time-series data preprocessing mechanism in an artificial neural network based process trend and abnormality detection scheme.

---

[*] Current address, Southwestern Bell Co., St. Louis, Missouri

1 Introduction

The behavior of a dynamic system is usually studied by observing the evolution of a set of measured or state variables (signals) over time. The signals are usually analyzed to yield information about the state of the dynamic system. The analysis is done either entirely in the time domain (e.g. by calculating moments) or entirely in the frequency (or Fourier) domain. Such analysis is used for various purposes, e.g. to analyze system stability, to identify the system model. Time-domain analysis is a popular method for residence-time distribution studies in reaction engineering. Linear control theory and practice depends heavily on pure frequency domain analysis of systems.

In many applications of interest to chemical engineers, a *joint* time and frequency domain analysis would be helpful. Such analysis is especially useful if the signals are non-stationary or change their frequency content with time. Such signals are encountered in the analysis of dynamic chemical engineering systems. They are the norm in geoexploratory analysis where signals which have passed through several layers of different properties (and have thus changed in frequency-content and also have delays associated with them) have to be analyzed. In vibration/acoustic analyses of structural components (e.g. polymer composite structures) the changes in the nature of the test signal are analyzed for diagnosing material defects.

This paper discusses a novel technique, the Frazier-Jawerth transform (FJT) (Frazier and Jawerth, 1985, 1988, Kumar et al., 1992) which provides a joint time and frequency description (or a time-frequency description) of a signal. The treatment here will only include one-dimensional signals (the case which is of primary interest to chemical engineers) although the technique can be extended to the general case of N-dimensional signals. The analysis of two-dimensional signals (images) is of interest to the disciplines such as telecommunications and medical imaging. For the process industries, possible areas of application include process and structural fault diagnosis, geoexploration, numerical solution of partial differential equations, process control and artificial neural network applications.

The primary aim of this chapter is to provide a self-contained introduction to the FJT and its computational algorithms and a new algorithm for its online implementation. That the FJT enables a multiresolution description of a signal is demonstrated with the help of an example. The relation between resolution at different scales and trend analysis stands out as well.

2 Background

In the time domain representation of a signal, information about the frequency-content of a signal is hidden while in the Fourier representation, the information about the evolution of a signal in time gets obscured. Figures 1 and 2 depict a non-stationary signal and its Discrete Fourier Transform (DFT) respectively (the phases of the DFT are not plotted). The signal in Fig. 1 is clearly non-stationary since its statistical properties change over time. From Fig. 1, one can obtain information such as the average magnitude of the signal and the times at which there is a sudden change in the magnitude. This information cannot be obtained directly from Fig. 2, which does however yield information about the dominant frequency components in the signal. information such as time-shifts is in the phases of the DFT. Both representations reveal complementary information in the signal.

Developing techniques to examine the changes in frequency-content over time has been an active area of research in the field of digital signal processing over the last 50 years. Gabor (1946) suggested that the Fourier transform of a signal when viewed through a window that slides over the signal in time would yield such a description. He recommended the Gaussian window because it has small essential supports in the time and the frequency domains (i.e., it is appreciably non zero over a small portion and so is its Fourier transform). However, there are problems associated with using a single constant-size window. If a window with narrow width (i.e., small essential support) is used, one cannot discriminate the low frequencies in the signal well. With a wider window, local phenomena (high frequency components) cannot be detected. Improvements in choosing the window functions cannot alleviate the problems due to the use of constant-width windows. Gabor's method has been extended to a set of methods known as 'short time Fourier analysis' (Allen and Rabiner, 1977). The methods use windows with shapes other than the Gaussian window of Gabor. These techniques have found wide applicability in speech analysis.

The second approach to time-frequency analysis is reflected in the research on joint time-frequency distributions and a good review can be found in Cohen (1989) and summarized in Chapter 1. These distributions have many properties which make them useful for certain applications. For example,

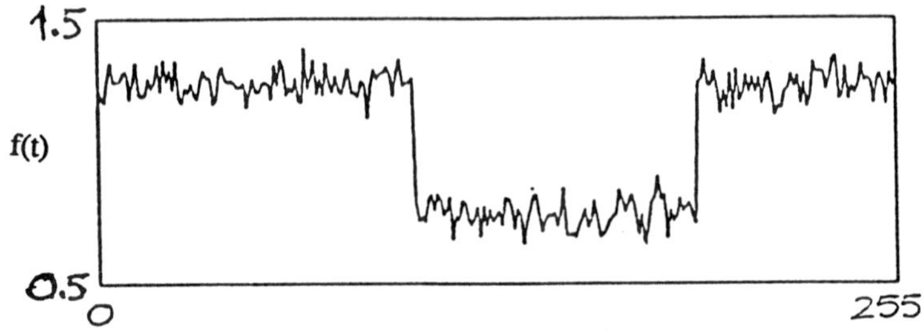

Figure 1. An Example Dynamic Response

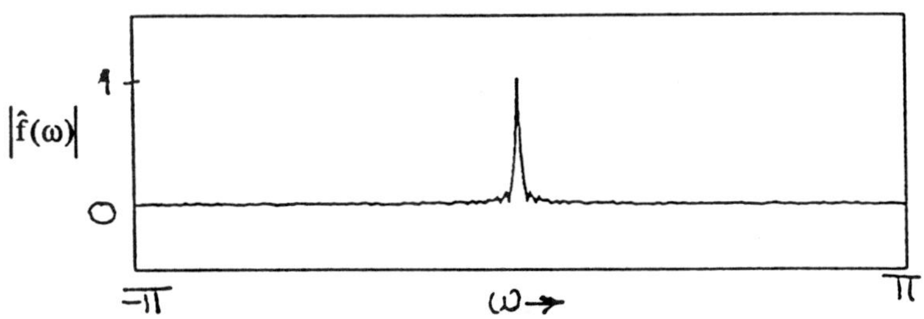

Figure 2. Fourier Spectrum of the Response in Fig. 1

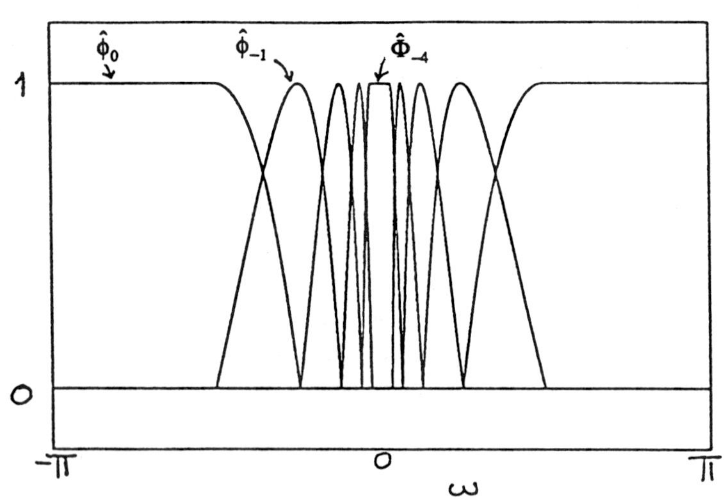

Figure 3. A Set of Analyzing Functions Covering the Frequency Line.

Boashash and co-workers (1986, 1987) have explored applications in seismic data processing and pattern recognition (1987). Forrester (1989) has shown that the Wigner-Ville distribution can be effectively used to discriminate between faulty/non-faulty helicopter components based on the vibration data

A simple technique for time-frequency analysis, the Frazier-Jawerth Transform (FJT), was recently proposed (Kumar 1990, Kumar and Fuhrmann 1990, Kumar et al. 1992). It was shown that any function belonging to the space of tempered distributions* can be represented as a weighted sum of a set of 'elementary' non-orthonormal functions called as the *synthesizing functions*. The weights of the sum are obtained by taking projections of the original function over a set of functions called the *analyzing functions*. The analyzing and the synthesizing functions are obtained by a simple construction; the process of dilation and translation. Further, if the analyzing functions are chosen so that they are appreciably non-zero only over small regions in both time and frequency domains, then the decomposition yields a time-frequency description. The advantage of using such analyzing and synthesizing functions to approximate a function is that the local regularity of the function can be studied with higher accuracy. Local changes in a function affect only the transform coefficients in the surrounding small region. A small change in the function characteristics results in only a small change in the transformed domain. As such, the FJ decomposition can be used to great advantage in numerical methods for solutions of partial differential equations. The Gibbs-phenomenon namely, the spurious oscillations observed with the spectral methods gets restricted to only a small region due to the localized approximations by the FJ decomposition.

3 Terminology

Let **Z, R** and **C**, denote the sets of integer, real and complex numbers and ℓ^2 denote the Hilbert space of absolutely square-summable sequences, i.e.

$$\ell^2(\mathbf{Z}) = \{f:\mathbf{Z} \rightarrow \mathbf{C}, \sum_{i \in \mathbf{Z}} |f_i|^2 < \infty\}$$

---

* All the dynamic responses that are typically encountered in chemical systems belong to this vector space.

All the functions in the vector space $\ell^2(\mathbf{Z})$ thus have finite energy in the above sense. The analysis in this paper holds for any signal belonging to ℓ^2 space. The *support* of a function $f : \mathbf{Z} \rightarrow \mathbf{C}$ is the topological closure of the set $\{x \in \mathbf{Z} | f(x) \neq 0\}$ i.e., the domain over which the function is non-zero. Let the set be denoted by *supp* f. Let $\hat{f}(n)$ denote the discrete Fourier transform (DFT) of the discrete signal f(n) defined by:

$$\hat{f}(k) = \left\langle f(n), e^{j(2\pi kn/N)} \right\rangle = \sum_{n=0}^{N-1} f(n) e^{-j(2\pi kn/N)} \tag{1}$$

N is the size of the sequence, f(x). Where $\left\langle f(n), e^{j(2\pi kn/N)} \right\rangle$ denotes the inner product of the function f(n) with the complex exponential function and the operator "^" denotes its discrete Fourier transform. The inner product of two functions f and g $\langle f, g \rangle$ in $\ell^2(\mathbf{Z})$ is usually defined by $\sum_{n \in \mathbf{Z}} f(n)\bar{g}(n)$ where the bar over g denotes its complex conjugation.

The inverse discrete Fourier transformation (IDFT) is given by:

$$f(n) = \frac{1}{N} \sum_{k=0}^{N-1} \hat{f}(k) e^{j(2\pi kn/N)} \tag{2}$$

which may be rewritten by using Eqn. 1:

$$f(n) = \frac{1}{N} \sum_{k=0}^{N-1} \left\langle f(n), e^{j(2\pi kn/N)} \right\rangle e^{j(2\pi kn/N)} \tag{3}$$

In the above equation a discrete function f(n) has been represented as a weighted sum of a set of basis functions (the complex exponentials). These functions are termed as the *synthesizing functions*. When a function is represented as a weighted sum of a set of some "elementary" functions, the expansion is called its *decomposition*. Thus the above equation represents the *Fourier decomposition* of a discrete signal f(n). The weights of this sum are obtained by taking projections of f(n) over some set of basis functions, called the analyzing functions. The *analyzing functions* in the Fourier decomposition are the same as the synthesizing functions. The weights of this decomposition are nothing but the Fourier transform coefficients.

4 The FJ Decomposition and the FJT

The basic idea behind the FJ decomposition can be best explained by referring first to the Fourier and the time representations. If any of the Fourier analyzing functions is plotted, it will be seen to extend all over in the time domain (they are global or *not localized* in the time domain). On the other hand, each represents one and only one frequency channel and thus is said to be perfectly localized in the frequency domain. The magnitude of a coefficient along a particular axis indicates the extent to which that particular frequency is present in the signal. On account of being localized in the frequency domain but being quite the opposite in the time domain, probing a signal with the Fourier analyzing functions obscures information about the time evolution of a signal whereas it makes the frequency characteristics explicit.

On the other hand, consider the representation of a signal in the time domain, i.e. the usual time-series data. Each time-point value may be thought of as the shadow of the signal from some abstract space over a unit pulse located at that time-point. The unit pulses may be thought of as the analyzing and the synthesizing functions of the time domain representation. Note that the unit pulse functions are perfectly localized in time, but are global in the frequency domain (because all frequencies show up with equal strength in its Fourier transform). This is quite opposite to the Fourier decomposition. The time domain representation thus makes the features of a signal pertaining to its time evolution explicit but hides the features regarding its frequency characteristics. Note that each representation in itself is *complete* in the sense that it does carry all the information. The difference lies in which information is explicit and which gets obscured.

The FJ decomposition aims to obtain a representation of a function that is intermediate to the time *and* the frequency domain representations and thus obtain time and frequency descriptions in a *single* representation. It obtains such representation by allowing the analyzing functions to have small supports in the time and the frequency domains.

4.1 Definitions

The Frazier-Jawerth decomposition of a signal in $\ell^2(\mathbf{Z})$ is given by:

$$f = \sum_{k \in Z} \langle f, \Phi_{Mk} \rangle \Psi_{Mk} + \sum_{v=M+1}^{0} \sum_{k \in Z} \langle f, \phi_{vk} \rangle \psi_{vk} . \tag{4}$$

Where M is any arbitrary negative integer. The FJT of the signal f is the sequence of inner-products given by $(\langle f, \Phi_{Mk} \rangle, \langle f, \phi_{vk} \rangle)_{v,k}$.

The above equation is not unlike Eqn. 3. The function f, just as in Eqn. 3, is represented as a weighted sum of certain elementary functions the weights of which are given by the projections of f on the elementary functions.

The functions Φ's and ϕ's are the analyzing functions the set of which will be denoted by **A,** i.e. $\mathbf{A} \underline{\underline{\Delta}} \{\Phi_{Mk}, \phi_{vk}\}_{v,k}$. Similarly, the synthesizing functions form a set $\mathbf{S} \underline{\underline{\Delta}} \{\Psi_{Mk}, \psi_{vk}\}_{v,k}$. ν (≤ 0) and k are termed the *dilation* and *translation* parameters respectively. Thus the set **S** is formed of functions Ψ and its translates and ψ and its dilates and translates. Similarly, **A** consists of the translates of Φ and the dilates and translates of ϕ. The sets **S** and **A** need not be different.

The *dilation* of an ℓ^2 function in the frequency domain is defined as:

$$\hat{\phi}_\nu(\omega) \underline{\underline{\Delta}} \hat{\phi}(2^{-\nu}\omega), 0 > \nu \in \mathbf{Z}. \tag{5}$$

The dilation in the time domain is defined as $\phi_\nu = (\hat{\phi}_\nu)^\vee$ where the operator "$^\vee$" denotes an inverse Fourier transformation. The dilation and translation of ϕ is defined as:

$$\phi_{\nu k}(n) \underline{\underline{\Delta}} 2^{-\nu/2} \phi(n - k2^{-\nu}), 0 > \nu \in \mathbf{Z} \tag{6}$$

ν and k are the dilation and translation parameters respectively. The function $\hat{\phi}_0$ is not a strict dilate of $\hat{\phi}$ in the interval $[-\pi,\pi]$ but rather spills over outside the interval. The translations $\phi_k(n)$ are defined simply as $\phi(n - k)$. The objective is to build these functions so that they cover the frequency space (which for discrete signals is bounded to the interval $[-\pi,\pi]$) and so that they have small essential supports in the time and frequency domains.

If the functions in the set **A** are chosen so that they have small essential supports in the time and frequency domains, then the FJT will give a localized time-frequency description. Since the FJ decomposition amounts to probing a signal with these functions by computing the projections of the signal on them (or computing the inner products), every inner-product will carry information about the behavior of the signal *only* in those regions of time and frequency domains where the corresponding analyzing function lives (i.e. over the support of the analyzing functions). The geometrical equivalent of this is to construct a non-orthonormal coordinate system such that each coordinate axis has only a small share of the frequency space and the time space and take projections of the signal along the axes. There is inevitably an overlap in the frequency space unlike the Fourier analyzing functions where each axis represents one and only one frequency.

The extent of time-frequency localization depends greatly on how small the supports for analyzing functions are in the time and frequency domains. The following section elaborates on the construction of the analyzing and the synthesizing functions and states conditions that these functions must satisfy.

The FJT is an injective and invertible transformation. It is really a family of transforms because the sets **S** and **A** can be freely chosen as long as the functions satisfy certain conditions (mentioned in the following section).

4.2 The Analyzing and the Synthesizing Functions for the FJT

Suppose a function ϕ is constructed such that it has a bounded support in the frequency domain. If $\hat{\phi}(\omega)$ is chosen to lie in the set of n times continuously differentiable functions, then $\phi(\tau)$ will decay at least as $1/(t-t_0)^n$ away from its peak at t_0. The function ϕ is then said to be *localized* in both time and frequency. The uncertainty principle (Gabor, 1946) dictates that no function can have a bounded support in both the time and the frequency domains. The analyzing functions for the FJT are chosen so that they have a compact support in the frequency domain and decay rapidly in the time domain or *vice versa*. Further, these analyzing functions are constructed from a single 'parent' function* using the process of *dilation* and *translation*. Thus instead of looking at a signal through a window of constant size (as in the case of the short-time Fourier transform), one looks at it through a sliding window of *changing width*. The amount by which this window slides also changes with its width. The shapes and the sizes of the windows are specified by the analyzing functions.

4.3 Construction by the Direct Method

There are several ways to construct the functions in **A** and **S**. The lemma stated below gives a prescriptive method for constructing any and all such functions. A direct method can also be followed where first the window functions, $\hat{\Theta}$ and $\hat{\theta}$, are selected so that they cover the frequency range and then factored out to yield the analyzing and synthesizing functions. If $\hat{\Phi}$ and $\hat{\phi}$ are n times continuously differentiable, then Φ and ϕ will decay rapidly

---

* The functions Φ and ϕ are inter-related.

(as $1/t^n$) away from their peak values and one will thus obtain small supports in both the domains.

Let $\hat{\Theta}$ and $\hat{\theta}$ be window functions such that *supp* $\hat{\Theta}(\omega)\subseteq\{\omega:|\omega| \leq\pi\}$ and *supp* $\hat{\theta}(\omega)\subseteq\{\omega:\pi/4\leq |\omega| \leq\pi\}$. Let the frequency interval $[-\pi,\pi]$ be covered with the dilates of $\hat{\Theta}$ and $\hat{\theta}$; so that:

$$\hat{\Theta}_M + \sum_{v=M+1}^{0} \hat{\theta}_v \equiv 1, \forall\omega \in [-\pi,\pi] \tag{7}$$

The choice of the supports is, in part, arbitrary. Any other choice that covers the frequency line will do as long as the above equation holds. Fig. 3 shows a set of windows covering the frequency line. By 'nicely covering' the frequency space thus, it is ensured that there is no loss of information. Eqn. 7 is used to construct the analyzing functions and the synthesizing functions for the decomposition.

The direct procedure for constructing the FJ analyzing function is then as follows:

(1) For a N-point discrete 1-D signal, choose the total number of window functions, W, such that $0 \leq W \leq \log_2(N)$.

(2) Fix $M = 1 - W(M \leq 0)$.

(3) Choose functions θ and Θ such that Eqn. 7 is satisfied.

(4) Factor Θ and θ:

$$\hat{\Theta} = \overline{\hat{\Phi}}\hat{\Psi} \tag{8}$$

$$\hat{\theta} = \overline{\hat{\phi}}\hat{\psi} \tag{9}$$

The 'bar' over a function denotes its complex conjugation. The following lemma by Kumar and Fuhrmann (1991) affirms the existence functions of $\hat{\Psi}$ and $\hat{\psi}$ when given $\hat{\Phi}$ and $\hat{\phi}$ covering the frequency range $[-\pi,\pi]$ such that equation (7) holds. The lemma is reproduced here for the sake of completeness:

Lemma *Given an m* $\in$ Z; *given* $\hat{\phi}(\omega)$ *such that the properties* $P1(\hat{\phi})$, $P2(\hat{\phi})$, *and* $P3(\hat{\phi})$, *below are true; given* $\hat{\Phi}(\omega)$ *such that* $P1(\hat{\Phi})$,

$P4(\hat{\Phi})$, *and* $P5(\hat{\Phi})$, *are true;* $\exists\hat{\psi}(\omega)$ *satisfying* $P1(\hat{\psi})$ *and* $P2(\hat{\psi})$; *and* $\exists\hat{\Psi}(\omega)$ *satisfying* $P1(\hat{\Psi})$ *and* $P4(\hat{\Psi})$; *such that* $\forall\omega\in\mathbf{R}$,

$$\overline{\hat{\Phi}}_M(\omega)\hat{\Psi}(\omega) + \sum_{v=M+1}^{0} \overline{\hat{\phi}}(\omega)\hat{\psi}(\omega) \equiv 1$$

where for some positive, real constants c and ε ,

$$P1(\hat{\phi}): \hat{\phi} \in \mathbf{C}^\infty$$

$$P2(\hat{\phi}): \textit{supp } \hat{\phi}(\omega) \subseteq \{\omega : \pi/4 \le |\omega| \le \pi\}$$

$$P3(\hat{\phi}): \left|\hat{\phi}(\omega)\right| \ge c, \omega \in \{\omega : (3\pi/8) - \varepsilon \le |\omega| \le (3\pi/4) + \varepsilon\}$$

$$P4(\hat{\Phi}): \textit{supp } \hat{\Phi}(\omega) \subseteq \{\omega : |\omega| \le \pi\}$$

$$P5(\hat{\Phi}): \left|\hat{\Phi}(\omega)\right| \ge c, \omega \in \{\omega : |\omega| \le (3\pi/4) + \varepsilon\}$$

Moreover, the functions Ψ *and* ψ *are rapidly-decreasing.*

The functions in **A** and **S** are each defined from **R** to **C**.

(5) The dilates and translates of Φ_M and Ψ_M are given by:

$$\Phi_{Mk}(n) \underline{\underline{\Delta}} 2^{-M/2} \Phi_M\left(n - k2^{-M}\right) \tag{10}$$

$$\Psi_{Mk}(n) \underline{\underline{\Delta}} 2^{-M/2} \Psi_M\left(n - k2^{-M}\right) \tag{11}$$

The dilates and translates of ϕ and ψ for v = M +1...0 are obtained from the definition given by Eqn. 6. This completes the construction of the sets **A** and **S**.

With these FJ analyzing functions and the synthesizing functions, the FJ decomposition is given by Eqn. 4 which can be restated as:

$$\begin{aligned} f(n) = &\sum_{k=0}^{N2^M-1} 2^{-M}\left(\sum_{i=0}^{N-1} f(i)\overline{\Phi}_M\left(i - k2^{-M}\right)\right)\Psi_M\left(n - k2^{-M}\right) + \\ &\sum_{v=M+1}^{0}\sum_{k=0}^{N2^v-1}\left(\sum_{i=0}^{N-1} f(i)\overline{\phi}_v\left(i - k2^{-v}\right)\right)\psi_M\left(n - k2^{-v}\right) \end{aligned} \tag{12}$$

4.4 An Example of Construction by the Direct Method

Let $\hat{\theta}(\omega)$, the "window function", be given by:

$$\hat{\theta}(\omega) \underline{\underline{\Delta}} \begin{cases} 1/2(1-\cos(\pi \log_2|\omega|)) & \pi/4 \le |\omega| \le \pi \\ 0 & \text{otherwise} \end{cases} \tag{13}$$

The sum of all the dilations of $\hat{\theta}$ is 1, for all $\omega \neq 0$. The functions $\hat{\phi}$ and $\hat{\psi}$ can now be chosen to be the square root of $\hat{\theta}$; $\hat{\phi} = \hat{\psi}$. The functions $\hat{\Phi}$ and $\hat{\Psi}$ can be constructed so that equation (7) is satisfied and again $\hat{\Phi} = \hat{\Psi}$. Figs. 4 and 5 depict the analyzing functions thus constructed in the frequency and time domain respectively. The functions have a small, bounded support in the frequency domain and die rapidly in the time domain. The analyzing functions and the synthesizing functions were chosen to be the same in this example. This is *not* a requirement, however. One may choose to weight certain frequencies in this signal more by choosing the analyzing functions appropriately. In that case, the synthesizing functions will not be the same as the analyzing functions.

From the plots of the Fourier transforms of the analyzing functions (Fig. 4, it is clear that the analyzing function corresponding to $v = 0$. acts like a *high pass filter*, one corresponding to M acts like a *low pass filter* and ones corresponding to intermediate values of v act like *band-pass filters*. Probing the signal with the analyzing function corresponding to v =-4 will capture the low frequency characteristics of the signal. Probing it with the analyzing function corresponding to v =-3 will capture the characteristics in the next higher frequency band and so on. Finally, the function corresponding to v =0 will capture the highest frequencies in the signal. Increasing the number of analyzing functions (or filters) results in increased resolution of lower frequencies. Note that there is a frequency-band overlap among filters.

The FJT of the signal f is given by the sequence of inner-products $\{\langle f, \phi_{M+1\,k}\rangle, \ldots, \langle f, \phi_{0k}\rangle\}_k$. The FJT coefficients at level v are thus computed as $\{\langle f, \phi_{vk}\rangle\}_k$. At every filter level (characterized by v) there are half as many coefficients as at the next filter level. The number of coefficients for level v for an N-point signal is given by $N2^v$.

5 Example

5.1 FJ Decomposition of the Dynamic Response in Fig. 1

Fig. 6 depicts the dynamic response in Fig. 1 with 256 points and its FJT coefficients at levels 0 through -4.

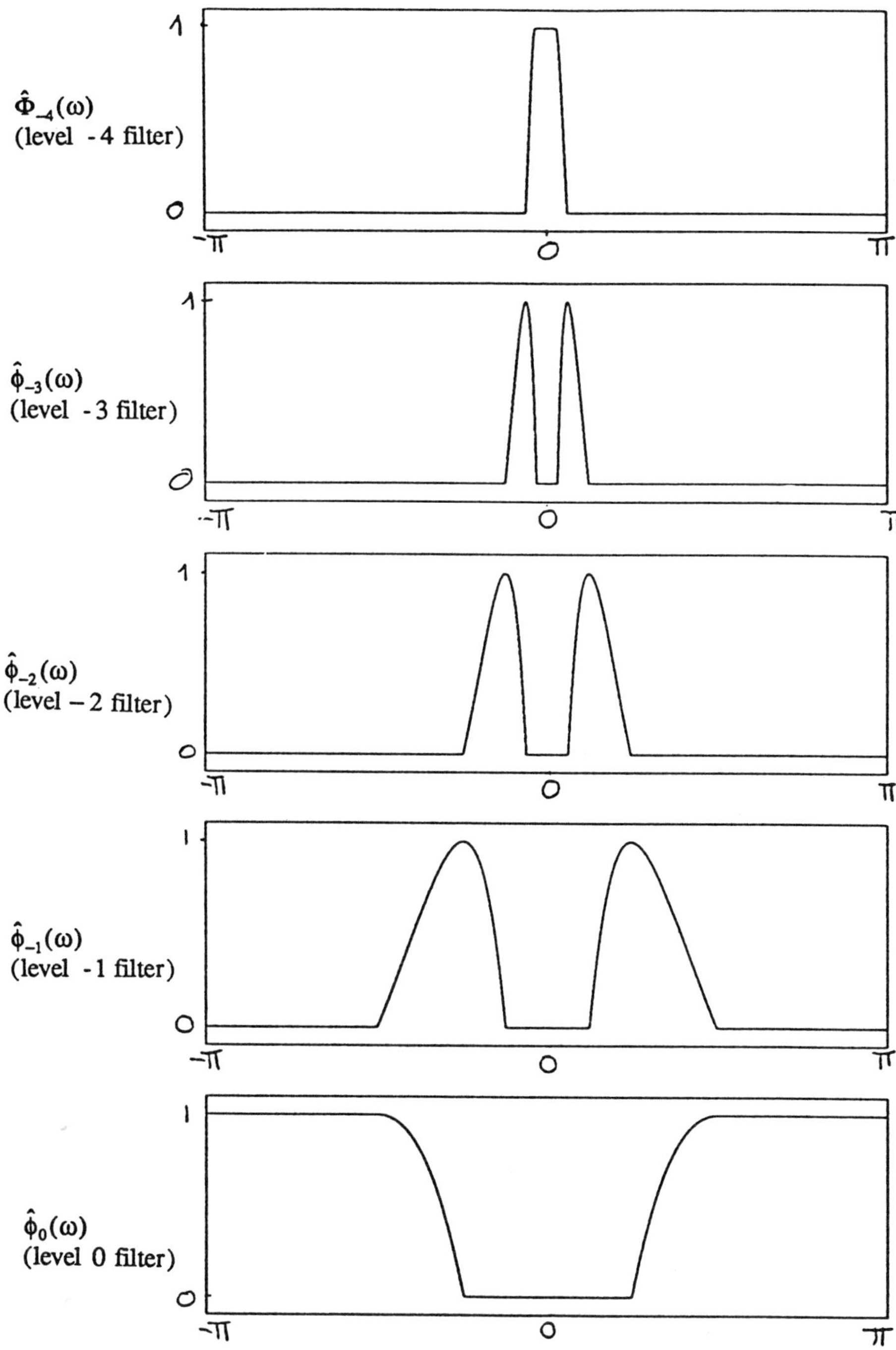

Figure 4. The Filters in the Frequency Domain.

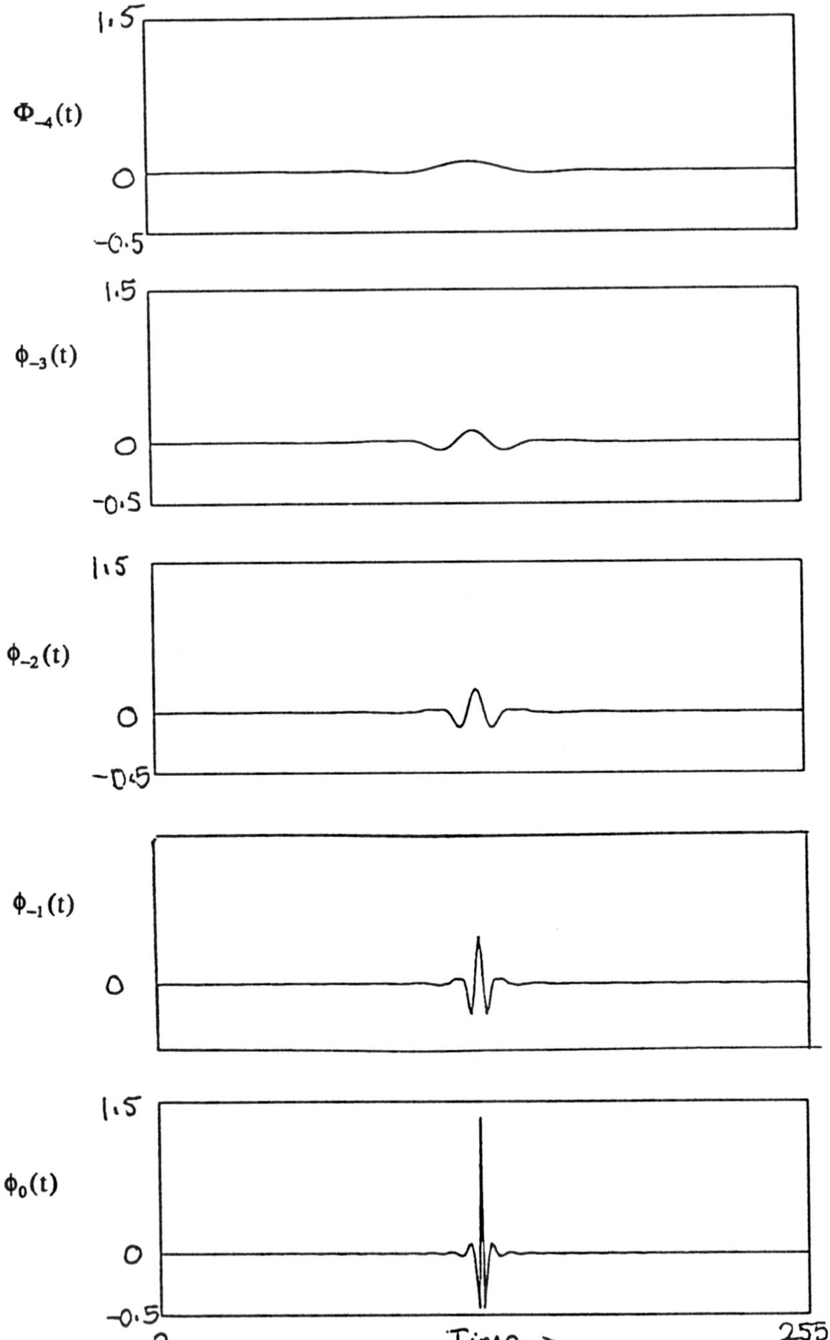

Figure 5. The Filters in the Time Domain.

The lowest level coefficients (level -4) capture the low frequency components in the signal. Level -4 coefficients ($256*2^{-4}$=16 in number) capture the shape of the curve and the actual value of the signal approximately. Coefficients at the next higher level capture additional details at that frequency level. The higher level coefficients can also be termed the *detail signals*. The higher level coefficients point out the presence of interesting phenomena. The largest level-3 coefficients are concentrated around the region where there are sudden changes in the signal. The largest level 0 coefficients are at the time point where there are sudden changes in the signal. Sudden changes such as first-order discontinuities in a signal are characterized by sudden changes in local frequency (from low to high). This is why the highest level coefficients are important around the regions of discontinuities. All the coefficients together (real and 496 in number) will reproduce the original signal exactly. However, considering that only a few coefficients at the higher levels are significantly non zero, an excellent approximation to the original signal can be obtained by including all the coefficients at level -4 and only a few coefficients at higher levels (indicating the time of occurrence of 'interesting' events --typically discontinuities or sudden changes in values) thus yielding 'data compression'. The simplest way to decide which coefficients to retain is to use 'absolute cut-off' at the detail signal levels. For an arbitrary cut-off parameter, η, a coefficient at levels above level -4 is retained if its absolute value is greater than η. The signal in Fig. 1 was generated by adding zero mean random noise with variance 0.1 (a significant amount of noise). Fig. 7 shows its reconstruction with η=0.15 which required only 41 coefficients (451 were discarded). Fig. 8 shows its reconstruction with η=0.2 which required only 26 coefficients (466 were discarded). From a pattern recognition perspective, this method of data compression is acceptable since it eliminates coefficients at higher frequencies. Sensitivity of the human eye tapers off rapidly at high frequencies does not deteriorate the ability of the human eye to recognize images, for example. Use can be made of this fact in a pattern recognition based methodology for trend detection.

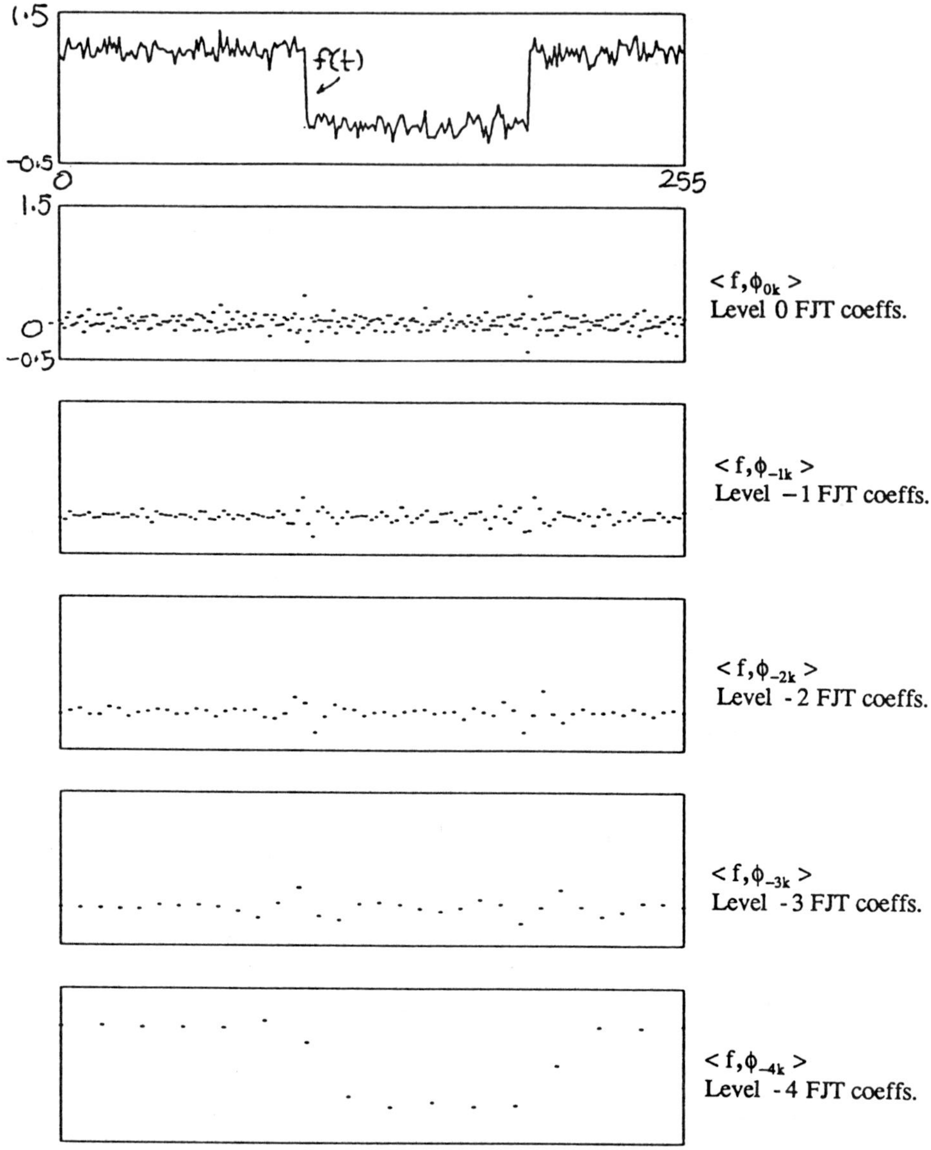

Figure 6. The FJ Decomposition of the Response of Fig. 1.

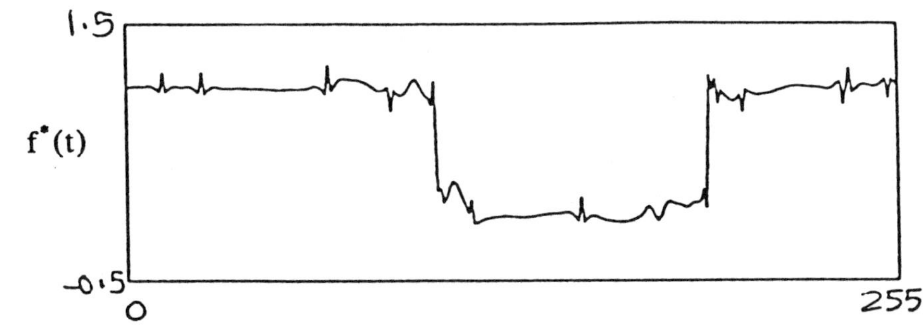

Figure 7. Reconstruction with $\eta = 0.15$ (41 coefficients).

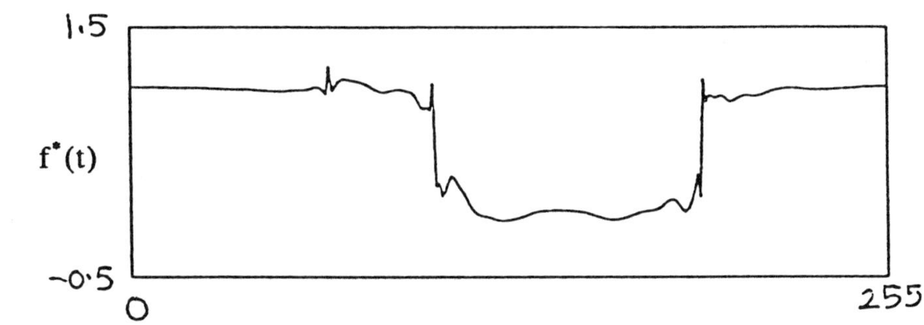

Figure 8. Reconstruction with $\eta = 0.20$ (26 coefficients).

At the highest level, the number of coefficients equal the number of signal points. Consequently, the total number of coefficients is actually *higher* than the number of signal points. This is not surprising in view of the fact that a signal originally represented with an orthonormal basis gets transformed into a space which the FJ analyzing functions span do not form a basis. Why would one want to transform the signal then? There are at least the following advantages to be gained:

- A joint time-frequency description is obtained.

- FJ analyzing functions that have small essential support in the frequency domain and that decay exponentially exist. Because of the functions are not required to be orthonormal, they are easy to construct. Time-frequency localization has a tangible advantage in signal analysis. For example, consider the FJ decomposition in (Fig. 6). The high-frequency (level 0) coefficient with the largest magnitude is at the point of sudden change. This is a direct consequence of the fact that the high frequency filter has very small essential support in the time domain. Such information that is present in the time-domain but gets obscured in the Fourier domain is still available.

- This property of the FJT is an asset when the FJT is used in pattern recognition. Further, it is possible to determine the *order of discontinuity* at a point from the asymptotic rate of decay of the FJT coefficients. This is a very useful property especially when the FJ analyzing and synthesizing functions are used to approximate a function space and solve partial differential equations.

- As is demonstrated in the following section, one can obtain a *multiresolution description* of a signal. The lowest resolution is obtained by using only the lowest level coefficients (i.e. v =M) and the highest by using all the coefficients (and which, of course, corresponds to the original signal). The use of the lowest level and the intermediate level coefficients yields descriptions at intermediate levels.

5.2 Multiresolution Reconstruction

Refer to the signal in Fig. 1. If its time constant is significantly greater than its sampling time, the fluctuations can be interpreted as noise and thus filtered out. On the other extreme, if the time constant is equal to the sampling time, then each of the fluctuations is relevant to its description and analysis and the fluctuations may no longer be filtered out. Time constants intermediate to these extremes will lead one to analyze signals at a different resolution. Essentially, the signal may be looked at different scales or being approximated at different resolutions. In many applications, it is useful to obtain such a multiresolution description of a signal. For example, in process trend analysis one is interested in looking at a signal at different scale (Cheung and Stephanopoulos, 1990). Multiresolution reconstructions are of importance in analyzing signals from processes where the variables being analyzed exhibit different time constants. A distillation column, for example, can exhibit time constants ranging from a few seconds (associated with the gas phase dynamics) to minutes (associated with tray hydraulics) or to hours (associated with composition responses). *FJ decomposition thus provides a neat and compact way of transient signal representation that can capture these widely varying transient characteristics.* Stanton (1990) underlines the importance of selective filtering of data in practical applications. It can be argued that process plant operators selectively filter out noise from trend displays and deduce conclusions about the process behavior by looking at *both* the overall behavior and the local behavior. The largest scale provides some sort of 'context' for the signal behavior. Decreasing the scale gradually reveals its localized behavior. The signal corresponding to the smallest scale is the original signal. For a theoretical treatise on the concept of scale, refer to the paper by Lindeberg (1990).

Typically, the largest scale descriptions of signals are of primary interest. However, some of the local phenomena may be equally important. The use of traditional filtering techniques will yield the largest scale description but will also smooth out any local phenomena--an undesirable characteristic. FJT reconstruction allows one to examine a process signal at many resolutions. One may choose the appropriate resolution at which to examine the signal or may choose to examine the signal at more than one. Depending upon the desired resolution, one can choose the number of coefficients required for analyzing the signal.

Multiresolution descriptions of signals are obtained from partial reconstructions, i.e. by using levels of coefficients incrementally in reconstruction. Fig. 9 shows the multiresolution reconstruction of the response of Fig. 1. It is obvious that if all the coefficients are used, the highest possible resolution reconstruction, the original signal, is obtained (shown in the top panel of Fig. 9). If the lowest resolution reconstruction is required then only the lowest level coefficients suffice. For obtaining signal reconstructions, the sets of coefficients at each level are used incrementally. The lowermost panel shows the reconstruction at the lowest level (level -4 reconstruction) which uses only level -4 coefficients. It captures the 'overall' trend of the signal. The next higher level or reconstruction, level -3, uses level -4 and level -3 coefficients and captures more details. Assuming that very high frequency-content of the signal is noise, level -1 reconstruction gives a sufficiently detailed approximation of the signal.

The resolution of the original signal will be denoted by the numeral 1. The resolution at level ν will then be given by 2^{ν}. In the above example, the lowest level of resolution is 1/16.

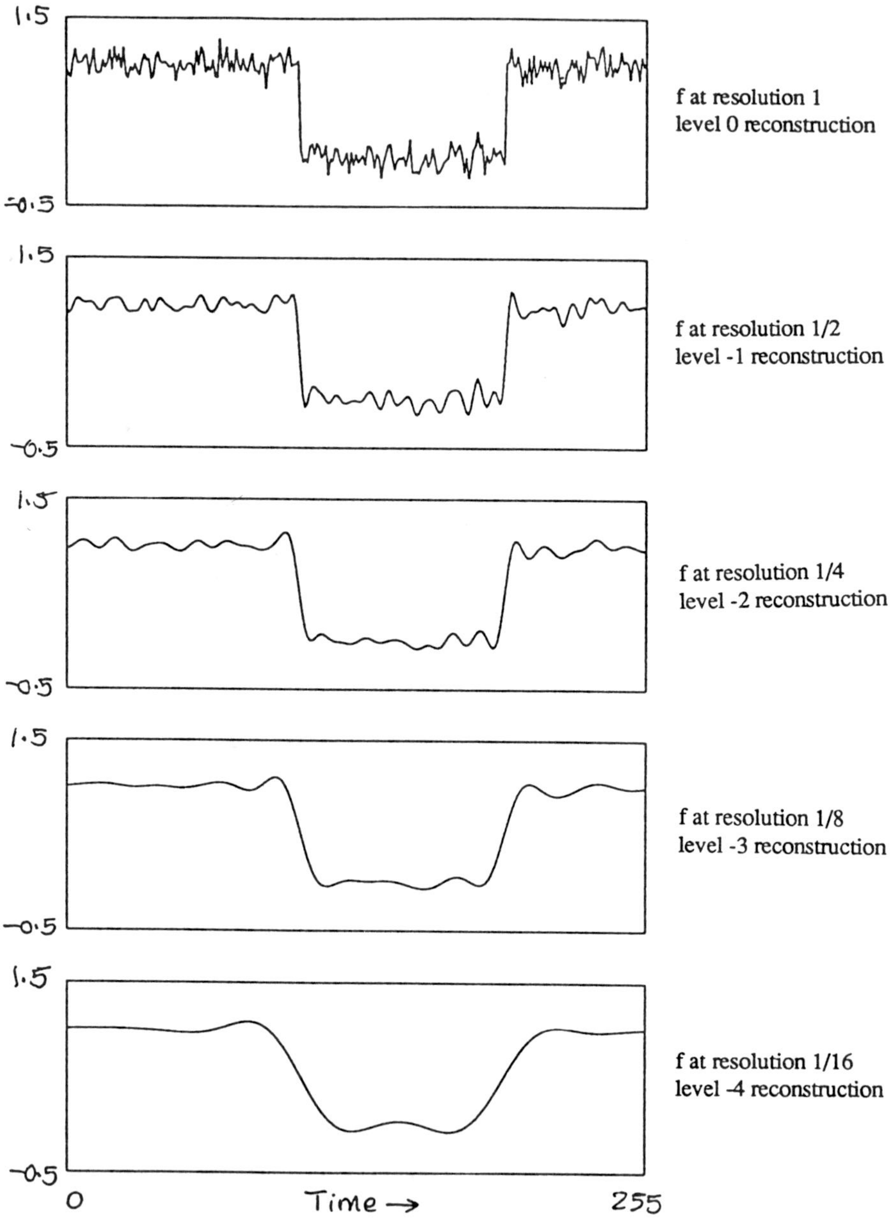

Figure 9. Multiresolution Reconstruction of the Signal of Fig. 1.

5.3 Choices

Two choices remain with the user in the application of the FJT:

- the analyzing functions, and
- the number of filters for the decomposition.

Both are dictated by the nature of application. Supports and magnitudes of the analyzing functions can be chosen to emphasize a particular frequency band. However, certain relationships must be obeyed (Kumar and Fuhrmann, 1990). Once the analyzing functions are selected, their number in the decomposition (the negative valued parameter M) should be chosen by examining the nature of signals that need to be analyzed. A lower value of M (more filters) will allow better separation of low- and mid-frequency components in the signal. Because the high-pass filter cannot be recursively decomposed in the FJT, in general, the high-frequency separation suffers. The other factor influencing the number of filters in the decomposition is the resolution at which the signals need to be analyzed. A lower value of M will mean that the lowest level of coefficients represent the signal at a lower resolution.

6 FJT Algorithms

Sec. 6.1 and Sec. 6.2 describe the O(N log N) algorithms for FJ decomposition and reconstruction respectively. For the detailed analysis of the computational complexity of these algorithms, the reader is referred to (Kumar, 1992). In Sec. 7.2 a method for implementing the FJT on-line is proposed.

6.1 Algorithm for FJ Decomposition

Theorem 1 $\langle f, \phi_{vk} \rangle = 2^{v/2}\left(\hat{f}^*\overline{\hat{\phi}_v^*}\right)^{\vee}(k)$, $k \in Z_{N2^v}$ where $\hat{\phi}_v^*(p)$ is defined as follows:

$$\hat{\phi}_v^*(p) \underline{\Delta} \begin{cases} \hat{\phi}_v(p) & \text{if } p \le N2^{v-1} \\ \hat{\phi}_v(p+N-N2^v & \text{otherwise} \end{cases} \tag{14}$$

The function $\hat{f}^*(p)$ similarly.

Proof

$$\langle f,\phi_{vk}\rangle=\sum_{n=0}^{N-1} f(n)\overline{\phi}_{vk}(n) \tag{15}$$

$$=\sum_{n=0}^{N-1} f(n)2^{-v/2}\overline{\phi}_v(n-k2^{-v}) \tag{16}$$

$$=2^{-v/2}\sum_{n=0}^{N-1}\left(\frac{1}{N}\sum_{m=0}^{N-1}\hat{f}(m)e^{j2\pi mn/N}\right)\overline{\phi}_v(n-k2^{-v}) \tag{17}$$

$$=\frac{2^{-v/2}}{N}\sum_{m=0}^{N-1}\hat{f}(m)\sum_{n=0}^{N-1}\overline{\phi}_v(n-k2^{-v})e^{j2\pi mn/N} \tag{18}$$

$$=\frac{2^{-v/2}}{N}\sum_{m=0}^{N-1}\hat{f}(m)\sum_{p=0}^{N-1}\overline{\phi}_v(p)e^{j2\pi m(p+k2^{-v})/N} \tag{19}$$

$$=\frac{2^{-v/2}}{N}\sum_{m=0}^{N-1}\hat{f}(m)e^{j2\pi mk/N2^{v}}\sum_{p=0}^{N-1}\overline{\phi}_v(p)e^{j2\pi mp/N} \tag{20}$$

$$=\frac{2^{-v/2}}{N}\sum_{m=0}^{N-1}\hat{f}(m)\overline{\hat{\phi}}_v(m)e^{j2\pi mk/N2^{v}} \tag{21}$$

The above development is valid since the arithmetic is (implicitly) being done in the ring $\mathbf{Z}_N$. Since *supp* $\hat{\phi}_v\subseteq\{\omega:|\omega|<\pi 2^v\}, \hat{\phi}_v(m)=0$ if $m \geq N2^{v-1}$ and $m \leq N - N2^{v-1}$, it follows that,

$$\langle f,\phi_{vk}\rangle=\frac{2^{v/2}}{N2^v}\left(\sum_{m=0}^{N2^{v-1}}\hat{f}(m)\overline{\hat{\phi}}_v(m)e^{\frac{j2\pi mk}{N2^v}}+\sum_{m=N-N2^{v-1}+1}^{N-1}\hat{f}(m)\overline{\hat{\phi}}_v(m)e^{\frac{j2\pi mk}{N2^v}}\right) \tag{22}$$

By definition in Eqn. 14, $\hat{\phi}_v^*, \hat{f}^* \in \mathbf{C}^{N2^v}$ and the above equation may be written as an $N2^v$-point DFT:

$$\langle f,\phi_{vk}\rangle=\frac{2^{v/2}}{N2^v}\left(\sum_{p=0}^{N2^{v-1}}\hat{f}^*(p)\overline{\hat{\phi}}_v^*(p)e^{\frac{j2\pi pk}{N2^v}}+\sum_{p=N2^{v-1}+1}^{N2^v-1}\hat{f}^*(p)\overline{\hat{\phi}}_v^*(p)e^{\frac{j2\pi(N-N2^v+p)k}{N2^v}}\right)$$

$$=\frac{2^{v/2}}{N2^v}\sum_{p=0}^{N2^v-1}\hat{f}^*(p)\overline{\hat{\phi}}_v^*(p)e^{j2\pi pk/N2^v}$$

$$=2^{-v/2}\left(\hat{f}^*(p)\overline{\hat{\phi}}_v^*(p)\right)^{\vee}(k) \tag{23}$$

Hence for any fixed $v\leq 0$, the FJT coefficients $\{\langle f,\phi_{vk}\rangle\}_{v,k}$ can be computed with an $N2^v$-point DFT. The DFT of f when convoluted with the conjugate of the DFT of a filter yields the DFT of the FJT coefficients at that level.

6.2. Algorithm for FJT Reconstruction

Define S_ν to be the first sum in the second term of the Eqn. 12, i.e. for $\nu \le 0$:

$$S_\nu(n) \triangleq \sum_{k=0}^{N2^\nu-1} \langle f, \phi_{\nu k} \rangle \psi_{\nu k}(n) \tag{24}$$

The algorithm follows from the following:

$$\hat{S}_\nu(m) = \sum_{m=0}^{N-1} \sum_{k=0}^{N2^\nu-1} \langle f, \phi_{\nu k} \rangle \psi_{\nu k}(n) e^{-j2\pi mn/N} \tag{25}$$

$$= \sum_{k=0}^{N2^\nu-1} \langle f, \phi_{\nu k} \rangle \sum_{n=0}^{N-1} 2^{-\nu/2} \psi_n(n - k2^{-\nu}) e^{-j2\pi mn/N} \tag{26}$$

$$= 2^{-\nu/2} \sum_{k=0}^{N2^\nu-1} \langle f, \phi_{\nu k} \rangle \sum_{p=0}^{N-1} \psi_\nu(p) e^{-j2\pi m(p+k2^{-\nu})/N} \tag{27}$$

$$= 2^{-\nu/2} \sum_{k=0}^{N2^\nu-1} \langle f, \phi_{\nu k} \rangle e^{j2\pi mk/N2^\nu} \hat{\psi}_\nu(m) \tag{28}$$

$$= 2^{-\nu/2} \langle f, \phi_{\nu k} \rangle^\wedge(m) \hat{\psi}_\nu(m) \tag{29}$$

Thus, the reconstruction is performed by taking the inverse DFT of the sum of the terms that arise at each filter level ν. The number of filters used in the reconstruction determines the level of reconstruction and thus the resolution of the reconstructed signal. The number of operations for reconstruction is about the same as required for the decomposition.

7 On-line Implementation of the FJT

7.1 The Wrap-around Problem

When an infinite signal is truncated to finite dimension to take its discrete Fourier transform, spurious 'end effects' are introduced (e.g. see (Press et. al., 1989)). Such truncation is necessary for on-line application. The discrete FJ transformation of such a chunk of a signal also introduces the same end effects. The DFT and the FJT treat a *discrete* finite dimensional signal as a periodic function with period *N*. This condition is hardly ever satisfied in reality. Not only must the sampled original signal be viewed as cyclic, but so must the analyzing functions. Therefore a portion at each end of the original function is erroneously wrapped around as a result of convolution with the analyzing functions. The DFT and the discrete FJT are thus said to be prone to the *problem of wraparound.* As a result, some of the coefficients at the two ends are actually spurious.

In the on-line implementation of the FJT, it is therefore necessary to exercise caution. A few coefficients at the ends are going to be spurious. It is necessary to determine just how many coefficients would be spurious given the set **A**. In the following section, the method for determining the number of spurious coefficients at each level is presented. Only thus can one obtain a 'true' FJT spectrum estimation on-line.

7.2 The Method for Implementation

The wraparound problem may be illustrated with the help of the *level* - 4 analyzing function in Fig. 5 reproduced in Fig. 10 on a larger scale. This is a symmetric function around t_0 over which the function is appreciably non-zero. The *effective radius,* r, of this function is the distance from t_0 over which the function is appreciably non-zero. In the FJ transformation, this function is translated along the horizontal axis. At distances less than r from the ends, this function will 'spill over' the end and will in fact be wrapped around, i.e., the spilled-over portion will appear at the other end. This wraparound will result in a false 'pollution' when a signal is convoluted with it. For every analyzing function (characterized by level v), the extent of this 'pollution' is determined by the effective radius r_v. At level v, coefficients within distances r_v from the two ends get polluted.

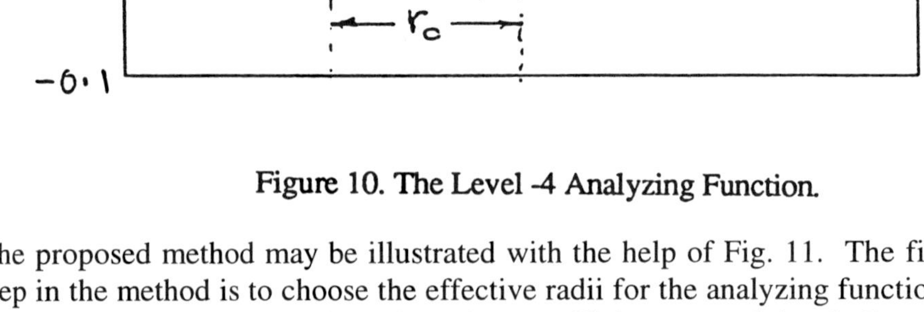

Figure 10. The Level -4 Analyzing Function.

The proposed method may be illustrated with the help of Fig. 11. The first step in the method is to choose the effective radii for the analyzing functions and thus determine the number of spurious coefficients at each level. It turns out that only the effective radius for the high-pass filter, r_o needs to be chosen. The reason for will be explained shortly. Once r_o is fixed, one can easily calculate the number of spurious coefficients at each end at each level. The method can be summarized as follows:

(1) Choose the initial chunk of signal EF (the size of which is typically chosen so that the number of signal points is power of two). Effectively, only the signal within the window EFGH is being inspected. Let the length of the segment EF be denoted by 1. Choose r_o.

(2) Take the FJT of the signal segment EF. The coefficients are shown in Fig. 11.

(3) Coefficients within distance r_o at each end at filter level 0 are spurious. At the filter level v, coefficients within distance $2^{-v} r_o$ at each end are spurious. In Fig. 11, the coefficients *outside* the region *abcd* will be thrown away. Effectively, FJT is known correctly for only a part of the original signal - the signal portion *ef*.

(4) The next step is to move the window EFGH along the horizontal axis over the signal by an amount equal to $1-(2^{1-M} r_o)$. The new rectangular window is labeled as E'F'G'H' in Fig. 11. Note that it overlaps the window EFGH and that the region *abcd* also gets changed to *a'b'c'd'*.

(5) Repeat steps (2), (3) and (4). The coefficients in the region of overlap between *abcd* and *a'b'c'd'* (the shaded area) are valid coefficients from the first FJT computation and these need not be recalculated. These are the coefficients that would have been spurious in the FJT of signal in the window E'F'G'H'.

Note that the FJT coefficients for a part of the signal at the very beginning are spurious and need to be discarded. This is its only apparent disadvantage with this method.

The above method requires specification of only one parameter, r_o, because the effective radii for the analyzing functions are related to each other. The functions are obtained by dilating a parent function. The function at level v + 1 is obtained by dilating the function at level v and scaling it (see Fig. 5 e.g.). If the scaling were not performed,

$$r_{v+1} = r_v / 2 \tag{30}$$

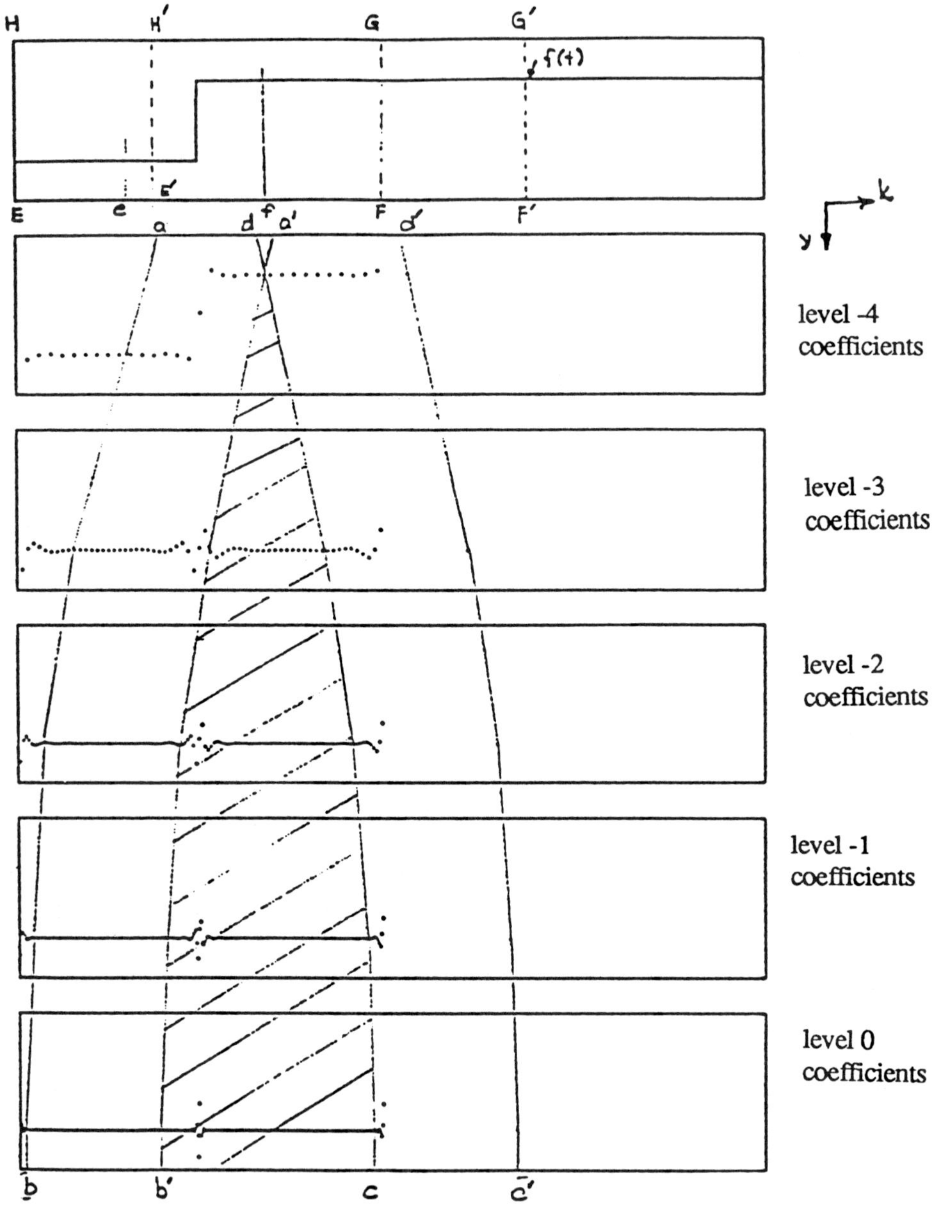

Figure 11. On-line Implementation of the FJT.

In that case, if r is fixed for any filter level *v*, then all other effective radii also get fixed. However, scaling *is* performed at each level and the equality of Eqn. 30 is not valid. It can be easily shown that $r_{v+1} > r_v / 2$. This inequality is the reason the base r value is chosen for level 0. The resulting effective radii that result from the use of Eqn. 30 then become conservative estimates. Using r_M for base value would yield erroneous results.

8 Conclusions

A new transform technique, the Frazier-Jawerth transform has been discussed. A new method for its on-line implementation has been proposed. The FJT would find use in applications where non-stationary signals have to be analyzed to extract information about a dynamic system. Such analyzes are required for example in residence-time distribution studies, fault diagnosis, process control. FJT yields *joint* time-frequency information which cannot be obtained from either the time domain values of a signal or its Fourier transform. When a signal varies smoothly, its lower frequency-level FJT coefficients will be most significant an will be few in number. When a signal varies rapidly, the higher frequency coefficients become significant. Many times this information is contained in only a few non-zero high frequency coefficients. Since most of the dynamic responses in chemical systems have predominantly low-frequency components, only a few FJT coefficients suffice to characterize a significant amount of information in the signal. Thus tremendous data compression is possible without loss of a great amount of information. Moreover, the FJT coefficients at one or more levels may be used together to approximate signals at different resolutions if needed. O(N log N) algorithms exist for the FJT so that the computations can be performed in real time. The FJT represents a mathematically rigorous method of treating data.

9 Acknowledgment

This work was partially supported by the National Science Foundation (Grant No. CTS-9108649).

References

"A New Wave in Applied Mathematics," Science, Vol. 249, pp. 858-859, August 1990.

Allen, J. B. and Rabiner, R., "A Unified Approach to Short-Time Fourier Analysis," *Proc. IEEE*, Vol. 65, no. 11, pp 1558-1564, Nov. 1977.

Boashash, B. and Whitehouse, H. J., "Seismic Applications of the Wigner-Ville Distribution," Proc. IEEE Int. Conf. Systems and Circuits, pp. 34-37, 1986.

Boashash, B., Lovell, B. and White, L., "Time Frequency Analysis and Patter Recognition Using Singular Value Decomposition of the Wigner-Ville Distribution," Advanced Algorithms and Architecture for Signal Processing, Proc. SPIE, Vol. 28, pp. 104-114, 1987.

Cheung, J. T.-Y. and Stephanopoulos, G., "Representation of Process Trends -- Part II. The Problem of Scale and Qualitative Scaling", Computers chem. Engng, Vol. 14, No. 4/5, pp. 511-399, 1990.

Cohen, L., "Time-Frequency Distributions --A Review," Proceedings of the IEEE, Vol. 77, No. 7, July 1989.

Forrester, B. D., "Use of the Wigner-Ville Distribution in Helicopter Fault Detection," Proc. ASSPA 89, pp. 78-82, 1989.

Frazier, M. and Jawerth, V., "Decomposition of Besov Spaces," Indiana U. Math. J., Vol. 34, No. 3, pp. 777-799, 1985.

Gabor, D., "Theory of Communication," J.III (London), Vol. 93, pp. 429-457, 1946.

Heil, C. E. and Walnut, D. F., "Continuous and Discrete Wavelet Transforms," SIAM Review, Vol. 31, no. 4, pp. 628-666, 1989.

Kumar, A., and Fuhrmann, D., "The Frazier-Jawerth Transform," Proc. of the IEEE Intl. Conf. on Acoustics, Speech, and Signal Processing, PP. 2483-2486, Albuquerque, NM., April 1990.

Kumar, A., Fuhrmann, D., Frazier, M. and Jawerth, B,. " A New Transform for Time-Frequency Analysis," IEEE Trans. on Signal Processing, Vol. 40, no. 7, pp 1697-1707, 1992.

Kumar, A., D.Sc. Dissertation, Dept. of Computer Science, Sever Institute of Technology, Washington University, St. Louis, Missouri, November 1990.

Lindeberg, T., "Scale-Space for Discrete Signals," IEEE Trans. on Pattern Analysis and Machine Intelligence, Vol. 12, No. 3, March 1990.

Press, W., Flannery, B., Teukolsky, S. and Vetterling, W., Numerical Recipes, Cambridge University Press, 1989.

Stanton, B. D., "Using Historical Data to Justify Controls," Hydrocarbon Processing, Vol. 69, No. 6, pp. 57-60, June 1990.

Chapter 4

Process Signal Features Analysis

Xue-dong Dai, B. Joseph and R. L. Motard
Washington University
Chemical Engineering Department
One Brookings Drive, Box 1198
St. Louis, Missouri 63130

Abstract

In this chapter, a phase plane analysis method has been proposed as a new way of analyzing nonstationary signals, based on wavelet theory. Some basic concepts are defined to form a theoretical framework of this method. Several possible applications are discussed. This approach offers the potential for building new intelligent process signal analysis systems to identifying the process "finger prints", i.e. the hidden time-frequency structure in signals.

1. Introduction

Signal features, especially the temporal signal features of the process measurements usually carry the most important information of the underlying process behaviors. One of the approaches to understand these behaviors is to analyze the output signal from the sensor directly. Traditional practice is to do the analysis in the time domain, or do the Fourier transform to look at the spectrum of the signal in the frequency domain. For a nonstationary signal, it is better to study the signal features in the two dimensional time-frequency domain. This leads to the possibility of discovering unique characteristics of the process which may be hidden otherwise. Recently, wavelet theory, including wavelet transform and its extension wavelet packets theory, provides a new time-frequency signal analysis method, and has attracted much attention from pure science to engineering fields (Jawerth and Sweldens, 1993, Daubechies et al eds., 1992).

In some applications, it is of interest to study phenomena occurring at different scales. The wavelet transform provides a multiscale decomposition of signals that offers a good possibility for such an analysis. Many wavelet-based applications take the advantage of this property of the transform. By using this multiscale analysis approach, we can scan and/or analyze the information scale by scale to obtain the complete results. If we just scan some pre-determined scales, it is hard to guarantee that the signal features to be analyzed are always included in these particular scales. In other words, there will be a problem in selecting the analyzing scales when we don't have enough knowledge of signals. Therefore, it will be nice to find a way to "look at" signals from a global point of view. Such a "global" representation is particularly useful in extracting the overall features of signals in a simple way. Actually, starting from time-frequency distributions, spectrograms for short-time Fourier transform and scalograms newly introduced for continuous wavelet transforms, these two-dimensional time-frequency representations have already shown their distinguishing power in dealing with the problem (Cohen, 1989, Rioul and Vetterli 1991, Hlawatsch and Boudreaux-Bartels, 1992).

Based on the wavelet theory, we have proposed a new transient signal analysis method named the phase plane analysis method to study nonstationary process signals, as part of an effort to develop an intelligent process signal analysis system (Dai, 1992). This presents a novel approach to apply the wavelet theory for solving engineering problems. More specifically, this approach offers the potential for building a new intelligent process signal analysis system to analyze the process "finger print", i.e. the hidden time-frequency structure of signals. There are number of possible applications, such as a visual sensor data analysis tool for studying the unknown time-frequency features in signals, new data filtering techniques for noise reduction and/or specific signal feature extraction and a technique for detecting difficult to measure process changes and abnormalities. The proposed method has been successfully applied to identify as well as classify the different phenomena in a corrosion process.

In Section 2, we begin with the discussion of phase plane definition introduced by Wickerhauser (1991), and then introduce a concept of signal patterns along with other supporting concepts with their properties in a new framework. Another important concept, the signal dependent basis, is presented in Section 3. In Section 4, we show that the proposed framework can be implemented to form useful analysis tools for visual data analysis as well as sensor data filtering and/or feature extraction. Finally Section 5 concludes the chapter.

2. Signal patterns on the phase plane

It is obvious that a compact and reliable representation of signal features is the first step for developing successful process signal analysis tools. In this section, we will define one such representation, and discuss its stability.

2.1. Phase plane definition

By extending the theory of wavelet transform, a library of the orthonormal bases called the "wavelet packets library" can be constructed (See Chapter 1, Section 4.3 for details). It covers the wavelet basis as well as the other orthonormal bases and thus provides a more flexible scheme to analyze signals (Coifman, Meyer, Wickerhauser, 1992a,b).

A two-dimensional time-frequency representation, the phase plane, is defined for the wavelet packets by Wickerhauser (1991). In this representation, each wavelet packet transform coefficient is associated with a time t and a frequency f, with their time and frequency uncertainty amounts Δt and Δf respectively. The result is interpreted as a rectangular patch of dimensions Δt by Δf, located around (t, f) on the phase plane. The smallest area (Δt by Δf) of the patch is limited by the Heisenberg uncertainty principle. The patch is assigned a color or gray scale in proportion to the amplitude of the corresponding transform coefficient (Figure 2.1).

Certain bases have been characterized in terms of the configuration of the patches presented on the phase plane. As shown in the Figure 2.2, we can clearly see that the standard basis has optimal time localization and no frequency localization, while the Fourier basis has optimal frequency localization, but no time localization. A windowed Fourier transform with a fixed window size has fixed time-frequency resolutions, and the orthonormal

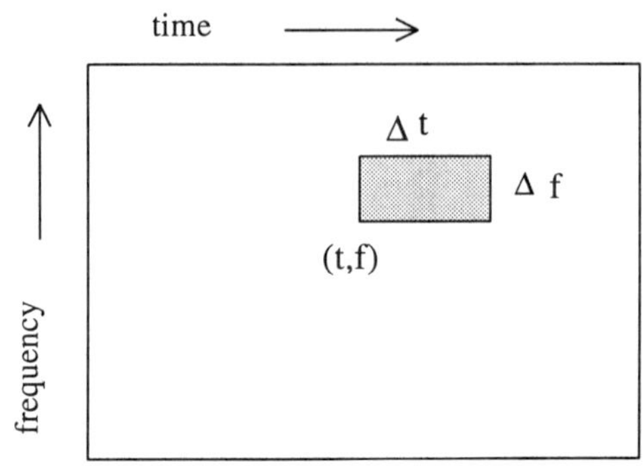

Fig. 2.1 The definition of the phase plane

wavelet basis has good time resolution in the high frequency band and good frequency resolution in the low frequency band.

The representation of the whole transform coefficient sequence corresponding to an orthonormal basis exactly is a disjoint cover of the entire phase plane by the patches. Thus, the phase plane shows the complete information of the transformed signal within the analyzing window, which is the "global" view of the signal we expected. For example, for a pair of chirps

$$S(t) = \sin(250\pi t^2) + \sin(190\pi t^3)$$

where $0 < t < 1$, sampled 512 times (Fig. 2.3 (a)). The transform results are shown in the Figure 2.3 (b) and (c). From the figure, it is easy to see that all signal features are shown in one picture (Fig. 2.3 (b)). The best basis of wavelet packets represents the analyzed signal with the least information, i.e. it minimizes the number of grey patches on the phase plane. Indeed, the best basis provides less dark patches and clearer patterns in which the superposed chirps can be seen (Fig. 2.3 (c)).

This implies a more important observation that the wavelet packets can represent the signal features in a more flexible way than the wavelet basis, which not only benefits signal and image compression as proposed by Wickerhauser and his colleagues (1992a), but holds promise for further

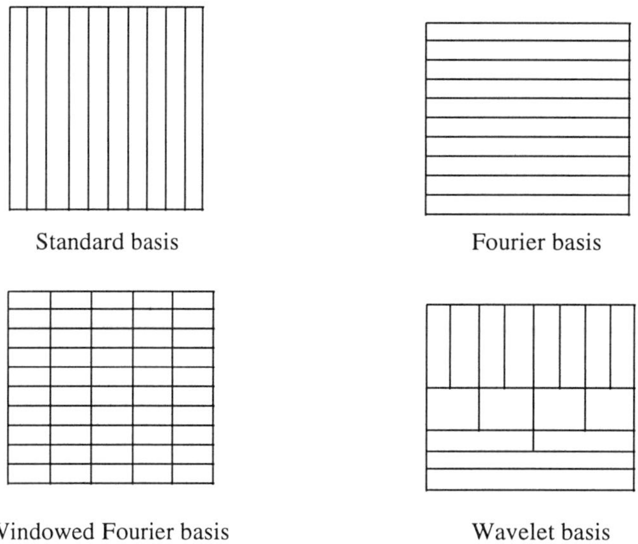

Fig. 2.2. The configuration of some bases on the phase plane

development of some new analysis methods based on the above phase plane definition, the global view of signals.

We use this definition as the basic starting point for the development of a new signal analysis method. In Figure 2.4, we describe the relationship of the proposed method and other existing wavelet theoretical fundamentals and methods.

2.2. Signal pattern representation

2.2.1. Signal pattern definition

Suppose we have a finite sequence $x = \{x_1, ... x_n\} \in X^n$, e.g. n sampled values of a signal. The orthonormal wavelet packet transform[1] *WP* is a linear mapping

$$WP : \begin{cases} X^n \to W^n \\ x \to w = WP(x) = \langle x, B \rangle \end{cases} \tag{2.1}$$

[1] We can use the *WP* transform on intervals or the periodic wavelets (Daubechies, 1992).

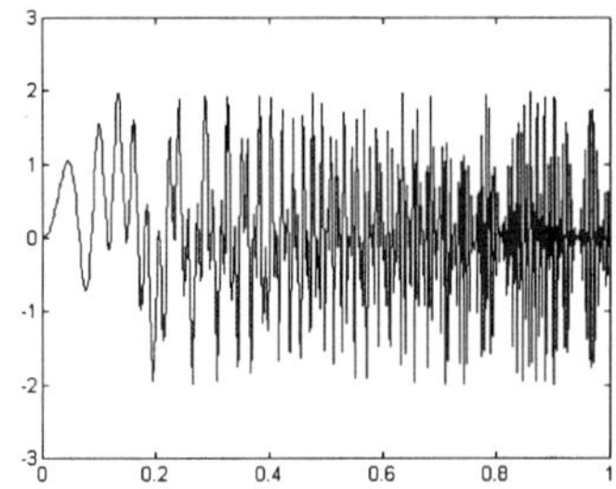

(a) The original signal in time domain

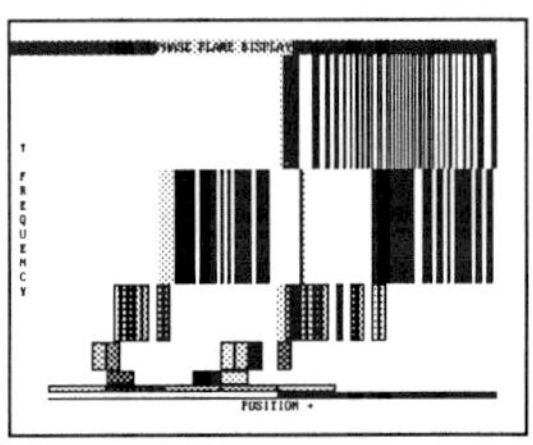

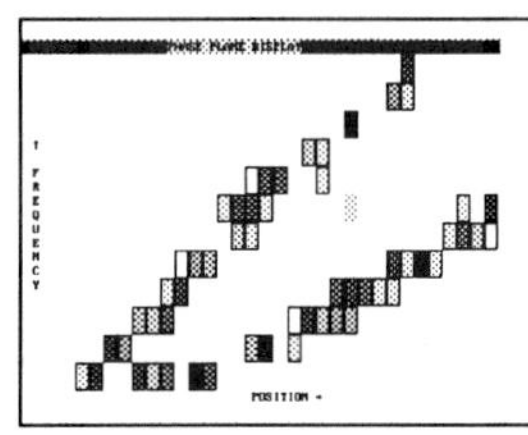

(b) Wavelet basis (c) Best basis

Fig. 2.3 The superposed chirps on the phase plane

where $w = \{w_1, ...w_n\} \in W^n$, B is an orthonormal basis selected from the wavelet packets library, and <*> denotes the operation of inner product. When the wavelet basis is chosen in the transform, the mapping represents the orthonormal wavelet transform.

There exists a unique inverse transform, i.e. the inverse mapping

$$WP^{-1} : \begin{cases} W^n \rightarrow X^n \\ w \rightarrow x = WP^{-1}(w) \end{cases} \tag{2.2}$$

In transform domain, each transform coefficient w_i of the sequence w actually carries the time-frequency structure information $(t, f)_i$, the index in the time-frequency domain. When the sequence w is organized by its coefficients' time-frequency indices $(t, f)_i$, it represents the time-frequency

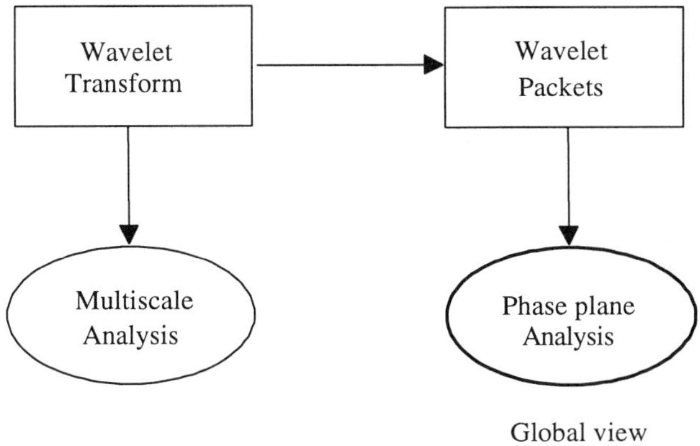

Fig. 2.4. The phase plane analysis and other existing wavelet methods

structure of the original sequence x. This time-frequency structure can further be decomposed into sub-structures if their time-frequency features are clearly differentiable. According to the time-frequency localization principle of the transform: if a local event is essentially localized in the time interval [-T, T] and its Fourier transform is essentially localized in [-Ω, Ω], then only those coefficients corresponding to time-frequency points within or close to the region $[-T,T]\times[-\Omega,\Omega]$ are needed to reconstruct the local event approximately, we can use these time-frequency sub-structures to represent the local events in x. This implies that a new analysis unit can be established, based on the time-frequency structure and/or sub-structure. This leads to the concept of *signal patterns*:

<u>Definition 2.1(a)</u> Given a finite sequence $x \in X^n$ with its *WP* mapping sequence $w \in W^n$, a *signal pattern P* is w organized in its time-frequency structure. A *sub-pattern* $S \subset P$ is then a time-frequency sub-structure in P, which can be related to a local event in x.

In parallel, by presenting w on the phase plane defined in the previous section, we have the following equivalent definition for the signal pattern on the phase plane:

Definition 2.1(b) A *signal pattern P* is formed by the coefficients of wavelet packet transform appearing as the configuration of time-frequency patches on the phase plane. A *sub-pattern S* of the signal refers to parts of the pattern associated with parts of the localized signal features, $S \subset P$.

Obviously, the signal pattern and the sub-pattern we defined here may spread to several scales in the frequency band as well as a period of time and appear as a time-frequency region which presents the localized time-frequency structure. Therefore, this representation provides the "global" view of the signal features as expected. Also, the signal pattern is visualized in the analysis by the definition 2.1(b). For example, Figure 2.5 displays a time domain signal in (a) and its corresponding approximation signal pattern on the phase plane in (b). It is easy to further identify a localized sub-pattern,

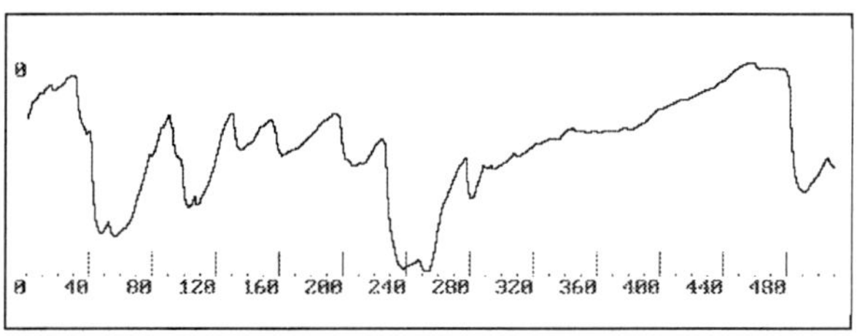

(a) The original signal x

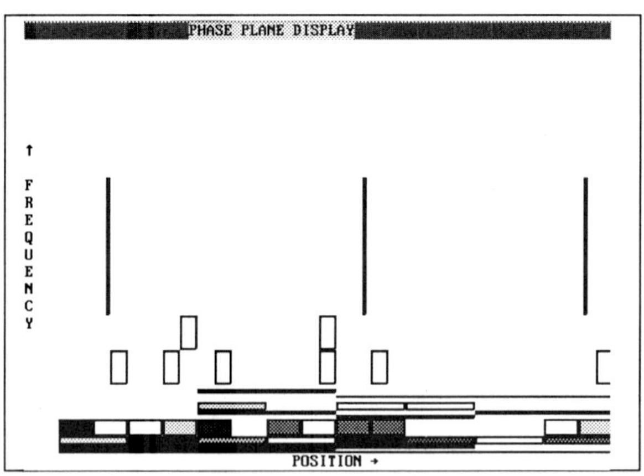

(b) Signal pattern $P*$

Fig. 2.5 Signal pattern on the phase plane

the long vertical black patch, from the signal pattern, which can be associated to a local event, the big jump in the original time domain signal. In fact, there are three similar sub-patterns in the picture. Now we can simply process the signal pattern (sub-pattern) to finish the tasks such as extraction, identification, approximation and classification of the signal features we expected.

2.2.2. Signal pattern approximation

Since the same orthonormal basis selected from the wavelet packet library is used in both the mapping (2.1) and its inverse (2.2), the results w and x are considered to be unique and complete. However, we often use approximation w^* instead of w in practice. Then, the inverse mapping becomes:

$$WP^{-1}: \begin{cases} W^n \to X^n \\ w^* \to x^* = WP^{-1}(w^*) \end{cases} \tag{2.3}$$

where $w^* = \{w_1, ...w_j, 0, ...0\}$ is a subset of w. Accordingly, we have the definition for *an approximation signal pattern* P^* as follows

Definition 2.2 Given a finite *WP* mapping sequence w corresponding to the original sequence x, if its approximation sequence w^* is selected from w in such a way that the retained coefficients always keep the main time-frequency structure of w, then P^* corresponding to w^* is said to be the *approximation signal pattern* of P. Similarly, S^* is defined for S, with $S^* \subset P^*$.

In fact, the approximation operation required by the definition can be achieved by simply applying a shrinking algorithm, using a signal pattern (sub-pattern) as the processing unit without being limited by the scales. There are several approximation criteria designated for different applications. For example consider the simple threshold criterion. For a given threshold ε, it eliminates the coefficients of absolute value less than ε. The energy threshold, on the other hand, it retains those large value coefficients which exceed a given energy level (represented by the square of the coefficients).

The accuracy and retaining ratio (j/n, the number of retained coefficients over the total number of the transform coefficients in the sequence w)

depend on the sparseness of the mapping result w. We try to maximize sparseness by choosing an appropriate basis from the library.

Donoho and Johnstone (1992) proposed a principle for spatially-adaptive estimation, call selective wavelet reconstruction, which indicates the significance of the wavelet approximation scheme. The principle provides reconstructions by selecting only a subset of the wavelet transform coefficients. This is the case of approximation inverse mapping Eq. (2.3) in using the orthonormal wavelet basis in the transform. Their selected wavelet reconstruction can perform as good as the other ideal spatial adaptation methods. This ideal selected wavelet reconstruction can guarantee accuracy of the approximation depended on the data alone.

Their ideal selected reconstruction is based on the time-frequency structure of the original sequence, i.e. the approximation signal pattern P^* (sub-pattern S^*) defined above. This leads to an improvement of their implementation results in applications.

2.3. Stability of signal patterns

Due to the nature of the transform, the signal pattern will show some variations when the original signal shifts in time. For the same process behavior, its "finger print" may be corrupted by process and sensor noise, and also show some variations in its signal pattern. So, it is important to consider the stability of the signal pattern defined above. Roughly speaking, a necessary and sufficient condition for a signal pattern to be a stable signal pattern is that the variation of itself should be always less than the difference between itself and the other patterns. This guarantees that we can still perform the recognition and classification correctly.

To measure the similarity and difference between the two signal patterns, we introduce the concept of *the distance of the signal patterns.*

Let P_1^*, P_2^* be two approximation signal pattern corresponding to two *WP* mapping sequence $w^*_1=\{w_{11}, \dots w_{1j}\}$, $w^*_2=\{w_{21}, \dots w_{2m}\}$ respectively, where j and m may or may not be equal. Consider the following index mapping for P^* with corresponding sequence w^*:

$$In : \begin{cases} P^* \to IP^* \\ w^* = \{w_1, \dots w_m\} \to In(w^*) = \{e_1, \dots e_m\} \end{cases} \tag{2.4}$$

where

$$e_i = \begin{cases} 1 & \text{if } w_i \neq 0 \\ 0 & \text{if } w_i = 0 \end{cases} \tag{2.5}$$

The index set IP^* represents the time-frequency structure information of P^*. Thus, we have the following definition of distance:

<u>Definition 2.3</u> Let IP_1^* and IP_2^* be two index sets defined by Eq. (2.4) corresponding to P_1^* and P_2^*, the *set distance* of P_1^* and P_2^* is that

$$D_1(P_1^*, P_2^*) = n(IP_1^* \oplus IP_2^*) = n[(IP_1^* \cup IP_2^*) - (IP_1^* \cap IP_2^*)] \quad \in N \tag{2.6}$$

where $n(A)$ denotes the number of elements in set A.

In the case of set distance $D_1(P_1^*, P_2^*) = 0$, we further define the second pattern distance measure by w^* directly:

<u>Definition 2.4</u> Let P_1^*, P_2^* be two signal patterns, their *weight distance* is defined as

$$D_2(P_1, P_2) = \sum_{i=1}^{m} (w_{1i} - w_{2i})^2 \quad \in R \tag{2.7}$$

The set distance D_1 is designated to count the major difference, the time-frequency structure difference, between the signal patterns, while the weight distance D_2 measures further details, the intensity difference.

Due to the nature of the *WP* transform, for each signal pattern P^* shifting along the time axis, P^* corresponding to signal x will become $P^{*\prime}$. This variation can also be measured by the distance measure either $D_1(P^*, P^{*\prime})$ or $D_2(P^*, P^{*\prime})$. For example, weight distance D_2 measure should be

$$D_2(P^*, P^{*\prime}) = \sum_{i=1}^{m} (w_i - w_i')^2 \quad \in R \tag{2.8}$$

Thus, we have a *necessary and sufficient condition* for a signal pattern P^* to be *a stable signal pattern*:

<u>Definition 2.5</u> A signal pattern P^* is *a stable signal pattern* if and only if P^* satisfies the following condition:

During an observation interval Δt for both distance measures D_1 and D_2 if

$$D_k(P^*, P^{*\prime}) < D_k(P^*, P_1^*) \quad k = 1, 2 \qquad (2.9)$$

where $P^{*\prime}$ is the variation pattern of P^* in Δt corresponding to a signal x, and P^*_1 is another arbitrary signal pattern representing a signal x_1 other than x. Clearly, condition (2.9) guarantees that the pattern recognition mechanism can differentiate the different signal patterns although each pattern itself may have variation when it shifts in time.

There are several factors which affect the stability of signal patterns. The first and most important one is the effect of changing basis. From the earlier discussion for the phase plane, we know that the shape of the time-frequency patches is determined by the basis, i.e. a different basis has a different shape and configuration of the patches on the phase plane which can be clearly seen in Fig. 2.3. It means that the signal pattern changes with the basis used in the transform. Therefore, the basis should be fixed in some way in the applications to avoid this problem. We will discuss this topic in detail in the next section.

The second factor affecting the stability of signal patterns is referred to as the lack of shiftability of orthonormal wavelet transform (Simoncelli et al. 1992, Mallat, 1989). The problem is that the wavelet transform coefficients do not shift as the input signal shifts in time. From observing the results of phase plane analysis, this problem has been identified as being caused by the periodic effect of the basic wavelet packet function (Dai, 1992).

In general, the period of the basic wavelet packet function can be easily calculated. Suppose the basis function belongs to scale s in the wavelet packets, then its period is 2^s sample points, where s denotes the scale as 0 to L from top to bottom in the wavelet packet library which is organized as a tree. This problem can also be explained as the uncertainty effect of the basic function whose uncertainty in position exactly equals 2^s. The corresponding time-frequency patches on the phase plane are 2^s periodic as well. According to the above observation, we have the following definition for *the period of a signal pattern* on the phase plane.

Definition 2.6 For a signal pattern P, its *period* is defined as the maximal period of the basis function (the dark time-frequency patch on the phase plane) in the pattern. The pattern will remain the same when the input signal

x is shifted in this period. *The period of a sub-pattern* is defined similarly as the maximal period of the basis function in the sub-pattern.

Hence, the signal pattern or sub-pattern can still be easily identified, if the signal x is shifted in its periods as defined above.

On the other hand, if we focus on the main time-frequency structure of P, the P^*, rather than its minor details (the individual basis function or the time-frequency patch), the stability condition (2.9) is much easier to satisfy.

3. Signal dependent basis

A stable signal pattern is a necessary condition for the application of pattern recognition. To obtain such a stable signal pattern, a new group of signal dependent orthonormal bases are desired. First, we need to determine the domain from which the basis can be selected. This leads to a concept of *shiftable basis* defined as follows

Definition 3.1 Consider the configuration of a basis (See Fig. 2.2) on the time-frequency domain. A *shiftable basis* is the basis whose elements always have the same scale along the time (or position) axis. The scale may be different along the frequency axis.

Each transform coefficient w_i actually contains three indices, a scale index s, a frequency index f and a time (or position) index p by the definition of wavelet packets transform. The time and frequency indices (t, f) determine the location of the coefficient in the two-dimensional time-frequency domain, and the scale index s decides the shape of the uncertainty area of this coefficient, i.e. the shape of the time-frequency patch on the phase plane. The same scale index means the same shape of the patch viewed on the phase plane. By the above definition, the basis functions of the shiftable basis, or the patches on the phase plane, have the desirable property of translation invariance[1] along the time (or position) axis. It guarantees that the input signal is treated the same wherever it appears in the analyzing window. This is a necessary condition for obtaining a stable signal pattern.

For the wavelet packets library, there are two ways to organize its basis functions, i.e. grouping by frequency or by time (position) (Wickerhauser,

---

[1] Here we consider the problem from the basis point of view, which differs from the discussion of stability of signal patterns in the previous section.

1991). In the former case, all the bases in the wavelet packet library are shiftable. But in the latter case, only those bases corresponding to a fixed scale decomposition are shiftable. We will use the wavelet packets library grouped by frequency in our studies.

A wavelet basis is shiftable and fixed in all times in the analysis. However, we have already found that a wavelet basis is well suited only for signals with high frequency components superposed on low frequency components. This is not always the situation in practice. Therefore, it may be difficult to obtain the compact signal patterns, which consist only a few large value coefficients. This can be seen in the example shown in Figure 2.3 (a).

The *best basis* method associated with the wavelet packets library proposed by Coifman and Wickerhauser (1992b) gives an adaptive decomposition scheme based on the signal features to be analyzed. A large number of orthonormal bases in the wavelet packets library provide this flexibility for choosing the most suitable basis, for a given signal. As observed in Figure 2.3 (b), the compact signal patterns can be obtained by using the *best basis* method. This is expected for further analysis. However, the best basis method only applies to the signal segment within the analyzing window. The selected "best basis" can be different as the input signal shifts in time (position). In other words, the shiftability of the basis can be destroyed by applying the best basis method directly.

Therefore, it is necessary to introduce another important new concept -- the *signal dependent basis*, whose basis functions are similar to the wavelet basis that preserves the time-frequency localization properties but adapts to the overall signal to be analyzed, rather than the signal segment within the analyzing window. This means that we fix this customized basis for the entire analysis of the signal instead of calculating the best basis for every analyzing window, to avoid the problems mentioned above (Fig. 3.1). Also, the analyzing basis is customized for every signal in analysis, to achieve better results, rather than using the same wavelet basis everywhere. This concept is particularly meaningful for the applications in process engineering, e.g. we can select the signal dependent basis for each sensor output in terms of the signal features of the individual parameter to be measured. More accurately, we have,

Definition 3.2 Given a time series signal y, *the signal dependent basis* B_y is selected from a collections of the shiftable bases Σ, e.g. the library of wavelet packets grouped in frequency, based on the time-frequency feature of y. Then B_y is fixed in the analysis of the signal y.

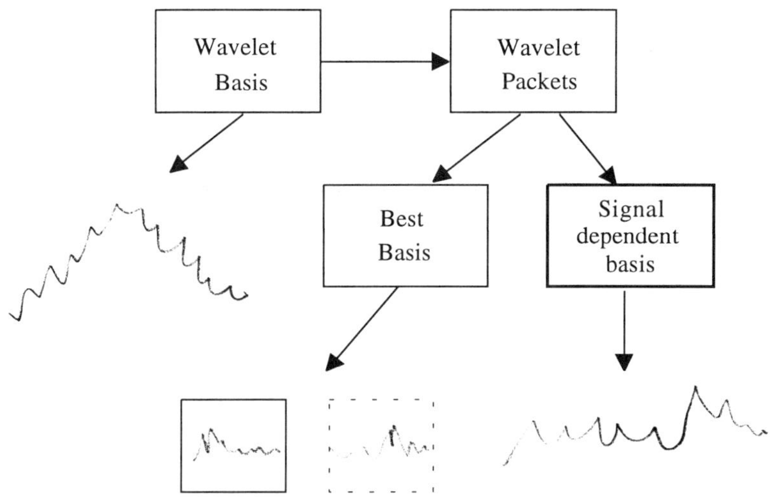

Fig. 3.1 The features of the different bases

In practice, a pre-analysis stage is proposed to determine B_y for a signal y. We believe that the performance of the signal dependent basis, pre-determined based on the signal features, will be better than that of the wavelet basis in many situations.

4. Application examples

In the practice of process monitoring, diagnosis and control in process industry, we often need to identify the underlying physical and/or chemical phenomena and process conditions. These phenomena can be process faults, senor and actuator failures, product quality, process physical changes, noise and interference in measurement, etc., which may be difficult to measure directly. One of the most direct and simple solutions of the problem is to find the unique hidden signal features, or in other words, to discover the process "finger print" from the temporal signal features of measurable parameters in the process.

The signal pattern we defined above actually represents the time-frequency structure information of the signal. It will be much easier to discover the hidden signal features inherent in the physical and/or chemical phenomena in the process, which is what we meant the term process "finger print". Having expressed the "finger print" by signal patterns, we can simply process the signal pattern to finish the tasks such as extraction, identification, approximation and classification of these signal features under studies.

This global view analysis method will be of great benefit for many applications. In the following paragraphs we consider a few applications areas.

4.1. Visual analysis tool

Visualization analysis is known to be a powerful tool for processing complex phenomena in nature. This is particularly helpful for understanding the features of many unknown phenomena around us. Hence, many different kinds of visual representations have been proposed and utilized in different fields.

By implementing the basic concepts of the method proposed in the previous section with some other necessary details, such as the signal sampling frequency, the lowest decomposition scale and the length of analyzing window requirements, phase plane end effects conditioning, mother wavelet selections, etc., we can construct a visual analysis tool for one dimensional signals. In this case, the process "finger print" represented by the signal pattern and/or sub-pattern can be seen directly on the visual phase plane. The human experts can then compare these visual signal patterns (sub-patterns) to extract their similarity and difference, discover the link between the signal pattern (sub-pattern) and the specific phenomenon, and finally recognize the inherent feature of the original phenomenon in the process. Therefore, this visual analysis tool will be very useful to identify the time-frequency feature of signals with unknown non-stationary phenomena embedded. Based on the signal feature identified by the visual analysis, tools can further be developed to process signals automatically.

Figure 4.1 shows the signal patterns of two basic functions: HeaviSine and Doppler (Donoho and Johnstone, 1992), which can be used to model spatially variable functions arising in signal processing. Figure 4.1 (a) shows the time domain function HeaviSine, and 4.1 (b) gives the time domain function Doppler. The signal pattern P_H^* in 4.1 (c) is corresponding to 4.1 (a), from which we can easily capture the time-frequency features, i.e. two peaks represent two singularity points in time and the darkest horizontal patches indicate the location of dominant frequency band in the signal. Figure 4.1(d), the signal pattern P_D^* of 4.1 (b), clearly shows the distinct non-stationary feature of the signal Doppler.

This visual analysis tool has been applied to the development of a new sensor for detecting and classifying signal features of chemical phenomena in a corrosion process. This process has been studied for decades, and no reliable methods can be used to associate the signal features observed from the indirect measurements to the phenomena occurring in the process. Many approaches have been tried unsuccessfully, including modelling, parameter correlation in either time domain or frequency domain, and so on. Actually, this is a multiple phenomena process. These phenomena can happen any time simultaneously and/or differently in the process. The objective of this application studies is to find a simple and reliable method to identify and classify two major types and intensity of the phenomena when they occur.

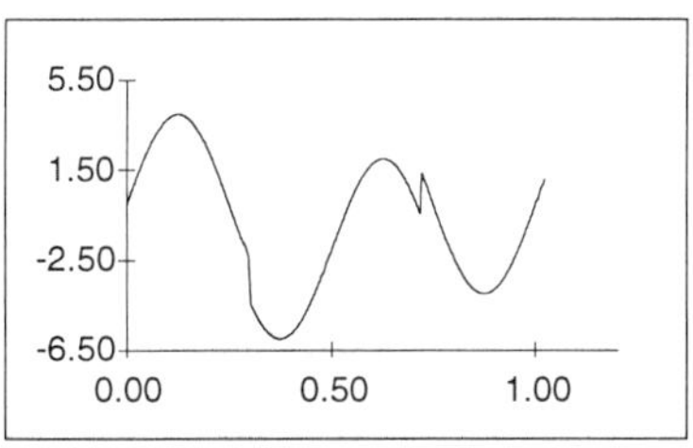

(a) Signal: HeaviSine

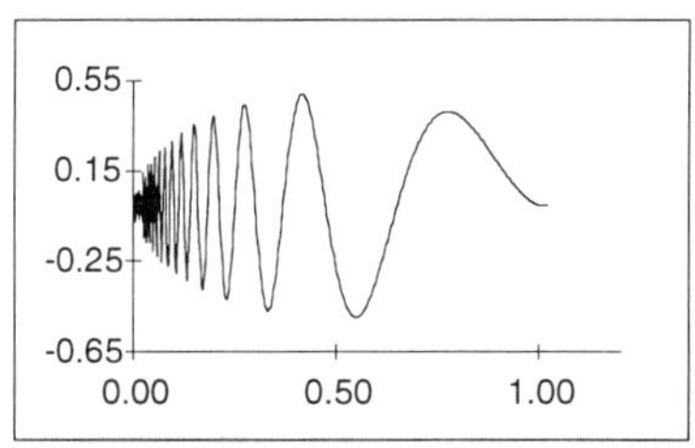

(b) Signal: Doppler

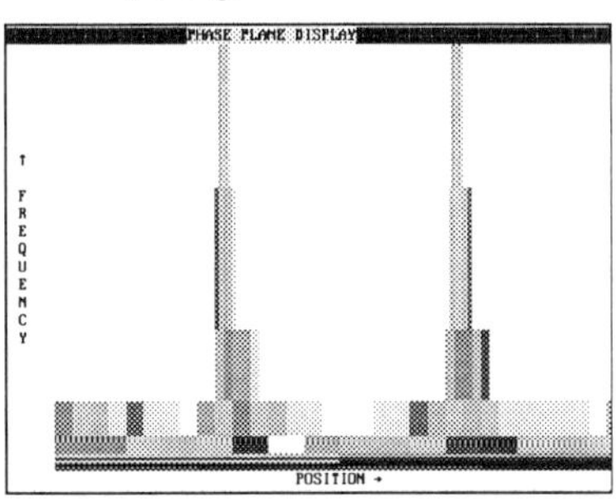

(c) Signal pattern: HeaviSine

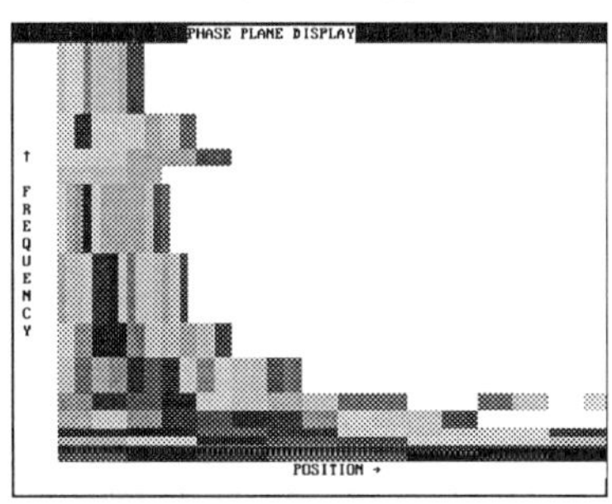

(d) Signal pattern: Doppler

Fig. 4.1 Visual representation examples

Over 1500 data samples have been collected in various process conditions from one measurable sensor output. By studying the signal patterns on the phase plane carefully, two specific signal patterns have been identified. That is A-type and B-type pattern (Figure 4.2). We hope that they can be used to represent the phenomenon A and B. These two patterns have been further divided into sub-types based on the intensity difference, to capture the intensity of the phenomena. The final results show a perfect match in type and intensity prediction for the case of strong and very weak activity present in the process.

4.2. Sensor data filtering for noise reduction and/or feature extraction

Many sensor data, from the laboratory instrument sensor outputs to process parameter measurements, are corrupted, sometimes severely, by noise. This is one of the major obstacles for obtaining accurate measurement data, and degrades the performance of downstream systems, such as parameter estimation from these measurements. Therefore, many efforts have been made to reduce or eliminate noise from measurements. However, these noise reduction techniques still retain some limitations for applications.

In the applications of wavelet theory, some researchers have already found the advantages for applying this newly emerging technique (Mallat and Hwang, 1992, Wornell and Oppenheim, 1992, Dai, 1992). Recently,

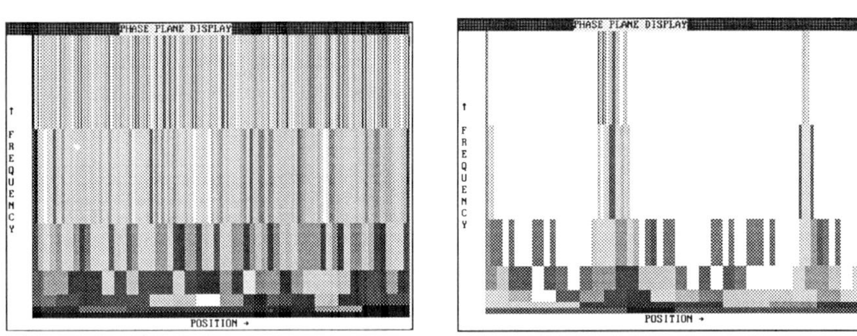

(a) A-type signal pattern (b) B-type signal pattern

Fig. 4.2 Two specific signal patterns identified in the process

Donoho and Johnstone (1992a, b) have further studied its mathematical fundamentals and proved that the "De-Noising" property of wavelet transform can give near-complete success in the area where many previous non-wavelets methods have met only partial success. Their implementations of the ideal "De-Noising" algorithms, however, are based on the concept of wavelet scales, which limits the performance of the algorithm in practice.

The ideal selected wavelet reconstruction principle they proposed actually is to shrink the wavelet transform coefficients according to the localized time-frequency structure without being limited by the wavelet scales. This is exactly the principle of signal pattern approximation presented in the definition 2.2 for P^*. Due to the flexibility of the wavelet packets library for the transform basis selection, a more adaptive basis can be selected for the signal to be analyzed by the signal dependent basis algorithm.

Moreover, a special case of sub-pattern, a noise pattern, can be defined based on the feature of the noise involved. Based on the ability of time-frequency two dimensional representation, we assume that the signal and noise pattern are not superposed together on the entire phase plane, or in other words, the non-superposed noise pattern can be identified in some portion of the phase plane, except for the cases that both the signal and noise features are really similar. As we observed, it is true for many practical situations we are dealing with. Then, the noise statistical features can be easily estimated based on the noise pattern identified.

Using these concepts, a new sensor data filtering technique for noise reduction can be obtained. This noise filtering technique is basically adaptive in nature since the noise feature can be estimated (updated) in every analyzing window. Thus, it will be useful in dealing with signals that have varying frequency features.

The following example illustrates the ability of this approach. Two original signals are taken from a sensor output of a mass spectrometer connected to a TAP reactor (Fig. 4.3 (a1) and (a2)). By doing the orthonormal WP transform with the selected basis for each signal, the transform results are presented on the phase plane (Fig. 4.3 (c1) and (c2)). On the phasc plane, the signal energy is highly concentrated on a few time-frequency patches, highlighted in white to represent the patterns for the original signals. On the other hand, the noise pattern can be identified as near uniformly distributed, very low signal energy intensity time-frequency patches that spread on the large portion of the phase plane. By removing the identified noise pattern, the de-noising signals are reconstructed directly from the highlighted signal patterns. Fig. 4.3 (b1) and (b2) show the de-noising reconstruction signals, from which we can clearly see that the high frequency details of signal (a1) are still retained while the noise interference is effectively removed. It is easy to notice that these signal patterns are obviously very different in their frequency nature, which is handled nicely by this signal pattern based de-noising approach.

On the other hand, for the application of process trend analysis (Cheung and Stephanopoulos, 1990, Joshi, 1992), a stable process trend is needed for further feature extraction and analysis. Fig. 4.3 (b1) and (b2) actually can be seen as the trend of the original signals in (a1) and (a2) respectively. Therefore, a "clean" and stable process trend can be obtained readily by applying the above de-noising procedure.

It is possible to extend this approach to the process and sensor fault detection, a more general topic of this aspect. The clues are that the symptoms of the process fault and sensor malfunction usually can be found in the output signals of the process sensors and their components with specific signal features. These features, of course, can be represented by the signal patterns in the time-frequency domain.

Case-based reasoning and neural networks are two promising candidates for constructing the complete intelligent processing systems, which can be linked to the proposed wavelet-based front end analysis. In building a case-based reasoning system, it is natural to consider the signal pattern to be identified as a case. Then the cases in the system can be indexed and stored in terms of the features of these patterns. As the system evolves, the cases are accumulated gradually and the ability to identify different signal patterns are enhanced at the same time, which is one of the most desirable properties of this scheme. Neural nets, on the other hand, can be connected to the front end directly when the input nodes and network topology are carefully arranged. It is possible to form a rapid processing unit by using this scheme. But the configuration could be application dependent. In the simplest case, the intelligent system can manage the task of classification in a straightforward way. However, in the complex case the system can be expanded to include some other intelligent components to fulfill the more complicated tasks such as diagnostics, monitoring, etc.

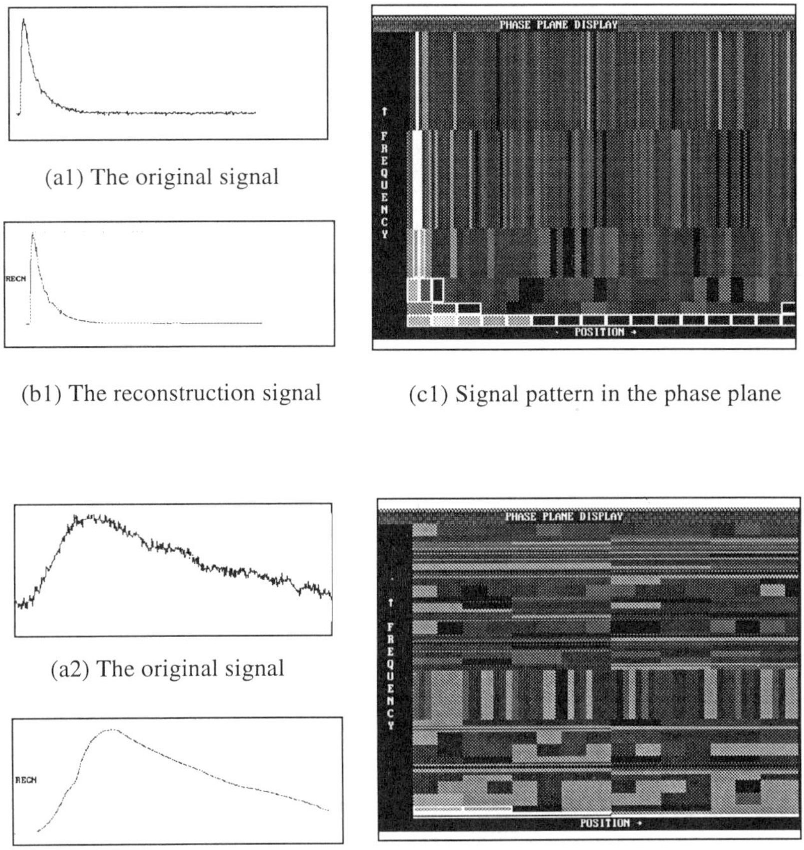

(a1) The original signal

(b1) The reconstruction signal

(c1) Signal pattern in the phase plane

(a2) The original signal

(b2) The reconstruction signal

(c2) Signal pattern in the phase plane

Fig. 4.3 The examples for the noise reduction effects

5. Conclusion

The framework of the phase plane analysis method has been presented in this chapter, based on the wavelet packet theory, an extension of wavelet transform theory. It provides a "global" view analysis approach which differs from the existing widely used multiscale analysis method in the application of wavelet theory. The signal pattern (sub-pattern) can be used naturally as a basic unit in analysis. The signal dependent basis, on the other hand, not only is an important supporting condition for obtaining the stable signal pattern in the proposed method, but also presents a novel concept for the wavelet theory applications in general. Finally, we highlight two application approaches of this framework. The visual implementation of this analysis method results in a new visual analysis tool for studying process signals with unknown time-frequency features. One of the practical applications of this visual analysis tool verifies the effectiveness of this tool. The application for signal noise reduction and feature extraction, on the other hand, illustrates that this signal pattern based algorithm actually can approach the existing problems from a different angle. In other words, the proposed methodology provides a new platform for process signal feature analysis, and can lead to novel solutions for many existing difficult-to-solve problems in the area of process data analysis.

Acknowledgements

The authors would like to acknowledge the support of NSF grant # CTS - 9108649.

6. References

Cheung, J. T.-Y and G. Stephanopoulos "Representation of process trends -- Part I. A formal representation framework." *Computers & chemical Engineering* 14, 4/5 (1990) : 495-510.

Cohen, L. "Time-frequency distributions -- A review." *Proceedings of the IEEE,* 77, 7 (1989) : 941-981.

Coifman, R. R., Y. Meyer, S. Quake and M. V. Wickerhauser "Signal processing and compression with wavelet packets." *Progress in Wavelet Analysis and Applications*, Meyer and Roques eds., Editions Frontieres, Toulouse, France, (1992a) : 77-93.

Coifman, R. R. and M. V. Wickerhauser "Entropy-based algorithms for best basis selection." *IEEE Trans. Information Theory* 38, 2 (1992b) : 713-718.

Dai, Xue-dong "Intelligent process signal analysis system." D.Sc research proposal, Department of Chemical Engineering, Washington University in St. Louis, (1992).

Daubechies, I. *Ten Lectures in Wavelets* Society for Industrial and Applied Mathematics, Philadelphia, Pennsylvania, 1992.

Daubechies, I., S. Mallat and A. S. Willsky eds. *Special Issue on Wavelet Transforms and Multiresolution Signal Analysis, IEEE Trans. Information Theory* 38, 2 (1992).

Donoho, David L. and Iain M. Johnstone "Ideal Spatial Adaptation by Wavelet Shrinkage." Preprint, Department of Statistics, Stanford University, (1992a).

Donoho, David L. "De-Noising by Soft-Thresholding." Preprint, Department of Statistics, Stanford University, (1992b).

Hlawatsch, F. and G. F. Boudreaux-Bartels "Linear and quadratic time-frequency signal representations." *IEEE Signal Processing Magazine,* (Apr. 1992) : 21-67.

Jawerth, B. and Wim Sweldens "An overview of wavelet based multiresolution analysis." Research report 1993:1, Department of Mathematics, University of South Carolina (1993).

Joshi, A. P. *Extraction of Information from Time-series Process Operational Data* D.Sc. Dissertation, Department of Chemical Engineering, Washington University in St. Louis, (1992).

Mallat, S. and W. L. Hwang "Singularity detection and processing with wavelets." *IEEE Trans. Information Theory* 38, 2 (1992) : 617-643.

Mallat, S. "Multiresolution approach to wavelets in computer vision." *Wavelets, Time-Frequency Methods and Phase Space* Combes, J. M. et al. eds. Springer, 1989) : 313-327.

Rioul, O. and M. Vetterli "Wavelets and signal processing." *IEEE Signal Processing Magazine,* (Oct. 1991) : 14-38.

Simoncelli, E. P., W. T. Freeman, E. H. Adelson and D. J. Heeger "Shiftable multiscale transform." *IEEE Trans. Information Theory* 38, 2 (1992) : 587-607.

Wickerhauser, M. V. *Lectures on Wavelet Packet Algorithms* Department of Mathematics, Washington University in St. Louis, (1991).

Wornell, G. W. and A. V. Oppenheim "Estimation of fractal signal from noisy measurements using wavelets." *IEEE Trans. Signal Processing* 40, 3 (1992) : 611-623.

Chapter 5

Learning at Multiple Resolutions: Wavelets as Basis Functions in Artificial Neural Networks, and Inductive Decision Trees

Bhavik R. Bakshi
Department of Chemical Engineering
Ohio State University, Columbus, OH 43210

Alexandros Koulouris and George Stephanopoulos
Department of Chemical Engineering
Massachusetts Institute of Technology, Cambridge, MA 02139

Abstract

Learning at multiple resolutions provides a fast, hierarchical and efficient technique for extracting models from empirical data. In this chapter we describe the application of wavelets for multi-resolution learning in artificial neural networks and inductive decision trees, and show how wavelets may provide a unifying framework for various supervised learning techniques. A *Wave-Net* is an artificial neural network with activation functions derived from the class of wavelets. *Wave-Nets* combine the mathematically rigorous, multi-resolution character of wavelets with the adaptive learning of artificial neural networks. Learning with *Wave-Nets* is efficient, and is explicitly based on the local or global error of approximation. The advantages of *Wave-Net* learning over other artificial neural learning techniques are highlighted, and learning methods for minimizing the L^2 or L^∞ norms are described. The reduced black box character of *Wave-Nets* is demonstrated by the explicit relationship between *Wave-Net* parameters and the quality of learning, and by the ability to extract if-then rules from a Haar *Wave-Net*. The relationship between Haar *Wave-Nets* and other rule-extraction techniques such as decision trees is described.

1. Introduction

Modeling the relationship between input and output variables is an important part of several chemical engineering tasks. In the operation and control of chemical processes, it is necessary to derive models relating measured and controlled variables to process behavior and conditions. Tasks such as adaptive control and system identification rely on the induction of input-output models from historical data. Fault diagnosis and supervisory control of chemical processes may require pattern recognition of measured process data and determination of the mapping between features in process data and process conditions. The simulation of ill-understood systems also relies on modeling from measured data.

The two broad classes of methods used for learning input-output relationships from historical and empirical data are based on, (i) statistics and (ii) artificial intelligence. Statistical techniques, such as principal least squares regression and principal components analysis (Mardia et al., 1980; Ljung, 1987), predate artificial intelligence techniques. Statistical methods usually require an assumed model and model a relationship that may not be easily adaptable when new data become available. Various machine learning techniques have become increasingly popular in recent years. These artificial intelligence-based techniques involve learning the mapping between inputs and outputs, and adapting it continually, upon availability of new data.

Artificial Neural Networks (ANN) are a powerful and general technique for solving machine learning and pattern recognition problems. A typical ANN consists of a set of inter-connected nodes arranged in multiple layers. Each node has several inputs and a single output. Depending on the type of node and network, an operator, defined by an activation function, is applied to the inputs to compute the node output. The ANN has several adjustable parameters, such as the weights on the edges between nodes, and the type and parameters of operators in the node, which are adjusted during the network training stage. Adaptation of the network to new examples is also accomplished by modifying the network parameters. Several attractive properties of ANNs, such as the simplicity of their structure and its inherent parallel nature, their ability to approximate a wide variety of mappings, the universality of their approximation ability, and their truly adaptive nature have made them extremely popular for solving engineering problems that require machine learning.

There have been numerous applications of ANN in chemical engineering, primarily for process operation and control (Venkatasubramaniam and McAvoy, 1992). ANNs have been explored for performing fault diagnosis (Hoskins and Himmelblau, 1988; Watanabe et. al., 1989; Kramer and Leonard, 1990; Venkatasubramaniam et. al., 1990; Leonard and Kramer, 1991), system identification and adaptive control (Ungar et. al., 1990; Ydstie, 1990; Bhat et. al., 1990; Hernandez and Arkun, 1992; Haesloop and Holt, 1990; Narendra and Parthasarthy, 1989; Barto, 1991), process trends analysis and pattern recognition (Whiteley and Davis, 1992), and modeling of complex chemical systems (Adomaitis et al., 1990).

Depending on the nature of the activation functions, several different types of ANN may be designed. Typical activation functions include global functions such as sigmoids, which are used in Backpropagation Networks (BPN), and local functions such as Gaussians, used in Radial Basis Function Networks (RBFN). Despite their popularity, ANN suffer from some significant disadvantages. No formal or rigorous techniques are available for designing an ANN. Decisions such as the number of nodes required for learning a mapping, initialization of network parameters and criteria for stopping the learning are at best arbitrary or heuristic. The model learned by the network is not physically interpretable, and is a black box. As a result, there is significant need for novel empirical learning techniques, and for improvements in the current methods.

Usually, the training data for learning a mapping are spread unevenly in the input space. The data may be sparse in some regions, and dense in others. Also, the hypersurface representing the input-output map may be slow-changing in some regions and fast-changing in others. An efficient technique for learning input-output relationships from data that are unevenly distributed in the input space is to learn at multiple resolutions. Learning a mapping at a single coarse resolution may not capture relevant details of the mapping at finer scales. On the other hand, learning at a finer resolution in the entire input space may capture unnecessary detail from some parts of the input space, leading to memorization of the training data, and a lack of generalization of the input-output model.

Another popular technique for learning input-output mappings from historical data, is induction via *decision trees*. Inductive decision trees approximate the input-output mapping by a piece-wise constant approximation of the hypersurface. Learning by decision trees overcomes some of the disadvantages of ANNs. The model learned is easily expressible as simple *"if-then"* rules, may even be physically interpretable, and is not a black box. Redundant or irrelevant variables are easily identified. Also, the construction of decision trees is fast and efficient. Applications in chemical engineering include pattern recognition for fault diagnosis and supervisory control (Bakshi, 1992) and for continuous process improvement (Saraiva and Stephanopoulos, 1992). One shortcoming of inductive decision trees is that the adaptation of decision trees with new examples is inefficient and may require reconstruction of the entire tree. Also, techniques for terminating the construction of a decision tree are often ad hoc.

The development of wavelet theory has provided novel techniques and solutions to several engineering problems. The derivation of wavelets from a formal and rigorous mathematical framework, combined with their relationship with approximation theory have also resulted in the development of a novel ANN. The Wavelet Network, or *Wave-Net* overcomes some of the disadvantages of conventional ANN, and have been developed in parallel by Bakshi and Stephanopoulos (1993) and Zhang and Benveniste (1992). The activation functions for *Wave-Nets* are derived from the class of wavelets. Learning a mapping from empirical observations proceeds at multiple resolutions, depending on the density and distribution of the training data. A coarse approximation is learned at first, using scaling functions that capture the hypersurface only at the lowest frequency. Additional wavelets are then introduced to capture details, and to improve the quality of the learned mapping in the input space, as necessary. Due to the multi-resolution nature of the basis functions, and the hierarchical mode of training, the model learned by a *Wave-Net* combines the benefits of global as well as local approximation, often leading to better performance. Batch and continuous training and adaptation procedures have been devised. The network parameters, such as the edge weights have a clear physical meaning, and are proportional to the error of the approximation. Other parameters, such as the position (translation) and extent (dilation) of the basis functions are directly related to the distribution of training data in the input space. Consequently, the model learned by *Wave-Nets* is less of a black box, as compared to other ANN. The training and adaptation procedures may be based on the desired quality of learning, and not on some heuristic criteria. Some disadvantages of other ANN such as unstable behavior and problems with local extrema on the error surface, are eliminated in *Wave-Nets* due to their rigorous and physically intuitive nature. Given the variety of basis functions in the class of wavelets, it is possible to derive the relationship between conventional ANN, such as perceptrons, BPN and RBFNs, and *Wave-Nets*. This provides greater insight into the relationship of the ANN learning paradigm with *Wave-Nets.*

The model induced by *Wave-Nets* using Haar basis functions may be easily expressed as simple if-then rules. The ability to extract rules from *Wave-Nets* further reduces the black box character of ANN. Haar wavelets readily provide a piece-wise constant hypersurface. Therefore, Haar *Wave-Nets* are best suited for solving classification problems, or problems where the outputs have only discrete values, such as class membership. The mapping learned has a simple relationship with that induced using decision trees. If adequate domain-specific knowledge is available, the induced rules even may be physically interpretable.

In this chapter, we describe the paradigm of learning at multiple resolutions as derived from wavelet analysis. The development of *Wave-Nets*, their properties and training procedures, the extraction of rules from *Wave-Nets* and their connection to other inductive rule extraction methods are described. Section 2 describes the relationship of artificial neural learning with mathematical principles of approximation theory. Section 3 contains a very brief introduction to wavelet theory and the notation used in this chapter, followed by the theory of multi-resolution approximation using wavelets, which forms the theoretical framework for *Wave-Nets.* Section 4 describes the development of *Wave-Nets* and various procedures for their training and adaptation. Illustrative examples are used to demonstrate the performance and properties of *Wave-Nets*. Section 5 describes the extraction of if-then rules from *Wave-Nets* using Haar wavelets as basis functions, and their relationship with inductive decision trees.

2. Learning and Approximation Theory

Learning is the process of deriving generalized models of some input-output functional dependence based on discrete observations of that dependence. Several techniques have been developed for inducing the model from input-output data in a variety of fields such as mathematics, statistics and artificial intelligence. Artificial Neural Networks (ANN) have been used extensively for solving learning problems in a variety of domains. An important feature of learning by ANN, as opposed to other approximation methods, is that the model can be continually adapted as new data become available. The construction of an ANN model involves the selection of a type of basis function, and the determination of the network structure, that is, how many nodes and how to interconnect them. These decisions are either made arbitrarily by the user or performed with the aid of empirical heuristics, giving ANNs a "black box" character. Finally, the network parameters are computed based on the available data, using the training and adaptation algorithm. Typical choices for the basis functions are global functions such as the linear threshold or its continuous analog, the sigmoid function, and local functions such as radial basis functions like the Gaussian.

Improved understanding of learning from empirical data has led to the realization that the techniques from all the disciplines may be unified as approximation problems. The tasks of curve or surface fitting, creating models from experimental data, or simulating the cause-effect behavior of a dynamical system require the approximation of the input-output function based on discrete data, and is equivalent to a functional estimation or approximation problem. Recent realization that learning in ANNs is equivalent to functional approximation by expansion on a set of basis functions (Poggio and Girosi, 1989) has provided significant insight into artificial neural learning.

Mathematically, the solution to an approximation problem involves the selection of a space of approximating functions which are linear or nonlinear combinations of known basis functions and are characterized by a finite set of adjustable parameters or weights. The approximated function, F(x) may be expressed as,

$$F(x) = \sum_{i=1}^{\infty} w_i \, \theta_i(x, k_i) \qquad (1)$$

where, θ_i $(i=1, \infty)$ represents the set of basis functions, and w_i and k_i, the corresponding set of adjustable weights. The modeling or learning task is to derive a set of values for the unknown parameters, w_i and k_i by minimizing some function of the empirical error on the data. All types of ANN may be expressed in a form given by equation (1).

The objective of the learning is to use the derived model to predict the output for unseen inputs, with reasonable accuracy. The critical question in learning an appropriate model is how close is the model to the underlying unknown function, or what is the magnitude of its *generalization error*. The generalization error is the normed difference between the data and the actual model, and consists of two components,

- *Approximation error*, arising from the finiteness of the function space. This refers to the distance in some functional norm of the model in the space of approximating functions from the real function. The best model is defined not with respect to the available data but for the entire input domain.
- *Estimation error*, arising from finiteness of the data. This measures the difference between the selected model based on limited information and the best approximating function in the space.

Therefore, the critical decisions related to the modeling scheme used are,

1) the selection of the approximating space where the model is sought and,
2) the criterion based on which a model within that space is chosen.

Since the real model is unknown in practical situations, the approximation error cannot be measured. However, the different learning techniques can be judged based on the a priori guarantees that they offer in minimizing the magnitude of both errors. The learning task consists of obtaining models with good global as well as local approximation properties, out of partial, localized and often incorrect information. Recent mathematical proofs have stated that any function can be approximated arbitrarily well by different types of ANNs. That is, ANNs are universal approximators. Unfortunately, achieving arbitrarily good approximation often requires arbitrarily large number of units. The limitations imposed by the finiteness of data and the empiricism involved in selecting the number of basis functions defy in practice, the reassurances of the universal approximation property.

Selecting the proper network and its architecture to obtain an accurate approximation involves a fair amount of guess work. Even if a good approximation is possible, no guarantees for attaining such a solution (obtaining solutions with small estimation error) with a given network are offered. The absence of any mathematical reassurance about the goodness of the approximating model can be attributed to the ill-posed nature of the learning problem. As long as the available observations can be equivalently interpreted by different general models, the question of how to select one of them needs convincing answers. All data regression schemes, including ANNs, postulate a model structure and search within that structure for the best model. The selection, however, of the particular model set can hardly be justified. Currently, ANNs are being subjected to significant mathematical scrutiny, and results on improving their performance and other theoretical aspects are bound to appear.

The quality of an empirical model depends on its ability to optimize the trade-off between memorization of the data and generality of the learned model. The underlying intuitive principle is that the simplest (smoothest) model that can approximate the data well should be found. Excessive emphasis on smoothness can result in poor generalization, and a large approximation error. On the other hand, pushing the approximation error towards zero may cause memorization of the training data resulting in anomalous behavior of the interpolating function. This trade-off is stated mathematically as the regularization problem. Approximation of data via regularization involves finding the function F(x) which minimizes the following functional

$$H[F(x)] = \sum_{i=1}^{N} (y_i - F(x_i))^2 + \lambda \, \|PF(x)\|^2 \tag{2}$$

where, y_i are the measured values of the unknown function, F(x) is the approximation of the unknown function, and λ is a positive real number called the regularization parameter. P is a constraint operator, and is normally selected to be a differential operator so that it reflects the desired degree of smoothness of the unknown function. Therefore, H[F(x)] expresses the compromise between the error of approximation and degree of smoothness of the unknown function. Poggio and Girosi (1989) have shown that radial basis functions provide a solution to the regularization problem, thus providing a mathematical basis for learning with RBFNs. This insight has resulted in better understanding of neural learning.

Both smoothness and accuracy are local properties. In some region of the input space, the underlying function can be coarse and void of any distinguished features, whereas in other regions, it can present significant details. Similarly, the goodness of the approximation achieved or sought is variable for the different regions in the input space. A modeling scheme that can efficiently capture the coarse as well as the detailed characteristics of the hypersurface is provided by learning at multiple resolutions. A coarse resolution may be used in regions where the data are sparse, or fine detail need not be captured, and a finer resolution where details are important. Techniques for learning at multiple resolutions in ANNs have been developed by Moody (1989) based on radial basis functions. Recently, wavelets have provided a novel method for learning at multiple resolutions under the framework of *Wave-Nets*. *Wave-Nets*, not only provide an appropriate framework for multiresolution learning problem, but also reduce the black box character of ANNs by generating physically interpretable models, and allowing straightforward characterization of its properties as an approximation scheme. Relationships may also be derived between *Wave-Nets*, other inductive learning methods, as well as mathematical approximation techniques.

3. Multi-Resolution Functional Approximation Using Wavelets

A family of wavelets is derived from the translations and dilations of a single function, and is given by,

$$\frac{1}{\sqrt{s}}\psi\left(\frac{x-u}{s}\right) \qquad \text{for } (s, u) \in R^2$$

The parameter, s, indicates the *dilation*, and u, the *translation* of the mother wavelet, $\psi(x)$. The wavelet belongs to the set of square-integrable functions, i.e. $\psi(x) \in L^2(R)$, and satisfies an admissibility condition, which implies that the mean value of $\psi(x)$ be zero,

$$\int_{-\infty}^{+\infty} \psi(x)\, dx = 0 \tag{3}$$

The resolution of wavelets at different scales, s, varies in the input and frequency domains as governed by the uncertainty principle. At large scale, the resolution is coarse in the input domain and fine in the frequency domain. As the scale, s, decreases, the resolution in the input space becomes finer while that in the frequency space becomes coarser.

Any function, F(x), belonging to the set of square-integrable functions, $L^2(R)$, may be represented as a weighted sum of a family of wavelets,

$$F(x) = \sum_{s=-\infty}^{\infty} \sum_{u=-\infty}^{\infty} c_{su}\psi_{su}(x)) \tag{4}$$

If frame properties are satisfied, then the coefficients of the expansion are related to the function F(x) and the family of basis functions, $\theta_i(x)$, (see equation (1)) as shown below (Daubechies, 1990). The set of basis functions, $\theta_i(x)$ constitute a frame if,

$$A\|F\|^2 \leq \sum_i |<F,\theta_i>|^2 \leq B\|F\|^2 \tag{5}$$

The numbers, A and B are positive and are called frame bounds. Under certain special conditions, the computation of the coefficients c_i in the expansion of equation (1) is straightforward. In general, we can write,

$$F = \sum \tilde{\theta}_i <\theta_i, F>, \quad \text{or} \quad F = \sum \theta_i <\tilde{\theta}_i, F>$$

where, $\tilde{\theta}_i$ is the dual frame. If the frame is tight, i.e. A=B, the reconstruction formulae become simpler, i.e.,

$$F = A^{-1} \sum_i \theta_i <\theta_i, F>$$

If A = B = 1, we have an orthonormal basis.

The wavelet parameters, s and u, have to be discretized for practical applications. Usually, the scale parameter, s, is discretized as,

$$s = C^m, \qquad m \in Z$$

where, C is a constant, and m is the dilation parameter. For most applications, a dyadic sampling scheme is selected, by choosing the constant to be C = 2.

The translation parameter, u, may be discretized as,

$$u = \alpha D^m k, \qquad (m, k) \in Z^2$$

where, α is the sampling interval, m is the dilation parameter, k is the translation parameter, and D is a constant. Traditionally, the parameters, C and D are selected to be C = D = 2 resulting in a dyadic sampling scheme. A signal of length O(N) can then be described by O(N) coefficients. A significant advantage of this discretization scheme is that for some choices of $\psi(x)$, the family of functions, $\psi_{mk}(x)$ $(m, k) \in Z^2$, forms an orthonormal basis of $L^2(R)$. Setting C = 2 and D = 1 gives uniform sampling of the translation parameter, and dyadic sampling of the dilation parameter, and O(NlogN) coefficients for describing a signal of length O(N). Discretized wavelets are represented as $\psi_{mk}(x)$, where m and k provide the discrete equivalents of the continuous parameters, s and u respectively. Over the years, different types of wavelets have been devised possessing various properties such as orthonormality, biorthogonality, semi-orthogonality, compact support, and various degrees of smoothness.

Representation of a function, F(x), by equation (4) requires variation of the translation and dilation parameters in the range $[-\infty, \infty]$. Such variation is not practical due to the availability of a finite amount of data. A scaling function is defined for each wavelet to make the expansion by equation (4) practical. Information at the lowest frequencies (coarsest scale) is captured by a scaling function, $\phi_{Lk}(x)$, where L represents the coarsest scale. Thus,

$$F = \sum_{m=1}^{L} \sum_{k=-\infty}^{\infty} c_{mk}\psi_{mk} + \sum_{k=-\infty}^{\infty} d_{Lk}\phi_{Lk} \tag{6}$$

Equation (5) may then be written as,

$$A\|F\|^2 \le \sum_{k} |\langle F, \psi_k \rangle|^2 + |\langle F, \phi_L \rangle|^2 \le B\|F\|^2$$

Equation (6) indicates how the function F may be built hierarchically, by introducing details at higher frequencies. The approximation at resolution L is given by,

$$F \approx F_L = \sum_{k=-\infty}^{\infty} d_{Lk}\phi_{Lk}$$

The approximation at resolution L-1 is obtained by adding wavelet terms at resolution L,

$$F \approx F_{L-1} = \sum_{k=-\infty}^{\infty} d_{Lk}\phi_{Lk} + \sum_{k=-\infty}^{\infty} c_{Lk}\psi_{Lk} \tag{7}$$

This approximation may be made more accurate by adding more wavelet terms, until equation (6) is obtained, giving exact interpolation. Equations (6) and (7) characterize the multi-resolution, hierarchical learning that is performed by a *Wave-Net*.

Error relationships. One of the important benefits of the *Wave-Net* formulation is that the *Wave-Net* coefficients are physically meaningful due to their relationship to the empirical error. Equation (5) indicates that the wavelet coefficients are related to the 2-norm of the approximated function. An approximation of the original function may be obtained by using only some of the basis functions and neglecting others. Let G be the approximation of F obtained by neglecting some basis functions. The error of approximation is proportional to the coefficients of neglected basis functions, as shown below,

$$F(x) = \sum_{i=1}^{N} w_i \tilde{\theta}_i, \qquad w_i = \langle F, \theta_i \rangle$$

$$G(x) = \sum_{i=1}^{n} w_i \tilde{\theta}_i, \qquad w_i = \langle G, \theta_i \rangle, \qquad n \leq N$$

$$e^2 = \int (F - G)^2 \, dx$$

If the basis functions are orthonormal, the least-squares error of approximation is directly related to the coefficients of the neglected basis functions, i.e.,

$$e^2 = \sum_i w_i^2 \qquad \text{with } w_i \in S_- \tag{8}$$

where S_- is the set of neglected wavelet coefficients.

In general, the *square of the wavelet coefficients is proportional to the* L^2 *norm of the function being approximated,* as indicated by equation (5). Consequently, the error of approximation is proportional to the square of the neglected coefficients, provided that the frame conditions are satisfied, i.e.,

$$e^2 \leq A^{-1} \sum_i w_i^2 \qquad \text{with } w_i \in S_- \tag{9}$$

The smaller the coefficient, the smaller the contribution to reducing the L^2 approximation error of the function.

The error of approximation given by equations (8) and (9) is global in nature. Since the approximation by wavelets is localized in the input space, local errors of approximation may be calculated based on coefficients in the relevant region. The local least squares error around a point l in the input space is proportional to the coefficients in the interval $[l\text{-}q, l\text{+}q]$.

$$e_{local}^2 \propto \sum_{m=1}^{L} \sum_{k=l-q}^{l+q} c_{mk}^2 \tag{10}$$

As described above, wavelets that span the space of square-integrable functions can approximate any function in this space with arbitrary accuracy. Therefore, the property of universal approximation is inherent in approximation with wavelets.

Approximation in multiple dimensions. Learning problems are usually multi-dimensional due to the presence of several input variables. The theory and formulation described above is easily extended to multiple dimensions.

The most common technique for developing multi-dimensional wavelets is by taking a tensor product of the single-dimensional wavelet and scaling function. For two inputs, the scaling function is given by,

$$\Phi(x_1, x_2) = \phi(x_1)\phi(x_2)$$

and the three wavelets by,

$$\Psi^l_{mn}(x_1, x_2) = 2^{-m}\, \Psi^l(2^{-m}x_1 - n_1, 2^{-m}x_2 - n_2)$$

where $l = 1, 2, 3$, $n \in Z^2$ and

$$\Psi^1(x_1, x_2) = \phi(x_1)\, \psi(x_2)$$
$$\Psi^2(x_1, x_2) = \psi(x_1)\, \phi(x_2)$$
$$\Psi^3(x_1, x_2) = \psi(x_1)\, \psi(x_2)$$

If $\phi(x)$ and $\psi(x)$ constitute an orthonormal basis, then the Ψ^l_{mn} with $l = 1, 2, 3$, $m \in Z$, $n \in Z^2$, constitute an orthonormal basis of wavelets for the $L^2(R^2)$ space. Each Ψ^l_{mn} captures information at different orientations as shown for the two-dimensional Haar wavelet in Figure 1. The number of wavelets increases exponentially with the dimensions, giving $(2^d - 1)$ wavelets for d inputs. Therefore, for every grid point in the input space, the weights for a large number of wavelets need to be computed. The exponentially increasing complexity of tensor product wavelets is controlled by using other types of wavelets, or by sacrificing the completeness of the basis functions. Tensor product wavelets do have some advantages in ANN learning since they facilitate efficient extraction of rules from *Wave-Nets*, as described in Section 6.

In general, multi-dimensional wavelets are represented as,

$$\det(\mathbf{A})^{m/2}\, \psi(\mathbf{A}^m\mathbf{x} - \gamma) \qquad (11)$$

where, $\mathbf{A}$ is the dilation matrix, and γ is the translation matrix. The structure of the matrix, $\mathbf{A}$, determines the dilation and rotation of the family of wavelets that spans the input space. Multi-dimensional dyadic sampling is obtained if $\mathbf{A}$ is a diagonal matrix with determinant of 2. Wavelets that are represented by equation (11), and are dense in $L^2(R^d)$, do not lead to an exponentially increasing number of basis functions, as the number of dimensions increases. Examples include the mexican

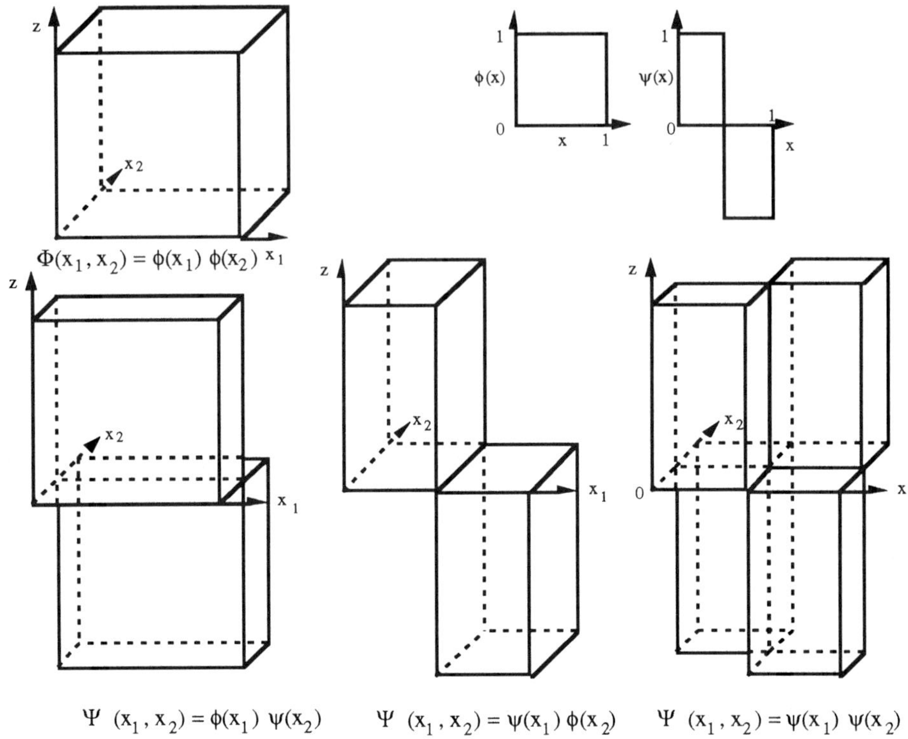

Figure 1. Orientation specific Haar wavelets

hat wavelet (Daubechies, 1992), and first and second derivatives of a Gaussian. In this chapter, we will be concerned with diagonal matrices only. The error equations and frame theory relationships may be directly extended to multi-dimensional problems (Zhang and Benveniste, 1992).

4. Theory of *Wave-Nets*

Learning a model relating inputs and outputs within the wavelet framework requires the estimation of network parameters from empirical data. The approximation of a function, F(x), and the adjustable parameters are given by equations (6) or (7). Representing this approximation problem as an Artificial Neural Network provides a convenient framework for adaptive approximation, and for comparison with other learning techniques. A typical *Wave-Net* is shown in Figure 2. As in a typical artificial neuron, each node has multiple inputs, an activation function and one output. The *Wave-Net* has a single hidden layer of nodes. Each node has one of two types of activation functions: scaling functions, ϕ_{Lk}, and wavelets, ψ_{mk}. The weights on the input edges to the hidden layer are equal to one. The weights on the output edges correspond to the wavelet coefficients c_{mk} and d_{Lk}, derived for representing the hypersurface. The scaling function nodes, or ϕ-nodes capture information at the coarsest scale, L, or lowest frequency, and is analogous to the "constant" or "bias value" that is present in other ANN. The scale L is determined such that a single scaling function ϕ_{Lk} cover the entire input space where training data are present. The ϕ-nodes approximate an average value of the training data. The hypersurface that remains to be approximated, after introduction of the ϕ-nodes, has output values distributed around a mean value of zero, and constitutes detail information that was not captured at the coarsest scale. These details are then captured by wavelets, ψ_{mk}, $m \leq L$, $-\infty < k < \infty$, constituting the ψ-nodes. The ψ-nodes are introduced to capture increasing amounts of detail, and facilitate hierarchical learning.

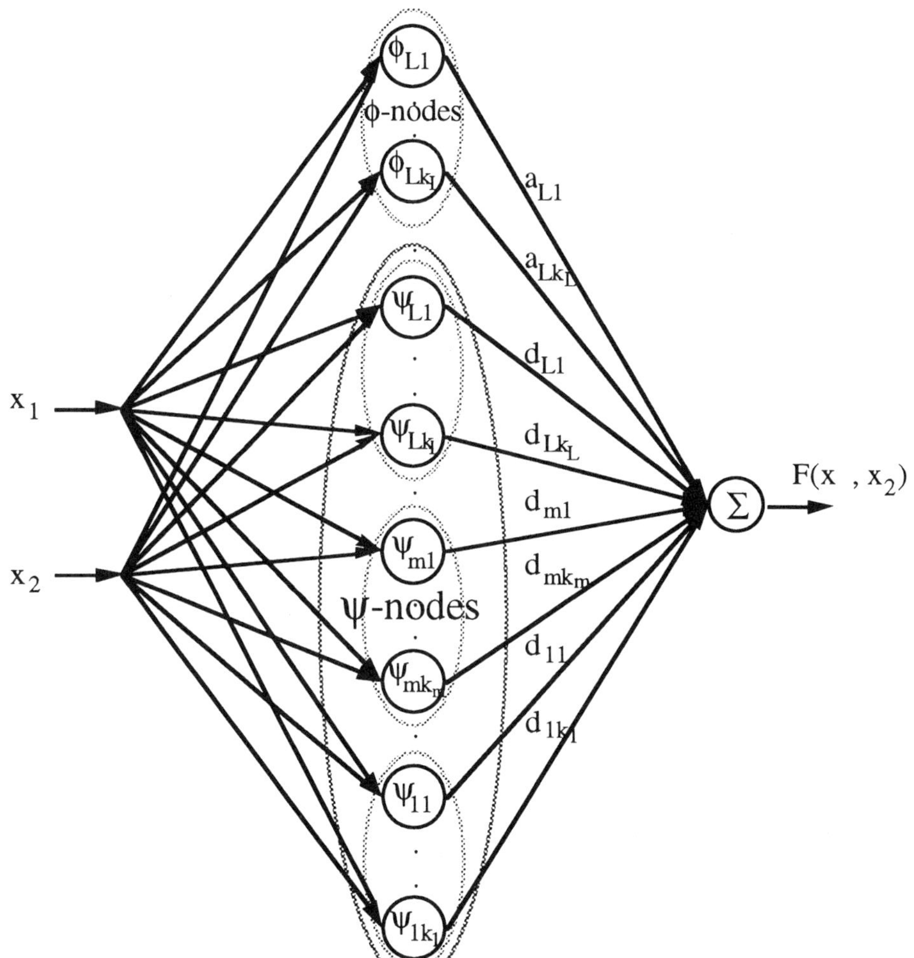

Figure 2. Structure of typical *Wave-Net*

4.1 Optimization Formulation of *Wave-Net* Learning

The model learned by a *Wave-Net* should provide good prediction for the training data as well as for new data that are unseen by the network. The construction of a *Wave-Net* requires estimation of several parameters such as the edge weights, and the translation and dilation parameters. The formulation of *Wave-Net* learning as an optimization problem provides insight into the decision variables, and their impact on the quality and complexity of the model learned. The optimization problem consisting of the objectives and the constraints is described below, followed by an analysis of the decision variables, and the simplifying assumptions needed to solve the problem in a computationally tractable manner.

Objectives. The ultimate objective of a learning method is to use the given data to capture the underlying model representing the problem. A theoretical analysis of the types of errors encountered during learning from empirical data was presented in Section 2. In practice, the quality of the learned model is usually measured by evaluating the empirical error between the actual outputs, and the outputs predicted by the learned model, as defined by a L^p $(0 < p \leq \infty)$ norm. A model that approximates the training data well may not be general enough to perform well for unseen data. A good model should be able to provide reasonably accurate prediction for both, the training data, as well as new, previously unseen data. Therefore, the primary objective of the learning technique is to minimize the L^p error for the training as well as testing data. For the input x_i, a model prediction of $F(x_i)$, and an actual value of y_i, these objectives are stated as,

$$\min \left\{ \left\| F\left(x_i^{trng}\right) - y_i^{trng} \right\|^p \right\} \tag{12a}$$

$$\min \left\{ \left\| F\left(x_i^{tst}\right) - y_i^{tst} \right\|^p \right\} \tag{12b}$$

where, $\left(x_i^{trng}, y_i^{trng}\right)$ represent the training data, and $\left(x_i^{tst}, y_i^{tst}\right)$ represents the testing data. The solutions to the minimizations (12a) and (12b) represent the trade-off between memorization and generalization of empirical data. The same trade-off may also be represented as the regularization problem defined in equation (2).

The computational complexity of learning, adaptation and prediction is directly proportional to the number of parameters in the learned model. Therefore, a simpler model consisting of fewer adjustable parameters, that gives the same error of approximation as a more complex model, allows faster training and prediction. The objective of obtaining the simplest *Wave-Net*, with the smallest number of adjustable parameters is stated as,

$$\min \{N_{param}\} \tag{12c}$$

where, N_{param} is the number of adjustable parameters in the *Wave-Net*.

Constraints. The model should be represented as a weighted sum of wavelets and scaling functions,

$$F(x_i) = \sum_{m=1}^{L} \sum_{k=-\infty}^{\infty} c_{mk} \psi_{mk}(x_i) + \sum_{k=-\infty}^{\infty} d_{Lk} \phi_{Lk}(x_i) \tag{6}$$

where, $\psi_{mk}(x)$ represents the discretized family of wavelets, and satisfies the wavelet admissibility conditions,

$$\int_{-\infty}^{+\infty} \psi(x)\, dx = 0 \tag{3}$$

as well as the frame conditions,

$$A\|F\|^2 \le \sum_k |\langle F, \psi_k \rangle|^2 + |\langle F, \phi_L \rangle|^2 \le B\|F\|^2, \qquad A, B > 0 \tag{5}$$

to ensure completeness and universal approximation.

Designing a *Wave-Net* by solving the optimization problem formulated above is a complicated and potentially intractable task due to the nonlinear nature of the problem. It is necessary to make assumptions to simplify the optimization problem and make it reasonably tractable. Reduced computational complexity is achieved at the cost of only partial satisfaction of the objectives and constraints. All empirical modeling and learning techniques require such assumptions. The attractive feature of *Wave-Nets* as opposed to other artificial neural learning techniques is that the simplifying assumptions are more explicit and physically meaningful, as described below.

4.2 Optimizing the Trade-off Between Memorization and Generalization

The objectives given by the minimizations (12a) and (12b) represent the need for learning a model that performs well for training as well as testing data. In practice, the trade-off between memorization and generalization is optimized by satisfying the objectives in (12a) and (12b) separately, by splitting the learning problem in two parts. First, several increasingly accurate models that satisfy equation (12a), are learned using only the training data. The model that best optimizes the memorization-generalization trade-off is then selected via crossvalidation with testing data. The objective given in equation (12b) is satisfied by crossvalidation. The performance of each model is evaluated for the previously unseen testing data, and the *Wave-Net* that gives the least error of approximation for the testing data is selected as the optimum model. For some problems, the maximum acceptable error of approximation may be known and crossvalidation may not be necessary. If the memorization-generalization trade-off is addressed by solving a regularization problem, it is necessary to determine an appropriate value for the smoothing parameter, λ, in equation (2).

4.3 Choosing the Wavelet Basis Functions

Ideally, the basis functions selected for approximation or learning should minimize the norm, using the least complex model, consisting of the minimum number of basis functions. It is unrealistic to expect that there exists a single set of basis functions that satisfies these criteria for a variety of models, since for each function to be learned, the best basis function is that function itself. Several qualitative and intuitive criteria are used for selecting the set of basis functions. The input-frequency localization of wavelets, and the multi-resolution character of *Wave-Net* learning enables efficient approximation of a wide variety of hypersurfaces. Also, the class of wavelets is quite extensive and offers several options and types of wavelets. Some criteria for selecting a class of wavelets are summarized here.

Information about the nature of the hypersurface to be learned is often available from the problem definition. For example, for classification problems, the output usually consists of a small number of discrete classes, and is represented by a piece-wise constant hypersurface. Therefore, using a piece-wise constant set of basis functions, such as the Haar wavelets, is likely to provide a simpler solution. An important requirement of the learned model is that it should be general and smooth enough to allow reasonable interpolation and extrapolation for unseen data. Highly oscillatory basis functions may result in undesirable oscillations in the learned model. In general, a high degree of smoothness is essential for obtaining good generalization for multi-dimensional problems with limited training data (Poggio and Girosi, 1989).

Several types of wavelets have been developed for the analysis of measured data. Properties such as orthonormality, biorthogonality, and compact support are useful for many applications, such as data compression, image processing and for solving partial differential equations. A significant advantage of orthonormal wavelets, in addition to the input-frequency localization, is that when new wavelets are introduced to improve the approximation, the values of previously computed wavelet coefficients remain unchanged, when the least-squares error of approximation is minimized. This property leads to very efficient and intuitive algorithms for data analysis, as well as machine learning. The benefits of learning with orthonormal wavelets are reaped only if the training data allow satisfaction of orthonormality conditions as described in Bakshi and Stephanopoulos (1993). For the wavelet analysis of a measured signal, orthonormality conditions are usually satisfied since the data are regularly spaced, and efficient algorithms (O(N)) have been devised (Mallat, 1989). Unfortunately, training data for machine learning applications are unlikely to be regularly spaced, and the benefits of orthonormality are usually quite limited. Wavelets that do not overlap, such as Haar wavelets, are orthonormal at the same scale (semi-orthonormal) even if the data are irregularly spaced.

Radial basis functions are known to possess good approximation properties, and provide a solution to the regularization problem, as described in Section 2. Several radial basis functions also satisfy the wavelet admissibility condition. Consequently, radial wavelets are likely to be well-suited for multi-resolution learning, within the *Wave-Net* framework. Radial wavelets are of the form,

$$\psi(\|x - t_j\|)$$

where, t_j is the center of the basis function. Several such wavelets may be devised, such as,

$$\psi(x) = -(x - t_j)\exp\left(-\frac{(x - t_j)^2}{2}\right)$$

which is the first derivative of a Gaussian. Using radial wavelets for functional approximation allows combining the regularization obtained by radial basis functions, with the hierarchical learning, explicit network construction, and physically meaningful network parameters of *Wave-Nets*, making them especially attractive as basis functions.

4.4 Minimizing an Error Norm

The quality of the approximation is measured by comparing the value predicted by the learned model with the actual or expected value. The error is usually measured as a norm, but it is not clear as to which norm represents the learning problem in the best manner. Traditionally, the L^2 norm has been used for learning. Wavelets and frame theory are also designed to span the space of L^2 functions. This is not a restriction, and wavelets that span other spaces may be designed. Orthonormal basis functions minimize the L^2 norm, and the squared coefficients are directly proportional to the L^2 norm for functions satisfying the frame conditions (equation (9)).

Wave-Net training procedure for minimizing the L^2 norm was developed in Bakshi and Stephanopoulos (1993). The L^2 norm provides a global measure of the quality of the approximation. In many problems, it is necessary to monitor the local error of approximation, in addition to the global. A local L^2 error of approximation may be calculated and monitored using the wavelet coefficients over a local region, as shown in equation (10). The training procedure, however does not explicitly minimize the local error of approximation at every point.

In many learning problems, especially those related to process control, it is important to obtain an approximation that keeps the error at each point below an acceptable maximum. The global L^2 error provides a measure of the error of approximation that is an average over the entire input space. If a point-wise error criterion is to be used, the L^∞ norm is more natural and convenient. *Wave-Net* training and adaptation procedures for minimizing the L^∞ error are described in (Koulouris et al., 1993), and are comparable to those for minimizing the L^2 error.

4.5 Translation and Dilation Scheme

The optimization problem formulated in the minimizations (12a), (12b) and (12c), and the constraints (6), (3) and (5) may be simplified by devising a scheme for varying the translation and dilation parameters. The parameters should be set such that the family of wavelets can span the entire input space and be dense in $L^2(R^d)$, that is, be able to approximate any d-dimensional hypersurface. The simplest and most extensively studied scheme for discretizing the translation and dilation parameters is via dyadic sampling, i.e.,

$$s = C^m, \quad u = \alpha D^m k, \quad (m, k) \in Z^2$$

where scale (dilation) parameter is represented by m, and translation parameter by k. For multi-dimensional problems, det(D) = 2 gives dyadic sampling. If D is a diagonal matrix, then the dilation along each dimension is controlled independently of the other dimensions. The matrix, D is then called *separable*. For example, the dilation matrix,

$$D = \begin{bmatrix} \sqrt{2} & 0 \\ 0 & \sqrt{2} \end{bmatrix}$$

gives the dilation scheme depicted in Figure 3. The relative dilation along each dimension may be different, and is dependent on the relative magnitudes of the diagonal elements of D. If D has non-diagonal elements, the basis functions are also rotated when dilated. The dilation matrix,

$$D = \begin{bmatrix} 1 & -1 \\ 1 & 1 \end{bmatrix}$$

rotates the grid by $\pi/4$ and stretches it by $\sqrt{2}$.

The learning scheme described in this chapter considers a dyadic, separable (no rotation) grid. *Wave-Net* training with other degrees of freedom is studied by Zhang and Benveniste (1992). The dyadic dilation scheme for a two input problem is depicted in Figure 3. Such a preset translation and dilation scheme greatly simplifies the optimization problem, at the expense of possible violation of the minimization given by (12c). The model learned by the *Wave-Net* may not be the simplest possible solution, but is certainly more practical and efficient.

4.6 Estimation of Wavelet Coefficients

Having made the assumptions described above, the optimization problem is greatly simplified. The method for estimation and adaptation of the *Wave-Net* coefficients by minimizing the L^2 and L^∞ norms is summarized below.

Minimizing the L^2 norm. The simplified optimization problem consists of two parts, learning several models based on the training data, and then selecting the best model via crossvalidation with testing data, i.e.,

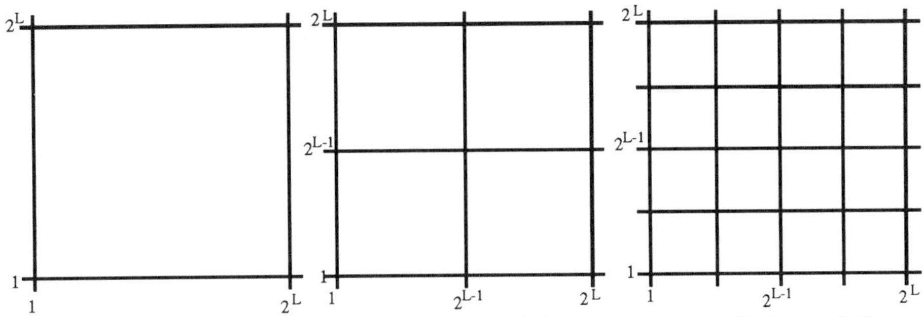

Scale, m = L   Scale, m = L-1   Scale, m = L-2

Figure 3. Evolution of the dyadic dilation scheme

$$\min\left\{\left\|F\left(x_k^{trng}\right) - y_k^{trng}\right\|^2\right\} \tag{13a}$$

such that,

$$\psi_{mk}(x) = 2^{-m/2}\,\psi(A^m x - k), \qquad (m,k) \in Z^d, \qquad \det(A) = 2 \tag{14}$$

$$F(x_k) = \sum_m \sum_k c_{mk}\psi_{mk}(x_k) + \sum_k d_{Lk}\phi_{Lk}(x_k) \tag{6}$$

where, $\psi_{mk}(x)$ satisfies the wavelet admissibility conditions, as well as the frame conditions. Solve the optimization problem given above for different numbers of basis functions, and then select the model that minimizes the error on testing data,

$$\min\left\{\left\|F\left(x_k^{tst}\right) - y_k^{tst}\right\|^2\right\} \tag{13b}$$

Equation (6) may be represented in matrix form as,

$$F = Ac$$

where, A is an $m \times n$ matrix for m training data, and n basis functions. Usually $n \leq m$. Computing the coefficients that minimize the L^2 norm involves finding the pseudo-inverse,

$$c = \left((A^TA)^{-1}A^T\right)F = A^+F$$

As shown in Bakshi and Stephanopoulos (1993) the matrix $(A^TA)^{-1}$ may have a convenient block diagonal structure that depends on the characteristics of the basis functions and the training data. If orthonormality conditions are satisfied, (A^TA) is an identity matrix, and the computation of wavelet coefficients is equivalent to using Mallat's algorithm (Mallat, 1989). But, orthonormality may not be satisfied for learning problems. In general, the structure of (A^TA) depends on the degree of overlap of the basis functions, and data density. The matrix is usually sparse and diagonally dominant allowing the use of efficient matrix manipulation routines.

As more data are obtained, the *Wave-Net* needs to be adapted by updating the weights. New data result in increasing the number of rows in the **A** matrix.

$$\mathbf{A}_{m+k,n} = \begin{bmatrix} \mathbf{A}_{m,n} \\ \text{--------} \\ \theta^T_{k,n} \end{bmatrix}$$

where, $\theta^T_{k,n}$ represents the rows due to the k new data points. For computing the pseudo-inverse of the matrix, $\mathbf{A}_{m+k,n}$, the pseudo-inverse of $\mathbf{A}_{m,n}$ that was computed previously may be used to provide an efficient technique for updating the wavelet coefficients.

The adequacy of the learned model, or the trade-off between memorization and generalization is tested via crossvalidation using testing data. As more training data are obtained, it may be necessary to add new basis functions to improve the quality of the model. This is necessitated when the error for training and testing data does not decrease upon adaptation of the coefficients. A new basis function introduces a new column in the **A** matrix, i.e.,

$$\mathbf{A}_{m,n+k} = \left[\mathbf{A}_{m,n} \mid \theta_{m,k}\right]$$

An incremental formula for computing the pseudo-inverse of this matrix may be derived as shown in the book by Cormen et al. (1990).

Minimizing the L^∞ norm. The problem for minimizing the L^∞ norm is also solved in two parts: learning and crossvalidation. The formulation is identical to that given for minimizing the L^2 norm, above, except that minimizations (13a) and (13b) are replaced by,

$$\min_c \max_k \left| F\left(x_k^{trng}\right) - y_k \right| \tag{15a}$$

$$\min_c \max_k \left| F\left(x_k^{tst}\right) - y_k \right| \tag{15b}$$

respectively. The minimax optimization problems described by (15a) and (15b), and the constraints in equations (14) and (6), may be solved as linear programming problems. As described by Koulouris et al. (1993), after specifying the type of basis functions, and the maximum error bound, the learning proceeds hierarchically. Starting with the basis functions for the coarsest approximation, the data are introduced point by point. The weights in the current *Wave-Net* are updated for the new data point by solving the minimax optimization problem. This forms the model derivation algorithm. If the error bound is not satisfied, then new basis functions at finer resolutions are introduced locally, until the error bound is satisfied. This forms the structural adaptation algorithm. The *Wave-Net* is constructed hierarchically such that the L^∞ error is less than a specified upper bound.

The adaptation techniques described above for both norms use computation done previously, such as the matrix inverse. Updating the weights for a new data point and/or for a new basis function requires storage of the A matrix obtained from the data so far. The A matrix can be stored in an efficient manner, taking advantage of its structure. For the least-squares error criterion, training and adaptation without requiring all the previously seen data may be performed by using gradient descent, or the LMS rule (Widrow and Hoff, 1960). Similar techniques for on-line training and adaptation for minimizing the L^∞ norm are being explored.

4.7 *Wave-Net* Training and Adaptation Procedure

The *Wave-Net* training procedure solves the optimization problem given by (13a), (14) and (6) for minimizing the L^2 norm, or (15a), (14) and (6) for minimizing the L^∞ norm. The simplifications and assumptions in solving this problem were described above. The *Wave-Net* obtained may not be the simplest model derivable within the general framework. The class of basis functions is selected based on knowledge about the learning problem. The translation and dilation parameters are set to obtain a dyadic grid for locating the wavelets. The network is then constructed hierarchically, and the coefficients are computed so as to minimize the L^2 or L^∞ norm. The learning algorithm proceeds as follows,

Given: Training data consisting of inputs, $\{\mathbf{x}\}$ and outputs, $\{\mathbf{y}\}$
Testing data and/or an acceptable L^2 or L^∞ approximation error.
1 Select a class of wavelets
2 Determine the largest scale, L such that the input space is covered by a single scaling function.
3 Construct the multi-resolution grid for selected translation and dilation parameters.
4 Select ϕ_{Lk} (coarsest scaling functions) with data in their support and compute *Wave-Net* coefficients
5 If error criterion, or crossvalidation with testing data is satisfied go to 13
6 else, for m = L
7   Select ψ_{mk} that have training data in the grid space, and approximation error > δ
8   Compute *Wave-Net* coefficients
9   If error criterion, or crossvalidation with testing data is satisfied go to 13
10   else, m = m - 1
11   go to 7
12   end
13 end

Upon receiving new data, the *Wave-Net* may be adapted by the following algorithm.

Given: Trained *Wave-Net*, and error (L^2 or L^∞) achieved
New data.

1 Compute error (L^2 or L^∞) for new data.
2 If error is greater than acceptable
3 until error criterion is satisfied, adapt current *Wave-Net* coefficients
4 else, implement steps 7 through 13 of training algorithm, selecting only basis functions that cover the new data
5 end

Simple examples illustrating the characteristics of *Wave-Nets* for one and two-dimensional problems, using orthonormal wavelets, and the L^2 norm are given in Bakshi and Stephanopoulos (1993).

4.8 Illustrative Examples

The training, adaptation and performance of *Wave-Nets* for multi-dimensional problems using non-orthonormal wavelets and using an L^2 or L^∞ error criterion are illustrated by the following examples. The methodology for minimizing the L^2 or L^∞ norm has been implemented using MatLab.

Example 1. This simple example illustrates the multi-resolution and adaptive character of the *Wave-Net* learning process. The model to be approximated is defined piece-wise as (Zhang and Benveniste, 1992),

$$f(x) = \begin{cases} -2.186x - 12.864 & -10 \le x < -2 \\ 4.246x & -2 \le x < 0 \\ 10e^{-0.05x-0.5}\sin[(0.03x + 0.7)x] & 0 \le x \le 10 \end{cases}$$

The *Wave-Net* was constructed by starting with 5 examples selected randomly from the input space, and adapting it until a total of 100 random examples were used. The network was constructed for a maximum L^∞ error of 0.7, following the methodology given above. The evolution of the model learned with increasing number of examples, and the location of the basis functions is shown in Figure 4. The mexican hat wavelet described in Daubechies (1992) is used for the approximation. Modeling the first order discontinuity at x=-2 requires a greater density of basis functions, as can be seen in Figure 4a to 4e. The final mapping learned with 100 training data requires 27 wavelet basis functions, with the corresponding weights being the adjustable parameters. The final model learned is close to the original function as shown in Figure 4e. The final L^∞ error of approximation is 0.6842.

Example 2. The Lorenz dynamical system provides a more interesting and complicated example to demonstrate the performance of the *Wave-Net* learning technique.

$$\frac{dx}{dt} = \sigma(y - x)$$

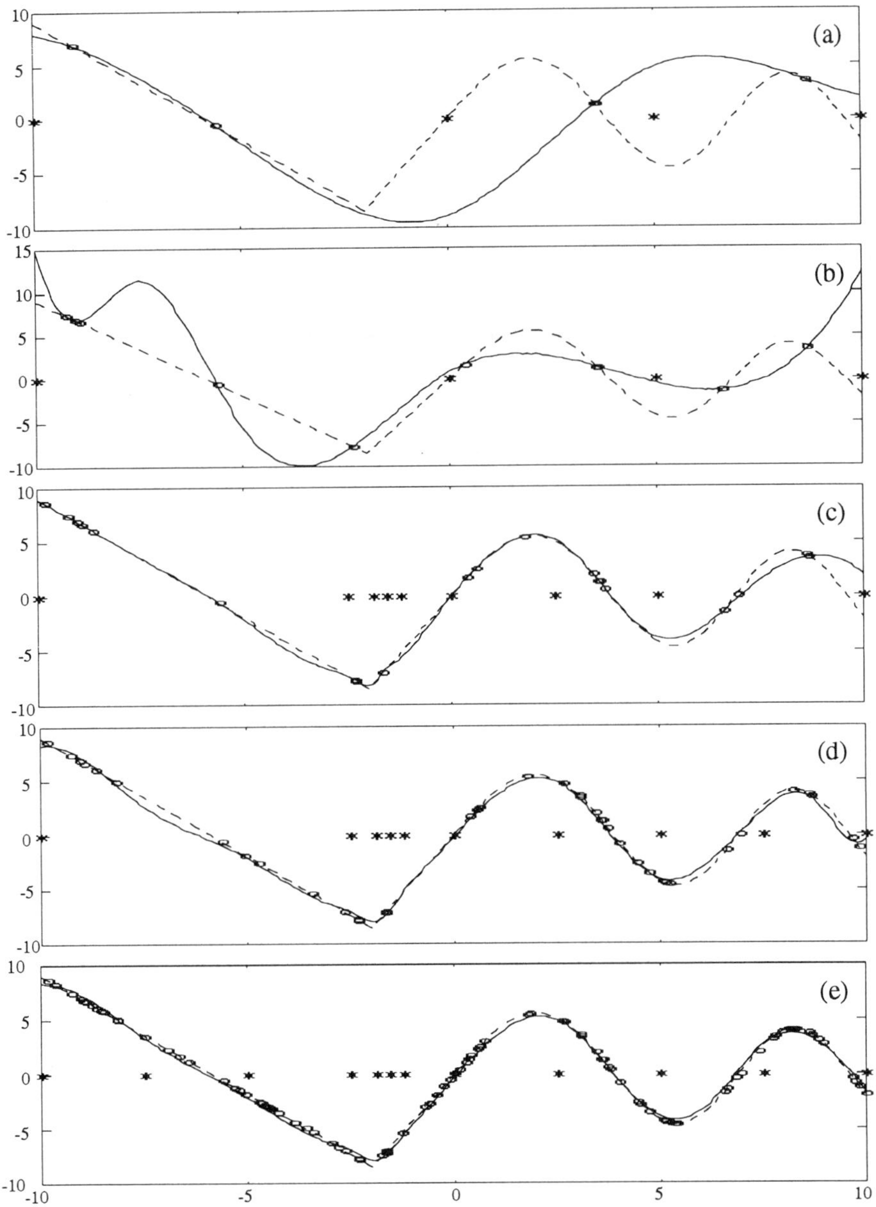

Figure 4. Multi-resolution learning for example1. Comparison of model (solid line) with real function (dashed line) after training with,

a) 5 points, 4 basis functions
b) 10 points, 9 basis functions
c) 20 points, 16 basis functions
d) 40 points, 23 basis functions
e) 100 points, 27 basis functions

Location of basis functions are indicated by asterisks.

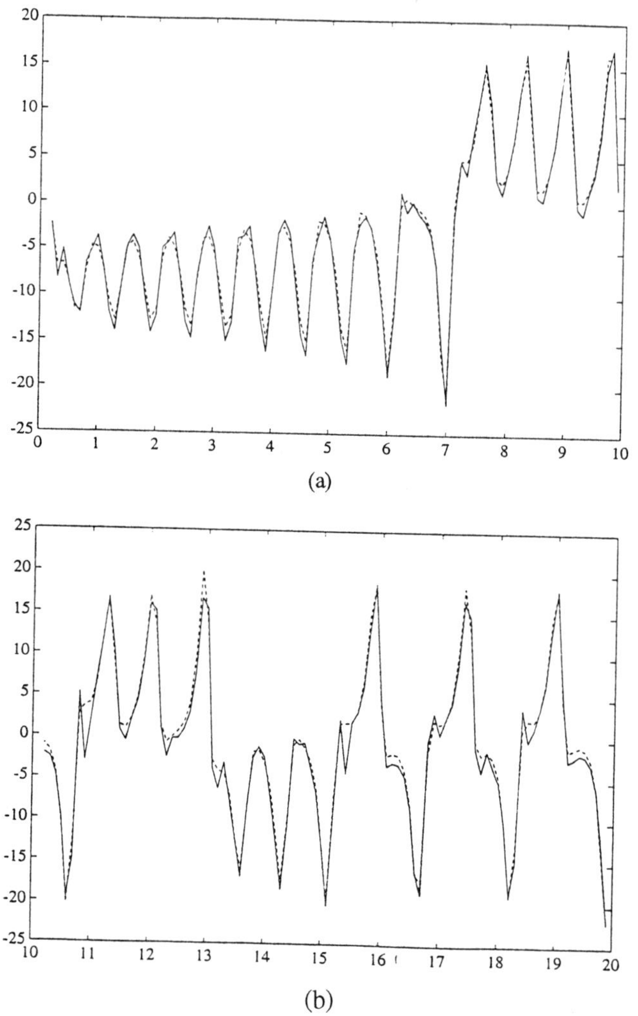

Figure 5. Multi-resolution learning using *Wave-Nets* for example 2. *Wave-Net* model: continuous line, actual values: dashed line.

a) Prediction of learned *Wave-Net* for training data

b) Prediction for testing data using training data in Figure 5a.

$$\frac{dy}{dt} = \rho x - y - xz$$

$$\frac{dz}{dt} = -\beta z + xy.$$

where, σ is the Prandtl number, ρ is the Rayleigh number and β corresponds to physical dimensions of the problem. This set of differential equations describes the heat convection in a two-dimensional fluid and was first introduced by Lorenz,

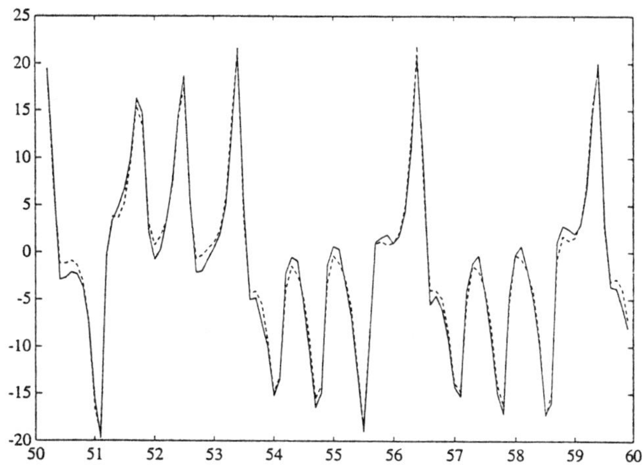

Figure 6. Prediction on testing data after training with 500 data points

(Ydstie, 1990). Following Ydstie the values of these parameters are set to 10, 28 and 8/3 respectively. The state vectors are given by x and y, but it is assumed that only y is observed at distinct intervals of 0.1 sec. The objective is to construct a model to predict the behavior of y one time step into the future based on the following one-step-ahead predictor model,

$$y(k + 1) = F(y(k), y(k - 1), y(k - 2))$$

where y(k) is the value of y at time 0.1k, k=0, 1, 2... and F is the unknown three-dimensional function which will be learned using *Wave-Nets*.

By solving the system of differential equations and sampling y, a time series y(k) is created and the data are fed to the *Wave-Net* algorithm. The mexican-hat wavelet is used as the basis function. This wavelet, introduced by Daubechies (1992) resembles the second derivative of the Gaussian, but is smooth, has compact support and, unlike most wavelets, has a simple algebraic representation. The L^{∞} error bound for the network training is set at 2 which is a tolerable error considering that the output y(k) takes values in the range between -30 and 30. The first 100 data points of the series are used for training the *Wave-Net* and the results are shown in Figure 5a. The resulting network consists of 15 basis functions distributed over three scales and exhibits a maximum pointwise error of 1.51. Another set of 100 points are used to test the prediction on unseen data as shown in Figure 5b. Although the overall prediction is quite satisfactory, the maximum empirical error for the testing data is 6.7, which exceeds the predefined bound of 2. Continuing the training with 400 more points, the error on testing data falls below the predefined bound of 2, without requiring additional basis functions. The prediction of this model on testing data is shown in Figure 6. The error bounds based on the training data are shown in Figure 7. It is remarkable that this accuracy is achieved without introducing additional complexity to the model. Even after training with 600 points the network still contains only 15 basis functions and equal number of adjustable

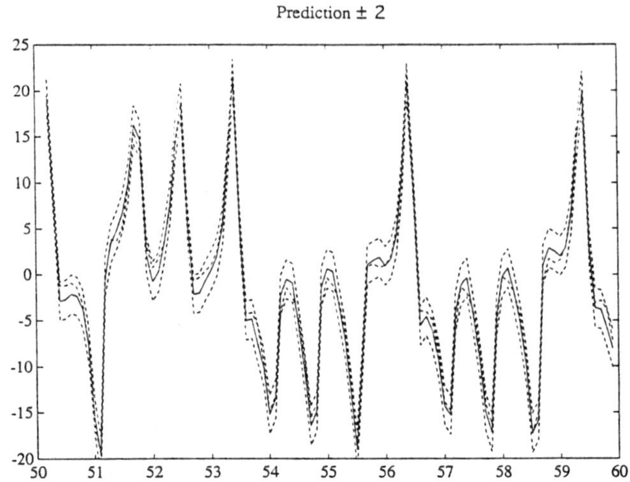

Figure 7. Error bounds on *Wave-Net* prediction for testing data

weights. For comparison, Ydstie uses a network of the backpropagation type with 44 hidden units and multiple number of unknown parameters. The prediction changes insignificantly if the L^2 error method was used instead, as shown in Figure 8 for the same sets of training and testing data.

This example demonstrates the practical merits of learning an unknown function at multiple resolutions using *Wave-Nets*. By monitoring the local approximation error modeling functions of low complexity and high accuracy can be constructed. The most remarkable advantage, however, is that the construction of such models does not require any arbitrary specifications on the form of the approximating function. Questions like how many basis functions to use and where in the input space to

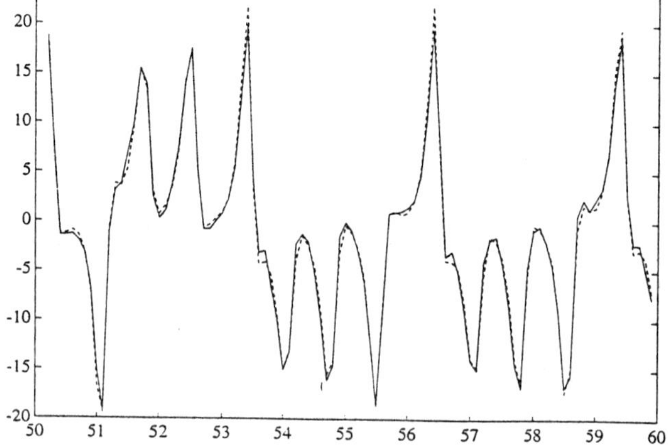

Figure 8. *Wave-Net* prediction on testing data for training with L^2 error criterion

place them are automatically handled by the *Wave-Net* algorithm without the need for any arbitrary or heuristic-based selection by the user.

5. Extracting Rules from *Wave-Nets*

The physically meaningful character of the input-output models learned by *Wave-Nets* is further enhanced by the ability to extract simple if-then rules from *Wave-Nets* using Haar wavelets as basis functions. The Haar wavelet is a natural choice for modeling problems where the output consists of a finite set of discrete values. For example, in classification problems, the output may consist of a finite number of distinct classes. The Haar basis functions may be represented in terms of discrete intervals as well as rules. For example, the Haar scaling function is,

$$\phi(x) = \begin{cases} 1 & 0 \le x < 1 \\ 0 & \text{otherwise} \end{cases}$$

which may also be expressed as,

If input, $x \in [0, 1)$ then, output, $\phi = 1$,
else, output, $\phi = 0$

The Haar wavelet is,

$$\psi(x) = \begin{cases} 1 & 0 \le x < 1/2 \\ -1 & 1/2 \le x < 1 \\ 0 & \text{otherwise} \end{cases}$$

which may also be expressed as,

If input, $x \in [0, \frac{1}{2})$ then, output, $\psi = 1$,

elseif input, $x \in [\frac{1}{2}, 1)$ then, output, $\psi = -1$,

else, output, $\psi = 0$

In multiple dimensions, the complete set of wavelets and scaling functions is computed by taking tensor products, as depicted in Figure 1. Multi-dimensional Haar wavelets may also be expressed as intervals or rules. For example, the wavelet, $\Psi^1(x_1, x_2)$ shown in Figure 1 is represented in interval form as,

$$\Psi^1(x_1, x_2) = \begin{cases} 1 & 0 \le x_1 < 1 \;\&\; 0 \le x_2 < 1/2 \\ -1 & 0 \le x_1 < 1 \;\&\; 1/2 \le x_2 < 1 \end{cases}$$

which may also be expressed as,

If input, $X = \left\{ [0, 1) \times [0, \frac{1}{2}) \right\}$ then, output, $\Psi^1 = 1$, elseif input, $X = \{ [0, 1) \times [\frac{1}{2}, 1) \}$ then output, $\Psi^1 = -1$,

else, output, $\Psi^1 = 0$

Similarly, other multi-dimensional wavelets may be easily represented in the form of both intervals and rules. The interval representation may also be used for expressing the result of adding Haar wavelets and scaling functions. For example, the result of the following weighted sum is,

$$4\Phi_{2,0}(x_1, x_2) + \Psi^1_{1,0}(x_1, x_2) = \begin{cases} 2 & 0 \le x_1 < 1/2 \ \& \ 0 \le x_2 < 1 \\ 0 & 1/2 \le x_1 < 1 \ \& \ 0 \le x_2 < 1 \\ 1 & \begin{cases} 1 \le x_1 < 2 \ \& \ 0 \le x_2 < 2 \\ 0 \le x_1 < 1 \ \& \ 1 \le x_2 < 2 \end{cases} \end{cases}$$

as depicted in Figure 9. Similarly, rules may be derived for weighted sums of higher dimensional Haar basis functions.

The generation of multi-dimensional wavelets by taking tensor products results in $(2^d - 1)$ wavelets for every grid point, where d is the number of input dimensions. This combinatorial explosion in the number of wavelets to be evaluated could result in expensive training and adaptation of such a *Wave-Net*. Of the one-dimensional Haar scaling function and wavelet, only the wavelet enables splitting the input space into different classes, and thus help classification. Taking the tensor product of a wavelet in one dimension with scaling functions in the other dimensions results in a multi-dimensional basis function that can classify only along the dimension spanned by the wavelet. As shown in Figure 1, the wavelet Ψ^1 can classify only along the x_1 dimension, and Ψ^2 along x_2. The wavelet Ψ^3 is constructed usingwavelets along both dimensions, and can classify along both x_1 and x_2. Thus, the hypersurface of the wavelet Ψ^3 is more complicated than that of Ψ^1 and Ψ^2. These characteristics of separable Haar wavelets are used to expedite the training. Some of the alternatives for dealing with the exponentially increasing complexity are as follows,

1) Tensor product basis functions may be evaluated for their classification ability in order of increasing complexity of the hypersurface. As described above, and illustrated in Figure 1, for the Haar wavelets, $\Psi^3 = \psi(x_1)\psi(x_2)$ is more complex and has greater classification ability than the wavelets, $\Psi^1 = \phi(x_1)\psi(x_2)$ and $\Psi^2 = \psi(x_1)\phi(x_2)$. Similarly, for a three input problem, the seven basis functions in order of increasing complexity, or classification ability are as follows,

$$\Psi^l(x_1, x_2, x_3) = \left\{ \begin{matrix} \phi(x_1)\phi(x_2)\psi(x_3) & \phi(x_1)\psi(x_2)\phi(x_3) & \psi(x_1)\phi(x_2)\phi(x_3) \\ \phi(x_1)\psi(x_2)\psi(x_3) & \psi(x_1)\phi(x_2)\psi(x_3) & \psi(x_1)\psi(x_2)\phi(x_3) \\ \psi(x_1)\psi(x_2)\psi(x_3) & & \end{matrix} \right\}$$

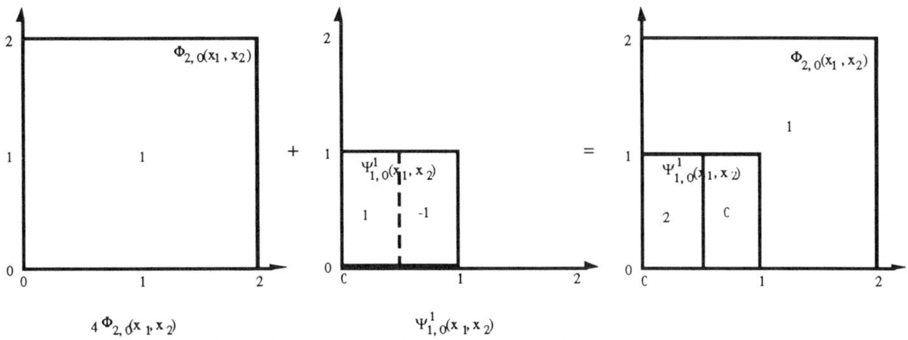

Figure 9. Addition of overlapping Haar basis functions for two inputs, x_1, and x_2

Instead of evaluating all the basis functions simultaneously, we evaluate the ability of the wavelets to approximate the hypersurface in the order of increasing classification ability as shown above. If there are irrelevant inputs or lack of variation in the data for some inputs, complete classification may be achieved without considering all the $(2^d - 1)$ wavelets, resulting in reduced complexity of the overall training procedure. If the wavelet along a chosen dimension results in no improvement in the approximation then the multi-dimensional form of this wavelet will also be irrelevant to the approximation and can be disregarded. These two observations may be used for exploring the basis functions in a step-wise and controlled manner, as shown in the illustration below.

2) Consider only those multi-dimensional wavelets that are formed by the tensor product of only one wavelet and (d-1) scaling functions. This results in d wavelets for d dimensions. For a three input problem, only the following wavelets are considered for classification,

$$\Psi^l(x_1, x_2, x_3) = \{\ \phi(x_1)\phi(x_2)\psi(x_3)\quad \phi(x_1)\psi(x_2)\phi(x_3)\quad \psi(x_1)\phi(x_2)\phi(x_3)\ \}$$

Each wavelet is oriented perpendicular to the axis covered by the wavelet, and split the data along that direction only.

3) Use non-separable wavelets, i.e. that are not created by taking a tensor product. This results in only one wavelet regardless of the dimension. Such wavelets were used in the examples in Section 4, but the approach is not relevant for obtaining piece-wise constant hypersurfaces.

The approach described in Item 2 above is the least complex, while maintaining the orthonormality of basis functions, but may compromise the completeness of approximation that is guaranteed by selecting $(2^d - 1)$ wavelets for every hyper rectangle in the grid. In practice following item 2, may require consideration of more scale levels for the same quality of approximation as that achieved by considering all $(2^d - 1)$ wavelets. For multi-dimensional Haar wavelets, the orientation specificity may be exploited quite effectively to detect irrelevant or redundant variables, and gain insight into the importance of measured variables.

Table 1. Data for illustrating induction using decision trees (H: High, L: Low, N: Normal, A: Abnormal)

	Input Features		Output
Example #	Pressure	Color	Quality
1	N (2)	N (3)	Good (1)
2	N (2)	A (1)	Bad (0)
3	H (3)	N (3)	Good (1)
4	L (1)	N (3)	Bad (0)
5	L (1)	N (3)	Bad (0)
6	N (2)	N (3)	Good (1)
7	L (1)	A (1)	Bad (0)
8	N (2)	A (1)	Bad (0)

5.1 Illustration

Consider the data shown in Table 1. We wish to learn a mapping for classifying the given examples using Haar *Wave-Nets*. The qualitative variables are given quantitative values as indicated by the numbers in the brackets in Table 1 to enable convenient application of the *Wave-Net* methodology. The distribution of the data in the input space is shown in Figure 10a. The multi-resolution evolution of the hypersurface is depicted in Figures 10b and c, and the values of the *Wave-Net* parameters are shown in Table 2. The two wavelets, Ψ^1, and Ψ^2, shown in Figure 1 are used to learn the mapping. First the scaling function, $\Phi(X)$ is introduced at scale m=1 resulting in an output of 0.375. The approximation of the hypersurface by wavelets at scale m=1 is shown in Figure 10b. Since the data are not approximated accurately, wavelets at scale m=0 are introduced resulting in the hypersurface shown in Figure 10c. In learning the mapping at scale, m = 0, the advantages of using orientation specific Haar wavelets is seen. For the data in the rectangle covered by the wavelet at coordinates (1, 1) the wavelet $\psi(x_1)$ does not decrease the error of approximation, as indicated by the zero coefficient. For the rectangle covered by the basis function at (1, 2), the wavelet $\psi(x_1)$ results in a zero error of approximation, and $\psi(x_2)$ need not be evaluated. Since there are no data in the region of the input space covered by the wavelet at (2, 1), no wavelets need to be evaluated in this region. The rules that may be derived from this surface are,

If Pressure is *Low* and Color is *(Normal or Abnormal)*
If Pressure is *(Low or Normal)* and Color is *Abnormal* } then Quality is BAD

If Pressure is *Normal* and Color is *Normal*
If Pressure is *High* and Color is *Normal* } then Quality is GOOD

This problem may also be solved using all the $(2^2 - 1)$ wavelets, and the rules derived from the resulting *Wave-Net* are identical to those listed above.

Table 2. Haar *Wave-Net* parameters for data in Table 1 and hypersurface in Figure 10

Figure	Scale	Coordinate	Basis function	Weight
10a	m=1	(1,1)	$\Phi(X)$	0.75
10b	m=1	(1,1)	$\Psi^1(X)$	-0.3125
			$\Psi^2(X)$	-0.5625
10c	m=0	(1,1)	$\Psi^1(X)$	0
			$\Psi^2(X)$	0.0625
		(1,2)	$\Psi^1(X)$	-0.5
			$\Psi^2(X)$	0
		(2,2)	$\Psi^1(X)$	-0.0937
			$\Psi^2(X)$	-0.0937

The simple example solved above illustrates the power of the *Wave-Net* learning technique. Unlike other neural learning methods, *Wave-Nets*, for the first time, provide physically interpretable models from empirical data. Generalization to multiple dimensions is straightforward.

5.2 Haar *Wave-Nets* and Decision Trees

Several techniques for inducing tree structured models from empirical data have been developed. Inductive learning using decision trees (Quinlan, 1986) is a popular technique for extracting rules from measured data. The mapping is learned by splitting the input space in a manner that provides the maximum split in the training data, as measured by an information theoretic entropy criterion. The algorithm is greedy, and continues developing the tree until the desired degree of classification is obtained. Several techniques, such as crossvalidation with testing data, are used for optimizing the trade-off between memorization and generalization. The decision tree developed for the data in Table 1, is shown in Figure 11. The rules derived from this tree are,

If Pressure is *Low*
If Pressure is *Normal* and Color is *Abnormal* } then Quality is BAD

If Pressure is *Normal* and Color is *Normal*
If Pressure is *High* } then Quality is GOOD

The rules derived above, and those obtained from the Haar *Wave-Net* are quite similar, but indicate some important differences between rule extraction from a *Wave-Net* and inductive decision trees.

Local versus global approximation. The model learned by a *Wave-Net* is localized in character, since the Haar wavelet is localized in the input space. Therefore, the *Wave-Net* allows only limited extrapolation beyond the region where training data are available. The model learned by a decision tree, on the other hand, is global, since it does not take into consideration the region of the input space

where training data are not available. As a result, prediction by decision trees can lead to large errors of extrapolation without warning. For the previously unseen example, with {Pressure=High, Color=Abnormal} the decision tree rule, "if pressure is high then quality is good" gives the potentially incorrect result that the quality is good. None of the *Wave-Net* rules applies to this example, indicating that the training data used are inadequate for proper classification of this example.

Optimization of learned mapping. The models learned by both *Wave-Nets* and inductive decision trees require crossvalidation to prevent overfitting. The greedy nature of the decision tree learning procedure may necessitate growing a large tree if the nodes selected early in the procedure are incorrect. Techniques used to counter this, such as pruning, require exhaustive generation of all possible trees, and are expensive. Similarly, adaptation to new data is also expensive. Learning within the framework of Artificial Neural Networks provides the advantage of fast adaptation, without losing the rule extraction capabilities of inductive decision trees.

The order of importance of the variables is clearly stated in the rules extracted by decision trees. This ordering is based on the entropy maximization principle, and is arbitrary. The rules extracted from Haar *Wave-Nets* do not have an explicit

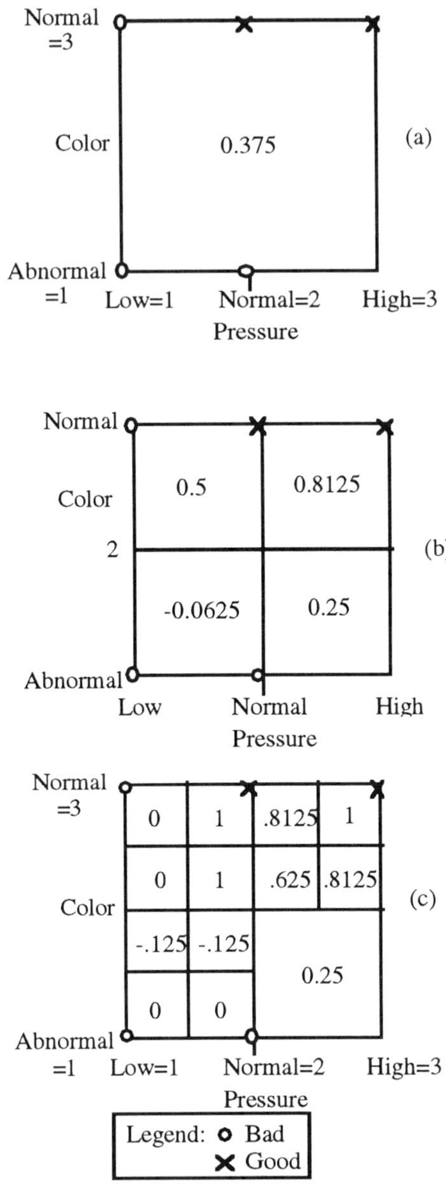

Figure 10. Evolution of hypersurface for data in Table 1. Haar *Wave-Net* parameters given in Table2.

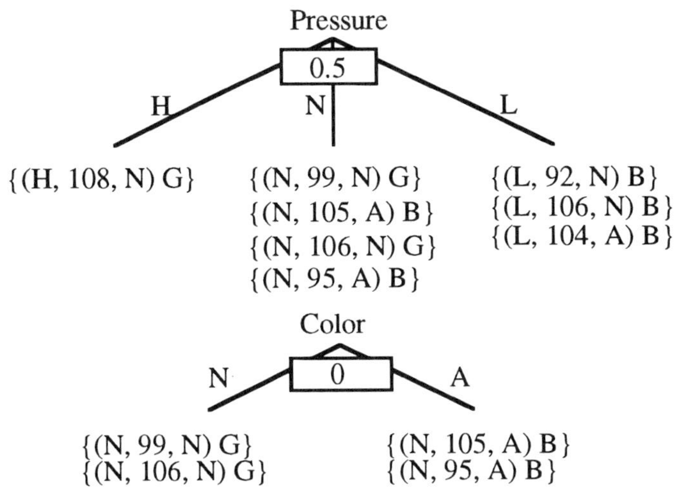

Figure 11. Resulting decision tree for data in Table 1.

ordering of the relative importance of measured variables.

The learning with Haar *Wave-Nets* considers sections of the input space in a hierarchical manner. For dyadic sampling, the input space is split into half as the resolution increases. Only those hyper rectangles are considered that contain training data. The rule extraction using Haar *Wave-Nets* is similar to many tree structured approximation and learning algorithms (Sanger, 1991).

6. Conclusions and Discussion

Learning at multiple resolutions provides a convenient and efficient framework for solving a variety of engineering problems. The theory of wavelets has provided impetus to the development of multi-resolution methods for learning input-output models from empirical data. *Wave-Nets* are a novel artificial neural network where the activation functions in the nodes are wavelets. *Wave-Nets* are based on the formal framework of wavelets and approximation theory, and possess several attractive properties.

- The model is learned at multiple resolutions, resulting in an efficient learning methodology.
- Decisions about the structure and construction of the network are explicitly based on the desired quality of approximation, and the availability of training data.
- The global as well as local error of approximation of the learned model is available as a meaningful criterion for its design.
- The model learned is physically meaningful. The network parameters such as the edge weights reflect the degree to which the corresponding basis

function reduces the least-squares error of approximation. The activation function parameters reflect the distribution and relevance of training data.
- Simple if-then rules may be derived from *Wave-Nets* used for solving classification problem.

Training and adaptation algorithms for minimizing the L^2 and the L^∞ error criteria are presented in this paper. Ongoing work on the extraction of rules from Haar *Wave-Nets* is also introduced.

Acknowledgments. Bhavik Bakshi acknowledges the post-doctoral assistantship at MIT and the resources available at Aware, Inc., Cambridge, MA.

References

Adomaitis, R. A., Farber, R. M., Hudson, J. L., Kevrekidis, I. G., Kube, M., Lapedes, A. S., Application of Neural Nets to System Identification and Bifurcation Analysis of Real World Experimental Data, *Los Alamos National Laboratory Technical Report*, LA-UR-90-515 (1990)

Bakshi, B. R., *Multi-Resolution Methods for Modeling, Analysis and Control of Chemical Process Operations*, Ph. D. thesis, Department of Chemical Engineering, Massachusetts Institute of Technology (1992)

Bakshi, B. R., Stephanopoulos, G., *Wave-Net*: a Multiresolution, Hierarchical Neural Network with Localized Learning, *AIChE J.*, 39, 1, 57-81 (1993)

Barto, A. G., Connectionist Learning for Control, in *Neural Networks for Control*, MIT Press, Cambridge (1991)

Bhat, N. V., Minderman, P. A., McAvoy, T., Wang, N. S., Modeling Chemical Process Systems via Neural Computation, *IEEE Control Systems Magazine*, 24-30, April (1990)

Cormen, T. H., Leiserson, C. E., Rivest, R. L., *Introduction to Algorithms*, The MIT Press, Cambridge, MA (1990)

Daubechies, I., *Ten Lectures on Wavelets*, SIAM (1992)

Daubechies, I., The Wavelet Transform, Time-Frequency Localization and Signal Analysis, *IEEE Trans. Inform. Theory*, 36, 5, 961-1005 (1990)

Haesloop, D., Holt, B. R., A Neural Network Structure for System Identification, *Proc. Amer. Control Conf.*, 2460-2465 (1990)

Hernandez, E., Arkun, Y., A Study of the Control Relevant Properties of Backpropagation Neural Net Models of Nonlinear Dynamical Systems, *Comp. Chem. Eng.*, 16, 227 (1992)

Hoskins, J. C., Himmelblau, D. M., Artificial Neural network Models of Knowledge Representation in Chemical Engineering, *Comp. Chem. Eng.*, 12, 881-890 (1988)

Koulouris, A., Bakshi, B. R., Stephanopoulos, G., Learning an unknown function from Examples: Analysis and Extension of the *Wave-Net* Solution, in preparation (1993)

Kramer, M. A., Leonard, J. A., Diagnosis Using Backpropagation Neural Networks - Analysis and Criticism, *Comp. Chem. Eng.*, 14, 12, 1323-1338 (1990)

Leonard, J. A., Kramer, M. A., Radial Basis Function Networks for Classifying Process Faults, *IEEE Control Systems*, 31-38, April (1991)

Ljung, L., *System Identification*, Prentice Hall (1987)

Mallat, S., A Theory for Multiresolution Signal Decomposition: The Wavelet Representation, *IEEE Trans. Pat. Anal. Mach. Intel.*, 11, 7, 674-693 (1989)

Mardia, K., Kent, J., Bibby, J., *Multivariate Analysis*, Academic Press, London (1980)

Moody, J., Fast Learning in Multi-Resolution Hierarchies, *Research Report*, Yale University, YALEU/DCS/RR-681 (1989)

Narendra, K., Parthasarthy, K., Identification and Control of Dynamical Systems using Neural Networks, *IEEE Trans. Neural Networks*, 1, 1 (1989)

Poggio, T., Girosi, F., A Theory of Networks for Approximation and Learning, *AI Lab., Memo no. 1140*, MIT AI Lab, July (1990)

Quinlan, J. R., Induction of Decision Trees, *Machine Learning*, 1, 1, 81-106 (1986)

Sanger, T. D., A Tree-Structured Algorithm for Reducing Computation in Networks with Separable Basis Functions, *Neural Computation*, 3, 67-78 (1991)

Saraiva, P., Stephanopoulos, G., Continuous Process Improvement Through Inductive and Analogical Learning, *AIChE J.*, 38, 2, 161-183 (1992)

Ungar, L. H., Powell, B. A., Kamens, S. N., Adaptive Networks for Fault Diagnosis and Process Control, *Comp. Chem. Eng.*, 14, 561-573 (1990)

Venkatasubramaniam, V., McAvoy, T. J., Computers and Chemical Engineering, special issue on Neural Networks in Chemical Engineering, 16, 4 (1992)

Venkatasubramaniam, V., Vaidyanathan, R., Yamamoto, Y., Process Fault Detection and Diagnosis Using Neural Networks -I. Steady-State Processes, *Comp. Chem. Eng.* 14, 7, 699-712 (1990)

Watanabe, K., Matsuura, I., Abe, M., Kubota, M., Himmelblau, D. M., Incipient Fault Diagnosis of Chemical Processes via Artificial Neural Networks, *AIChE J.*, 35, 1803-1812 (1989)

Whiteley, J. R., Davis, J. F., Knowledge-Based Interpretation of Sensor Patterns, *Comp. Chem. Eng.*, 16, 4, 329-347 (1992)

Widrow, B., Hoff, M., Adaptive Switching Circuits, *IRE WESCON Conv. Record, Part 4*, 96-104 (1960)

Ydstie, B. E., Forecasting and Control using Adaptive Connectionist Networks, *Comp. Chem. Eng.*, 14, 4/5, 583-599 (1990)

Zhang, Q., Benveniste, A., Wavelet Networks, *IEEE Trans. Neural Networks*, 3, 6, 889-898 (1992)

Chapter 6

Application of Wavelets in Process Control

Yugender Chikkula
Jay H. Lee*

Department of Chemical Engineering
Auburn University
Auburn, AL 36849-5127

Abstract

This chapter provides an overview of wavelets from a process control perspective. The time-frequency localization and multiresolution properties of wavelets are discussed in the context of control applications. We also discuss various control problems where wavelets could be particularly advantageous. We illustrate the benefits of wavelet formulations by presenting wavelet domain approaches to basis reduction and frequency domain tuning in model predictive control problems.

1 Introduction

Over the past few years, wavelets and wavelet based analysis have found their way into many different fields of science and engineering. Wavelets are new family of basis functions which are localized in both time and frequency domains. The wavelet transform is a tool that provides descriptions of functions or signals in the time-frequency plane (Daubechies, 1992). Wavelets form unconditional bases for important function spaces like $L^2(R)$ and enable signal descriptions at different resolutions. A basis which preserves the time-frequency information of the signal can be of significant advantage in control system synthesis and analysis. Wavelets, because of their time-frequency localization and multiresolution

*To whom all correspondence should be addressed: phone (205)844-2060, fax (205) 844-2063, e-mail JHL@ENG.AUBURN.EDU

properties, offer such an efficient framework for representation and characterization of signals, especially those non-stationary in nature. Wavelets can also be useful in obtaining process models which describe the process behavior on different time-scales. These properties are very attractive from process identification and controller design perspectives and recently have been shown to be of significant advantage for some of the process engineering problems (Elias-Juarez and Kantor,1992; Lee *et al*,1992a; Bakshi and Stephanopoulos,1992 and 1993; Carrier and Stephanopoulos,1992; Palavajjhala *et al*, 1993). Recent research results also emphasize the potential of wavelets as a suitable framework for simultaneously addressing the time-frequency considerations in controller design.

The purpose of this chapter is to discuss the wavelet transforms and their possible applications in process control which have already been studied or being investigated currently. One important aspect of these transformations is, they allow us to tailor the existing multi-variable control algorithms so that it is more computationally attractive and easier to design and tune for robust performance. We illustrate these benefits by considering a particular time-domain control technique known as Model Predictive Control (MPC). We show that formulation of this technique in the wavelet domain allows joint time-frequency descriptions of input and output behaviors. We discuss the time-frequency distributions and show how they can be useful in control design by presenting an exemplary application to the problem of basis reduction in MPC. In the following sections these aspects are highlighted through some of the existing results. We also provide motivations for some of the ongoing research.

2 Wavelets: A Process Control Perspective

The construction of orthonormal wavelet basis functions for $L^2(R)$ by Meyer (1985), followed by the realization that the construction of different wavelet bases can be addressed in the "Multiresolution Analysis" framework (Meyer, 1986; Mallat,1989a), has resulted in an explosion of interest in the general area of wavelets. These constructions have propelled wavelets into applications in signal processing (Coifman *et al*, 1992), numerical analysis (Beylkin *et al*,1991) and other allied fields. These developments also spurred significant interest in the chemical engineering community as well. The process signals are typically non-stationary and they represent the cumulative effects of the many underlying process phenomena such as the process dynamics, measurement noise, effects of external disturbances, faults, process degradation *etc.*, each of which may manifest on a different scale. Given this fact, wavelet decompositions which result in representations on different time scales, seem to be the natural framework for modeling process signals. From the control perspective the most appealing property of wavelets is their time-frequency localization. Basis functions which are localized in both the domains facilitate simultaneous incorporation of time and frequency considerations in control design.

2.1 Signal Representations with Wavelets

Process signal representations using different basis functions result in different manifestations of the inherent structure of the signal. The familiar pulse function and its translations in time, which is a natural choice for the zero-order hold generated signals of the sampled data systems, preserve the temporal structure because of their perfect localization in the time domain. But the same representation does not give any information about the frequency content of the signal due to the infinite support these basis functions have in the frequency domain. The pulse basis and its Fourier transform are shown in Figure 1. The other extreme is the sinusoid basis functions which are perfectly localized in the Fourier domain but have infinite support in time domain. Signal projections onto these bases (Fourier Transformation), result in a perfect decomposition in the frequency domain but do not retain information about the time evolution of the signal. A basis which can preserve both the time as well as the frequency domain information is advantageous in modeling process signals and the subsequent control system synthesis and analysis. As said earlier, the wavelets with their excellent time-frequency localization properties, offer such a framework for signal representation. In this section, we will look at wavelets and how they can be utilized in signal representations.

2.1.1 Wavelets and Multiresolution Analysis

Wavelets constitute a family of functions constructed from dilations and translations of a single function called the "mother wavelet". They are indexed by two parameters, a the dilation parameter and b the translation parameter.

$$\psi^{(a,b)}(t) = |a|^{-1/2}\psi\left(\frac{t-b}{a}\right) \qquad a, b \in R \quad (a \neq 0)$$

The translation parameter b governs the time localization center of the wavelet $\psi^{(a,b)}(t)$ and the dilation parameter a determines the support in the frequency domain. For $|a| < 1$, the function $\psi^{(a,b)}(t)$ is the compressed version of the basic wavelet $\psi(t)$ (alternately, it has a smaller support in time-domain) and consists mostly higher frequencies. Conversely for $|a| > 1, \psi^{(a,b)}(t)$ has a larger time-width than $\psi(t)$ and corresponds to lower frequencies. In other words the low frequency $\psi^{(a,b)}(t)$ has a broader time-width and the high frequency $\psi^{(a,b)}(t)$ are very narrow. Thus wavelets have time-widths adapted to their frequencies. This is the primary reason for the success of wavelets in time-frequency analysis and signal processing.

If the parameters a and b are varied continuously, then the wavelet trans-

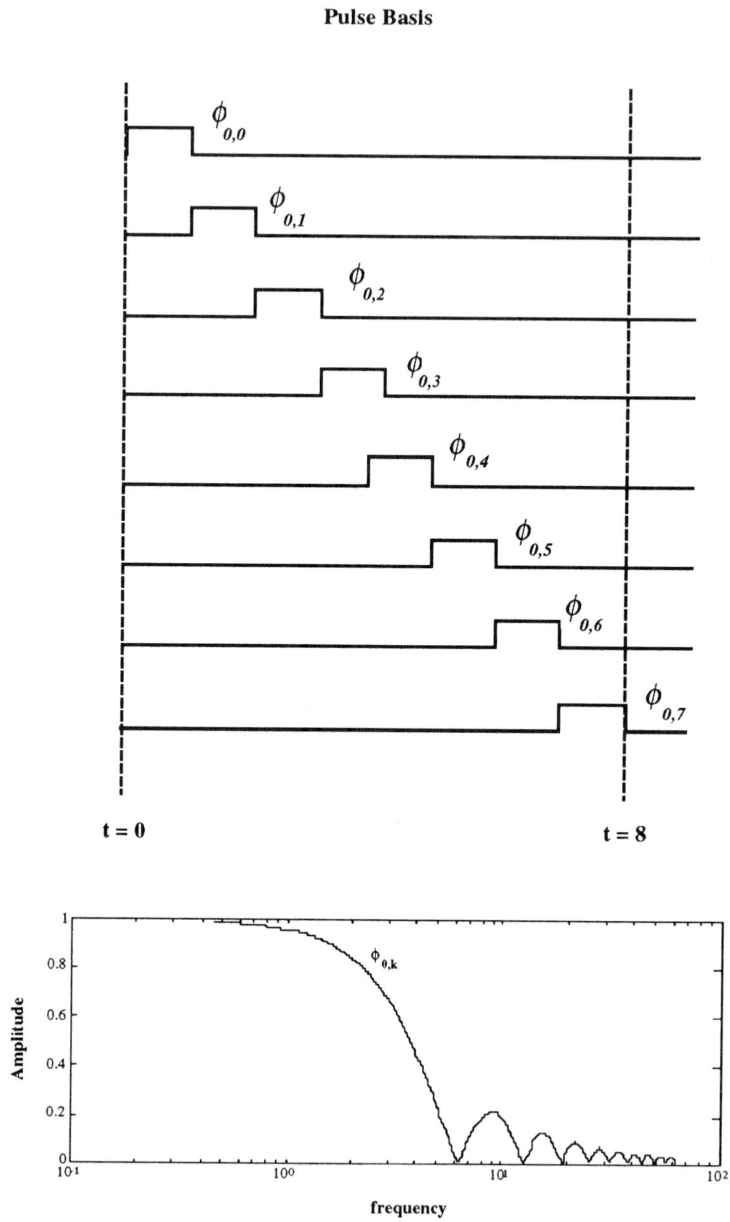

Figure 1: The Pulse Function $\phi_{0,k}$ and its Fourier Transform

form using this family of wavelet functions is known as the *Continuous Wavelet Transform* and is defined as follows:

$$(\Gamma f)(a,b) = \left\langle \psi^{(a,b)}, f(t) \right\rangle = |a|^{-1/2} \int f(t)\ \psi\left(\frac{t-b}{a}\right)\ dt \tag{1}$$

If we restrict the parameters a, b to discrete values as $a = {a_0}^m,\ b = nb_0{a_0}^m$ where $a_0 > 1,\ b_0 > 0$ and $m, n \in \mathbf{Z}$, we have the following family of discrete wavelets:

$$\psi_{m,n}(t) = |a_0|^{-m/2}\psi\left({a_0}^{-m}t - nb_0\right)$$

And the associated Discrete Wavelet Transform is defined as:

$$(\Gamma_{m,n}f) = \langle \psi_{m,n}, f(t) \rangle = |a_0|^{-m/2} \int f(t)\ \psi\left({a_0}^{-m}t - nb_0\right)\ dt \tag{2}$$

The inner products defined in equations (1) and (2) for various dilation and translation parameters are known as the *continuous* and *discrete* wavelet coefficients respectively. Any $f(t) \in L^2(R)$ can be represented in terms of these inner products, provided we choose the family of wavelets such that they form a basis for $L^2(R)$. For instance consider the following discrete wavelet expansion of an arbitrary $f(t) \in L^2(R)$:

$$f(t) = \sum_{m,n} c_{m,n}\psi_{m,n}(t) \tag{3}$$

where $\psi_{m,n}(t)$ forms a wavelet basis for $L^2(R)$. For the particular case of an orthonormal basis, the expansion coefficients can be evaluated from equation (2), *i.e.*, $c_{m,n} = \langle \psi_{m,n}, f(t) \rangle$.

The wavelet transformation of a function or signal provides an intermediate description between time and frequency domains. Each coefficient $c_{m,n}$ gives information about the function $f(t)$ around time and frequency corresponding to the region of localization of the basis function $\psi_{m,n}(t)$ in each of the domains respectively. This double localization facilitates the characterization of the local behavior of functions or signals based on these coefficients. This is one particular property of wavelet expansions of signals that can be exploited in control applications.

In general, we do not have explicit functional representations for $\psi_{m,n}(t)$. The construction of wavelets begins with the solution to a dilation equation of the following form (Daubechies,1988; Strang,1989):

$$\phi(t) = \sum_k c_k \phi(2t - k) \tag{4}$$

Here $\phi(t)$ is known as the *scaling function*. In the next stage, a wavelet $\psi(t)$ is constructed from the scaling function $\phi(t)$. The dilations and translations of the wavelet function $\psi(t)$ result in a wavelet basis. The choice of the coefficients c_k in equation (4) determine the approximation and orthogonality properties of the resulting wavelets. For more details on the methodologies and algorithms for constructing orthonormal bases of wavelets, readers are referred to Daubechies (1988,1992).

Now we introduce the Multiresolution Analysis of Mallat(1989a,b). Since in practice we deal with signals of finite length we present the multiresolution framework in the context of finite length zero order hold generated signals. We illustrate the construction of Haar basis which we use for the transformation of the MPC problem in the subsequent section.

2.1.2 Multiresolution Analysis

Multiresolution analysis is a convenient framework for hierarchical representation of functions or signals on different scales. The basic idea of multiresolution analysis is to represent a function as a limit of successive approximations. Each of these successive approximations is a smoother version of the original function with more and more of the finer "details" added. As an illustrative example, let us consider the following multiresolution approximation for $L^2(R)$, the space of all square integrable functions.

Let V_0 be the set of all square-integrable signals $x(t)$ that are constant on the unit interval $t = [i, i+1), \forall i$. It is the space of signals that can be generated from a discrete data set through zero-order hold of sampling time 1. V_0 can be expressed as

$$V_0 = \left\{ x \in L^2(R) : x(t) = \sum_{k=-\infty}^{\infty} x_k \phi(t-k), x_k \in R \right\}$$

where $\phi(t)$ is the pulse function

$$\phi(t) = \begin{cases} 1 \text{ if } t \in [0,1) \\ 0 \text{ otherwise} \end{cases}$$

$\{\phi_{0,k}\}_{-\infty<k<\infty}$ with $\phi_{0,k} \triangleq \phi(t-k)$ forms an orthonormal basis set for V_0. Each of these basis functions $\phi_{0,k}$, which is the familiar pulse function or its translate, is perfectly localized in time-domain, but has infinite support in the frequency domain as seen from its Fourier transform plotted in Figure 1. Moreover, all the basis functions have the same Fourier transform. Hence, the signal representations with the pulse basis do not carry any frequency information.

Next define V_1 as the subspace of functions in V_0 with half the original resolution. In other words, let V_1 be the space of signals that are constant over

$[2i, 2(i+1))$.

$$V_1 = \left\{ x \in L^2(R) : x(t) = \frac{1}{\sqrt{2}} \sum_{k=-\infty}^{\infty} x_k \phi\left(\frac{t}{2} - k\right), x_k \in R \right\}$$

$\{\phi_{1,k}\}_{-\infty<k<\infty}$ with $\phi_{1,k} \triangleq \frac{1}{\sqrt{2}}\phi(\frac{t}{2} - k)$ comprise an orthonormal basis set for V_1. We can generalize this idea to V_m, the space of signals with the resolution of $\frac{1}{2^m}$ (*i.e.*, signals that are constant over $[2^m i, 2^m(i+1))$. An orthonormal basis set for V_m is $\{\phi_{m,k}\}_{-\infty<k<\infty}$ where

$$\phi_{m,k} = \frac{1}{(\sqrt{2})^m} \phi\left(\frac{t}{2^m} - k\right)$$

Note that all these spaces are nested, *i.e.*, $\cdots \supset V_{-1} \supset V_0 \supset V_1 \supset \cdots \supset V_m \supset \cdots$.

We can think of the limiting case V_m as $m \to -\infty$ as the set of signals that are generated through zero-order hold of an increasingly smaller sampling time (2^m) and therefore signals of increasing resolution. Since this limit is dense in $L^2(R)$ any signal belonging to $L^2(R)$ can be approximated arbitrarily closely, in a L^2-norm sense, with a zero-order-hold-generated signal of sufficiently fine resolution. The complimentary case of V_m as $m \to +\infty$ represents signals on increasingly coarser scales. The set of spaces $(V_m)_{m\in\mathbf{Z}}$ form a multiresolution approximation for $L^2(R)$.

2.1.3 Representation of Finite-Length Zero-Order-Hold-Generated Signals in Different Resolutions

The multiresolution concept introduced above applies to infinite domain signals. It can be extended to signals of finite length by treating them as infinite-length signals belonging to V_0 with zero values outside the finite domain (we call this "zero-padding"). Let

$$V_0^\ell = \left\{ x \in L^2(R) : x(t) = \sum_{k=0}^{2^\ell - 1} x_{0,k}\phi_{0,k} \right\}$$

V_0^ℓ constitutes signals that are constant on the unit interval $t = [i, i+1)$ for $0 \leq i < 2^\ell - 1$ and zero-padded for $-\infty < t < 0$ and $2^\ell \leq t < \infty$. It can be thought of as the set of zero-order-hold generated signals (sampling time 1, without loss of generality) of finite length 2^ℓ. $\{\phi_{0,k}\}_{0\leq k \leq 2^\ell - 1}$ is the pulse basis set commonly used to represent such a signal in digital control. $x(t)$ is often represented by a 2^ℓ-dimensional vector $\left[x_{0,0}, \cdots, x_{0,2^\ell-1}\right]^T$. As before, we can define subsets of V_0^ℓ with different resolutions in the following manner:

$$
\begin{aligned}
V_1^\ell &= \left\{x \in L^2(R) : x(t) = \sum_{k=0}^{2^{\ell-1}-1} x_{1,k}\phi_{1,k}\right\} \\
V_2^\ell &= \left\{x \in L^2(R) : x(t) = \sum_{k=0}^{2^{\ell-2}-1} x_{2,k}\phi_{2,k}\right\} \\
\vdots \quad & \vdots \qquad\qquad\qquad\qquad \vdots \\
V_\ell^\ell &= \left\{x \in L^2(R) : x(t) = x_{\ell,0}\phi_{\ell,0}\right\}
\end{aligned}
$$

Once again, $\cdots \supset V_{-1}^\ell \supset V_0^\ell \supset V_1^\ell \supset \cdots \supset V_\ell^\ell$. In practice, however, we do not have signals of infinitely fine resolution. The fastest sampling rate determines the finest available resolution, *i.e.*, resolution of V_0^ℓ.

2.1.4 Haar Basis: Wavelets for Zero-Order-Hold Generated Signals

The above introduced multiresolution analysis can be used to construct an orthonormal wavelet basis for V_0^ℓ. Haar wavelet and its dilations and translations along with an appropriate scaling function $\phi_{0,\ell}$ can serve as a basis for V_0^ℓ. Its construction through the multiresolution decomposition of V_0^ℓ is demonstrated below.

Consider V_0^ℓ and express it as an "internal sum" of two orthogonal subspaces V_1^ℓ and W_1^ℓ:

$$V_0^\ell = V_1^\ell \oplus W_1^\ell$$

V_1^ℓ is the coarser approximation of V_0^ℓ with half its resolution. And W_1^ℓ represents the subspace containing the "lost" information as we go from V_0^ℓ to V_1^ℓ. As we saw earlier, $\{\phi_{0,k}\}_{0\le k\le 2^\ell-1}$ with $\phi_{0,k} \triangleq \phi(t-k)$ forms a basis for V_0^ℓ and similarly $\{\phi_{1,k}\}_{0\le k\le 2^{\ell-1}-1}$ with $\phi_{1,k} \triangleq \frac{1}{\sqrt{2}}\phi(\frac{t}{2}-k)$ is an orthogonal basis for V_1^ℓ. The basis functions for V_0^ℓ and its averaged version V_1^ℓ are related as follows:

$$\phi_{1,k} = \frac{1}{\sqrt{2}}\left[\phi_{0,2k} + \phi_{0,2k+1}\right]$$

Next define $\psi_{1,k}$ as:

$$\psi_{1,k} = \frac{1}{\sqrt{2}}\left[\phi_{0,2k} - \phi_{0,2k+1}\right]$$

Note that

$$\psi_{1,k} = \frac{1}{\sqrt{2}}\psi\left(\frac{1}{2}t - k\right)$$

where $\psi(t)$ is the well known Haar function defined as:

$$\psi(t) = \begin{cases} 1 \text{ if } t \in [0, \frac{1}{2}) \\ -1 \text{ if } t \in [\frac{1}{2}, 1) \\ 0 \text{ otherwise} \end{cases}$$

$\psi_{1,k}$ is orthogonal to $\phi_{1,k}$ and belongs to V_0^ℓ. Hence it lies in W_1^ℓ. In fact $\{\psi_{1,k}\}_{0 \leq k \leq 2^{\ell-1}-1}$ forms an orthogonal basis for W_1^ℓ. Any $x \in V_0^\ell$ can be represented in terms of $\phi_{1,k}$ and $\psi_{1,k}$ as:

$$x = \sum_{k=0}^{2^{\ell-1}-1} a_{1,k}\phi_{1,k} + \sum_{k=0}^{2^{\ell-1}-1} b_{1,k}\psi_{1,k}$$

The orthogonality ensures that $x_1 = \sum_{k=0}^{2^{\ell-1}-1} a_{1,k}\phi_{1,k}$ represents the "best" approximation (in L^2-norm sense) of x with a function half its resolution and $x_1^* = \sum_{k=0}^{2^{\ell-1}-1} b_{k,1}\psi_{k,1}$ is the "lost" information as we go from the finer to coarser resolution.

The above result extends to orthogonal decomposition of any V_m^ℓ to $V_{m+1}^\ell \oplus W_{m+1}^\ell$ and an orthonormal basis, $\{\psi_{m+1,k}\}_{0 \leq k \leq 2^{\ell-m-1}-1}$, for W_{m+1}^ℓ can be found.

$$\psi_{m+1,k} = \frac{1}{(\sqrt{2})^{m+1}}\psi\left(\frac{1}{2^{m+1}}t - k\right)$$

Applying this idea successively from V_0^ℓ, we can write

$$\begin{aligned} V_0^\ell &= V_1^\ell \oplus W_1^\ell \\ &= V_2^\ell \oplus W_2^\ell \oplus W_1^\ell \\ &\vdots \quad \vdots \\ &= V_\ell^\ell \oplus W_\ell^\ell \oplus W_{\ell-1}^\ell \oplus \cdots W_1^\ell \end{aligned}$$

Then, the 2^ℓ orthonormal basis functions for V_0^ℓ can be chosen as $\left\{\phi_{\ell,0}, \{\psi_{m,k}\}_{\substack{m=1,\cdots,\ell \\ k=0,\cdots,2^{\ell-m}-1}}\right\}$. For the purpose of the present exposition we refer to this set as Haar basis for V_0^ℓ. Figure 2 shows the Haar basis for V_0^3, $\left\{\phi_{3,0}, \psi_{3,0}, \{\psi_{2,k}\}_{k=0,1}, \{\psi_{1,k}\}_{k=0,1,2,3}\right\}$ along with the plots of their Fourier

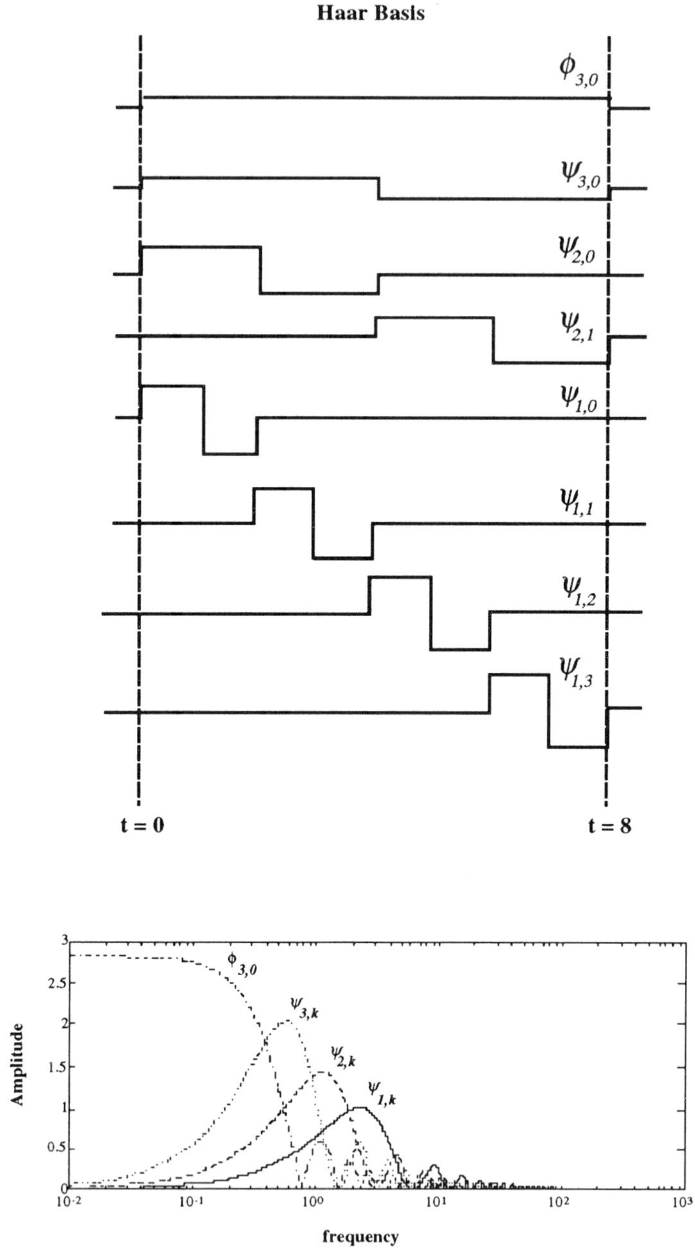

Figure 2: Haar Basis and Its Fourier Transforms

transforms. The significant improvement in frequency localization of the basis functions, compared to the earlier pulse basis, can be observed. We can also associate different frequency ranges with basis functions of different dilation parameters. The Fourier transform of $\psi_{m,k}$ shows its dominant energy shifting to an increasingly higher frequency range as $m \to 1$.

Haar Decomposition of a Signal

The Haar decomposition of a finite dimensional vector can be viewed as a change of basis from the pulse basis to the wavelet basis. And this change of basis is a unitary transformation as both are complete orthonormal basis sets for V_0^ℓ. Consider $x \in V_0^\ell$. Then

$$\begin{aligned} x &= \sum_{k=0}^{2^\ell-1} a_{0,k}\phi_{0,k} \\ x &= a_{\ell,0}\phi_{\ell,0} + \sum_{m=1}^{\ell}\sum_{k=0}^{2^{\ell-m}-1} b_{m,k}\psi_{m,k} \end{aligned}$$

are the two representations of x with respect to the pulse and the Haar basis respectively. The two representations are related through a unitary transformation $\tilde{x}^* = U\tilde{x}$ where

$$\begin{aligned} \tilde{x}^* &= \begin{bmatrix} a_{\ell,0} & b_{\ell,0} & b_{\ell-1,0} & b_{\ell-1,1} & \cdots & \cdots & b_{1,2^{\ell-1}-2} & b_{1,2^{\ell-1}-1} \end{bmatrix}^T \\ \tilde{x} &= \begin{bmatrix} a_{0,0} & a_{0,1} & a_{0,2} & a_{0,3} & \cdots & \cdots & a_{0,2^\ell-2} & a_{0,2^\ell-1} \end{bmatrix}^T \end{aligned}$$

and U is an appropriate orthonormal matrix. For example, for $x \in V_0^3$,

$$\begin{bmatrix} a_{3,0} \\ b_{3,0} \\ b_{2,0} \\ b_{2,1} \\ b_{1,0} \\ b_{1,1} \\ b_{1,2} \\ b_{1,3} \end{bmatrix} = \begin{bmatrix} \frac{1}{\sqrt{8}} & \frac{1}{\sqrt{8}} & \frac{1}{\sqrt{8}} & \frac{1}{\sqrt{8}} & \frac{1}{\sqrt{8}} & \frac{1}{\sqrt{8}} & \frac{1}{\sqrt{8}} & \frac{1}{\sqrt{8}} \\ \frac{1}{\sqrt{8}} & \frac{1}{\sqrt{8}} & \frac{1}{\sqrt{8}} & \frac{1}{\sqrt{8}} & \frac{-1}{\sqrt{8}} & \frac{-1}{\sqrt{8}} & \frac{-1}{\sqrt{8}} & \frac{-1}{\sqrt{8}} \\ \frac{1}{2} & \frac{1}{2} & \frac{-1}{2} & \frac{-1}{2} & 0 & 0 & 0 & 0 \\ 0 & 0 & 0 & 0 & \frac{1}{2} & \frac{1}{2} & \frac{-1}{2} & \frac{-1}{2} \\ \frac{1}{\sqrt{2}} & \frac{-1}{\sqrt{2}} & 0 & 0 & 0 & 0 & 0 & 0 \\ 0 & 0 & \frac{1}{\sqrt{2}} & \frac{-1}{\sqrt{2}} & 0 & 0 & 0 & 0 \\ 0 & 0 & 0 & 0 & \frac{1}{\sqrt{2}} & \frac{-1}{\sqrt{2}} & 0 & 0 \\ 0 & 0 & 0 & 0 & 0 & 0 & \frac{1}{\sqrt{2}} & \frac{-1}{\sqrt{2}} \end{bmatrix} \begin{bmatrix} a_{0,0} \\ a_{0,1} \\ a_{0,2} \\ a_{0,3} \\ a_{0,4} \\ a_{0,5} \\ a_{0,6} \\ a_{0,7} \end{bmatrix}$$

Note that U is an orthonormal wavelet matrix ($U^TU = I$) and each row in U, except for the first one, corresponds to a sampled version of a Haar wavelet of particular dilation and translation. The first row corresponds to the scaling function $\phi_{0,\ell}$.

Multiresolution Approximation: Example

Consider a function $x_0 \in V_0^5$ (with 32 data points) and its multiresolution approximations, $x_1 \in V_1^5$ through $x_5 \in V_5^5$ are shown in Figure 3. The lost information or the detail signal, as we move to successively coarser resolutions, expressed as $x_1^* \in W_1^5$ through $x_5^* \in W_5^5$, are also shown. The original function can be reconstructed as follows:

$$\begin{aligned} x_0 &= x_1 + x_1^* = x_2 + x_2^* + x_1^* \\ &= \cdots = x_5 + x_5^* + x_4^* + x_3^* + x_2^* + x_1^* \end{aligned}$$

Also note that $\arg\{\min_{x \in V_m^5} \|x_0 - x\|_2\} = x_m$ with the optimal error being $x_m^* + \cdots + x_1^*$.

2.2 Wavelets : Potential Applications

The most appealing properties of wavelets, as stated earlier, are their time-frequency localization and multiresolution approximation capabilities. As we have seen, projections of functions or signals onto the wavelet bases result in decompositions which provide time-frequency information together. These decompositions along with the attractive computational properties of wavelets can be exploited in process control applications. The following are some of the areas where wavelets can be of advantage.

1. Robust Control

Most of the robust control theory is developed in frequency domain where as the controller design schemes such as MPC are most attractive when formulated in time domain. For example, the time-domain MPC allows for explicit consideration of nonlinearity and constraints on operating variables. The frequency domain techniques, such as H_∞-norm minimization offer a convenient framework to address the model uncertainty issues in controller design, but can not explicitly consider time-domain considerations such as rise-time, process constraints, etc. On the other hand, time domain formulations are less amenable to convenient and explicit treatment of the model uncertainty issues. Thus there is a need for a framework which can address the time-frequency considerations simultaneously. The wavelet decompositions facilitate such joint time-frequency interpretations of process signals. These decompositions in turn could be used in formulating new design criteria for controller design incorporating both time and frequency domain considerations.

2. Adaptive Control

Most of the chemical engineering systems exhibit behaviors on different

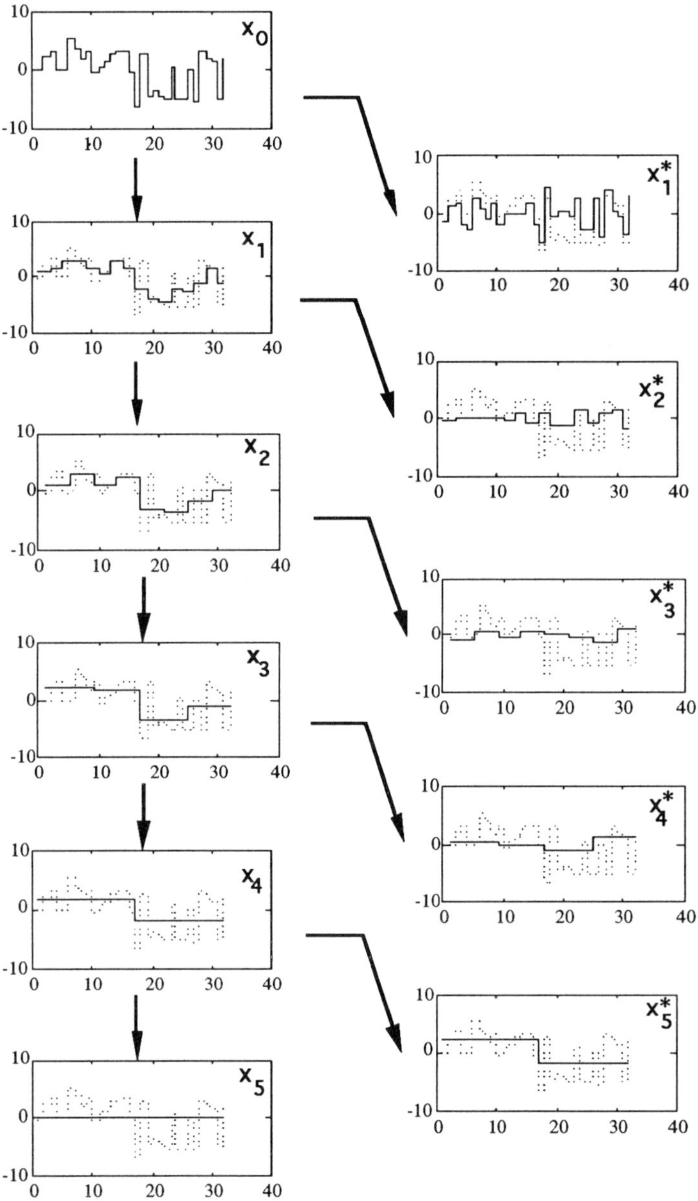

Figure 3: Multiresolution Approximation of a Signal $x_0 \in V_0^5$

time scales. Explicit consideration of these behaviors in adaptation strategies is desirable and wavelets with their multiresolution capabilities could be the right framework to reformulate the adaptation problems. For example, the data retention in adaptation, to a large extent is still an unresolved issue. Addressing the problem in time-frequency plane appears to be more natural and convenient than addressing in either domain alone.

3. Nonlinear Control

Solution of the nonlinear ODE/PDE's using wavelet basis functions has been the focus of many recent studies (Liandrat *et al*,1992; Nikolaou and You,1992). Multiresolution approximation properties of wavelet functions can be of advantage in localizing the approximation bias to desired regions of solution *e.g.* later parts of prediction horizon in nonlinear MPC. This could possibly have significant relevance for reducing the on-line computational requirements of nonlinear MPC implementations.

4. Process Identification

Most of the system identification techniques can be represented as projections onto a basis set (Carrier and Stephanopoulos,1992). The properties of this basis set to a large extent influence the range of validity of the identified model. Stephanopoulos and Carrier (1991) proposed control relevant model identification methodologies based on the multiresolution analysis of the input-output data. They use the orthogonal wavelet framework to construct reduced order models which are accurate in a given time-frequency range. They also utilize the time-frequency properties of wavelets in quantifying the associated model uncertainty and calculate error bounds for robust controller design.

Over the past few years, artificial neural networks have established themselves as a powerful paradigm for modeling complex, nonlinear systems. Some of the disadvantages of the neural network based system modeling have been the significant computational effort needed in training and adaptation of the networks and the lack of physical meaning of the resulting models. To a large extent, the computational complexity of the networks was due to the global nature of the node activation functions which in turn led to slow adaptation characteristics. Recent research efforts have been directed at using activation functions whose excitation characteristics are limited to a local region in the input space, such as radial basis functions (Leonard and Kramer,1991). Networks with local basis functions have been observed to have better training and adaptation properties. Because of their local nature, these functions are active only in a small range in the input space, and thus avoid significant extrapolation errors. In addition a reliability measure can also be associated with network predictions based on the input/output data used in network training

(Leonard *et al*, 1991). Neural Networks can be interpreted as function approximators. A function approximation such as a weighted sum of orthogonal basis functions, has a neural network representation provided each of the orthogonal basis functions is chosen to be one of the node activation functions. Such a choice of node activation functions facilitates the utilization of the powerful theory of orthogonal expansions in designing the networks. The orthogonality property is important from the network structure determination perspective. Thus node activation functions which are local and orthogonal are desirable in neural network applications and wavelets because of their local and orthogonal nature (we can construct orthogonal wavelet bases) are the right choice for such applications. Recently Bakshi and Stephanopoulos(1992) developed a neural network structure called "wave-net", in which the activation functions are drawn from a set of orthonormal wavelets. They exploit the multiresolution capabilities, localization as well as the orthogonal properties of wavelets in designing networks with better training and adaptation characteristics. Another important aspect of wave-nets is the explicit determination of the associated prediction error bounds, both global and local.

The recent work of Stephanopoulos and coworkers underline the fact that wavelets offer a powerful framework for addressing the linear/nonlinear identification problem from a control perspective. The relevance and significance of the generated models for the process control problem was also addressed in their work. The conceptual framework presented in their work has to be explored further, for instance the control relevant properties of wave-nets. The insights gained by the wavelet domain identification techniques can be incorporated into design schemes. The benefits could be reduced computational requirements and design of robust controllers based on the generated models.

The above are few of the applications the authors perceive at this juncture. Wavelets are also well suited for exploring pattern based solutions to most of the process automation problems. As the full power of wavelet analysis is realized, many more applications can be expected in near future. Wavelets have been proved to be very advantageous in addressing such higher level process automation tasks such as trend analysis and fault detection (Joshi and Motard,1991; Bakshi and Stephanopoulos,1992). Thus it is not a completely inconceivable scenario where many of the process automation strategies, high level tasks such as fault detection and diagnosis along with the lower level controls, are addressed simultaneously under the unified framework of wavelets.

3 Model Predictive Control: A Wavelet Domain Solution

As an exemplary application of wavelets in process control problems, next we present a wavelet transformation of the MPC problem. The purpose of the presentation is to illustrate how the time-frequency interpretations of the inputs and outputs can be of advantage in control design. The particular problem of basis reduction in MPC is addressed within the wavelet domain framework and the advantages of such an approach are discussed.

Process control problems often involve dynamics on mixed time-scales. A typical situation would be where certain manipulated variables may influence the process outputs through faster time-scale dynamics than others. Different outputs may also exhibit behaviors on different time-scales. In these cases, the sampling rate for each input and output would be chosen to be commensurate with the associated time-scale. In many cases, however, it is difficult to attribute specific time-scales to different inputs and outputs and their sample rates would have to be determined according to the fastest time-scale in order not to limit the achievable closed-loop performance. For the particular case of MPC, recent research results (Rawlings and Muske,1991; Lee and Yu,1992b) indicate that, while the choice for sample rates is governed primarily by the fastest time-scale, the prediction and control-move horizons should be chosen according to the slowest time-scale in order to ensure good closed-loop stability properties and performance. This is especially true when the constraints are important aspects of the control problem. The combination of a fast sample rate and large prediction/control move horizons leads to very demanding online optimization problem. In the next few pages, we present the technique of Lee *et al* (1992a), which addresses the problem of improving the computational efficiency of the MPC algorithm by reformulating the optimization problem in wavelet domain. This formulation offers an efficient and intuitive way of performing basis reduction through techniques called input *blocking* and output *condensation*. The following presentation is an adapted version of the original presentation of Lee *et al* (1992a).

3.1 Wavelet Transformation of the MPC Problem

In this section we demonstrate how the time-domain MPC can be transformed to wavelet domain. In order to facilitate the exposition, the treatment here is confined to the SISO case. The extension of the presented methodology to MIMO case is straightforward.

The typical time-domain optimization objective for MPC at time k can be written as:

$$\min_{\Delta\mathcal{U}(k)} \|\Lambda_y(\mathcal{R}(k) - \mathcal{Y}(k|k))\|_2^2 + \|\Lambda_u \Delta\mathcal{U}(k)\|_2^2 \tag{5}$$

such that

$$\mathcal{Y}(k) = \mathcal{S}^x \Delta x(k|k) + \mathcal{H}_c y(k|k) + \mathcal{S}^{\mathcal{U}} \Delta\mathcal{U}(k) \tag{6}$$

$$\mathcal{C}_u \Delta\mathcal{U}(k) \geq \mathcal{C}(k|k) \tag{7}$$

where $\mathcal{Y}(k|k) = [y(k+1|k), \cdots, \cdots, y(k+P|k)]^T$ is the predicted future output over the prediction horizon P, and $\mathcal{R}(k) = [r(k+1), \cdots, \cdots, r(k+P)]^T$ is the desired future output trajectory. $\Delta\mathcal{U}(k) = [\Delta u(k), \cdots, \cdots, \Delta u(k+M-1|k)]^T$ is the vector of optimal input moves to be computed so as to minimize the objective of equation (5) subject to the operating constraints on the inputs and outputs given by equation (7) where:

$$\mathcal{C}_u = \begin{bmatrix} -I_L \\ I_L \\ -I \\ I \\ -\mathcal{S}^{\mathcal{U}} \\ \mathcal{S}^{\mathcal{U}} \end{bmatrix}; \quad \mathcal{C}(k|k) = \begin{bmatrix} -\mathcal{U}_{high} \\ \mathcal{U}_{low} \\ -\Delta\mathcal{U}_{max} \\ -\Delta\mathcal{U}_{max} \\ \mathcal{S}^x \Delta x(k|k) + \mathcal{H}_c y(k|k) - \mathcal{Y}_{high}(k) \\ -\mathcal{S}^x \Delta x(k|k) - \mathcal{H}_c y(k|k) + \mathcal{Y}_{low}(k) \end{bmatrix} \tag{8}$$

$\mathcal{S}^{\mathcal{U}}, \mathcal{S}^x$ and $\mathcal{H}_c$ are constant matrices that are computed from the state-space model parameters. $\mathcal{Y}_{high}, \mathcal{Y}_{low}$ and $\mathcal{U}_{high}, \mathcal{U}_{low}$ define the operating bounds on the output and input variables respectively. There is a velocity constraint on the input move magnitudes as well, specified by $\Delta\mathcal{U}_{max}$. $y(k|k)$ is the filtered output using the measurements up to time k and $\Delta x(k|\ell)$ are the estimates of state increments, typically using a Kalman filter. Here the notation $\Delta x(k|\ell)$ stands for the optimal estimates of the state increments for $t = k$ using the measurements available up to $t = \ell$. Λ_y and Λ_u are the output and input weighting matrices and typically are constant diagonal matrices with the specified weights as the diagonal elements. For a detailed presentation of the state-space MPC technique, the readers are referred to Lee *et al* (1992c).

In a typical implementation, the Quadratic Programming problem given by equations (5-7) is solved on-line and the control moves are implemented in a receding horizon fashion, *i.e.*, only the first move of $\Delta\mathcal{U}(k)$ is implemented and the optimization is repeated for the next sampling time.

Now let us transform the above time-domain MPC objective and constraints into wavelet domain. Assume that the prediction horizon P and control horizon M are chosen to be $2^{\hat{P}}+1$ and $2^{\hat{M}}+1$ respectively. Hence, $\mathcal{Y}(k)$ and $\Delta\mathcal{U}(k)$ are vectors of size $(2^{\hat{P}}+1)$ and $(2^{\hat{M}}+1)$ respectively.

Let $\mathcal{U}_c(k,t)$ be the continuous future input vector defined as follows:

$$\mathcal{U}_c(k,t) = \begin{cases} 0 & \text{for } t < 0 \\ u(k+\ell) - u(k-1) & \text{for } \ell \le t < \ell+1, \quad 0 \le \ell \le 2^{\hat{M}}-1 \\ u(k+2^{\hat{M}}) - u(k-1) & \text{for } t \ge 2^{\hat{M}} \end{cases}$$

where $u(k-1)$ is the input implemented at the $(k-1)^{th}$ step. If we choose $\{\phi_{0,k}\}_{k=0,\cdots,2^{\hat{M}}-1}$ along with $\phi_{2^{\hat{M}},s}$, a unit step function starting at time $2^{\hat{M}}$, as a basis set, any $\mathcal{U}_c(k,t)$ can be expressed as a linear combination of these $2^{\hat{M}}+1$ basis functions. These functions are shown in Figure 4. The vector representation of $\mathcal{U}_c(k,t)$ in terms of the above basis (denoted as $\mathcal{U}(k)$) is as follows:

$$\mathcal{U}(k) = \begin{bmatrix} u(k) - u(k-1) \\ \vdots \\ u(k+2^{\hat{M}}) - u(k-1) \end{bmatrix}$$

$\Delta\mathcal{U}(k)$ and $\mathcal{U}(k)$ are related as $\Delta\mathcal{U}(k) = \Pi\ \mathcal{U}(k)$ where

$$\Pi = \begin{bmatrix} 1 & 0 & 0 & \cdots & 0 & 0 \\ -1 & 1 & 0 & \cdots & 0 & 0 \\ 0 & -1 & 1 & 0 & \cdots & 0 \\ \vdots & \ddots & \ddots & \ddots & \ddots & \vdots \\ 0 & 0 & \ddots & -1 & 1 & 0 \\ 0 & 0 & 0 & \cdots & -1 & 1 \end{bmatrix}$$

Defining the output tracking error, $\mathcal{E}(k) = \mathcal{R}(k) - \mathcal{Y}(k|k)$ and replacing $\Delta\mathcal{U}(k)$ with $\Pi\mathcal{U}(k)$, now we rewrite the objective and constraints (5)–(7) as:

$$\min_{\Delta\mathcal{U}(k)} \|\Lambda_y \mathcal{E}(k)\|_2^2 + \|\Lambda_u\ \Pi\ \mathcal{U}(k)\|_2^2 \tag{9}$$

such that

$$\mathcal{E}(k) = \mathcal{R}(k) - \mathcal{S}^x \Delta x(k|k) - \mathcal{H}y(k|k) - \mathcal{S}^{\mathcal{U}}\ \Pi\ \mathcal{U}(k) \tag{10}$$

$$\mathcal{C}_u\ \Pi\ \mathcal{U}(k) \geq \mathcal{C}(k|k) \tag{11}$$

Similarly the input constraints can also be expressed as $\mathcal{U}_{high}(k)$ and $\mathcal{U}_{low}(k)$ in terms of the above basis.

A similar interpretation can be extended to the predicted future output vector, $\mathcal{Y}_c(k,t)$. With an assumption of constant output between sampling intervals and also after the end of the prediction horizon, we have

$$\mathcal{Y}_c(k,t) = \begin{cases} 0 & \text{for } t < 0 \\ y(k+\ell|k) & \text{for } \ell - 1 \leq t < \ell,\ 1 \leq \ell \leq 2^{\hat{P}} \\ y(k+2^{\hat{P}}+1|k) & \text{for } t \geq 2^{\hat{P}} \end{cases}$$

Then $\mathcal{Y}(k|k)$ corresponds to the $2^{\hat{P}}+1$-dimensional coefficient vector for $\mathcal{Y}_c(k,t)$ with respect to the orthogonal basis set $\{\phi_{0,k}\}_{k=0,\cdots,2^{\hat{P}}-1}$ plus $\phi_{2^{\hat{P}},s}$, a unit step function starting at time $2^{\hat{P}}$. We can also think of the future reference

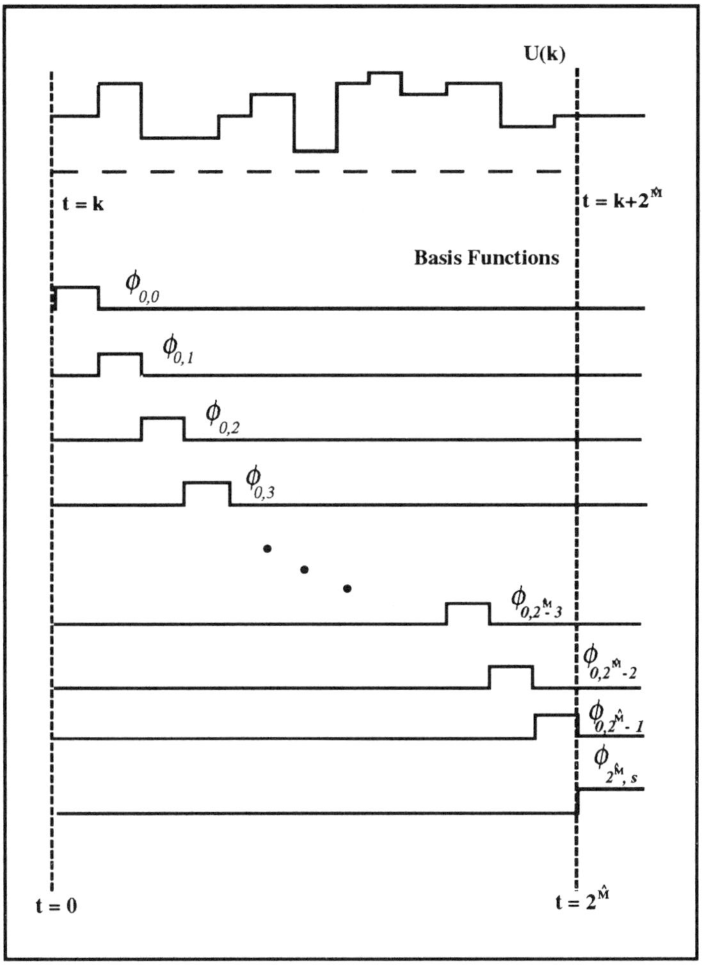

Figure 4: Basis for Representation of the Future Input Vector $\mathcal{U}(t)$

signal $\mathcal{R}(k)$ and output constraints $\mathcal{Y}_{high}(k)$ and $\mathcal{Y}_{low}(k)$ as representations with respect to the above orthonormal basis.

It was shown in the previous section that transformation of $\mathcal{U}(k)$ from the standard pulse basis into an orthonormal wavelet basis can be performed by multiplying it with an appropriate wavelet matrix. Recall that $\mathcal{U}_c(k,t)$ can be projected onto the wavelet basis as follows:

$$\begin{aligned}\mathcal{U}_c(k,t) &= \sum_{\ell=0}^{2^{\hat{M}}-1} \left(u(k+\ell) - u(k-1)\right)\phi_{0,\ell} + \left(u(k+2^{\hat{M}}) - u(k-1)\right)\phi_{2^{\hat{M}},s} \\ &= u^*_{\hat{M},+}(k)\phi_{\hat{M},0} + \sum_{\ell=1}^{\hat{M}} \sum_{j=0}^{2^{\hat{M}-\ell}-1} u^*_{\ell,j}(k)\psi_{\ell,j} + u^*_f(k)\phi_{2^{\hat{M}},s}\end{aligned}$$

Defining

$$\begin{aligned}\mathcal{U}^*(k) &= \Big[u^*_{\hat{M},+}(k), u^*_{\hat{M},0}(k), u^*_{\hat{M}-1,0}(k), u^*_{\hat{M}-1,1}(k), \cdots, \cdots, \cdots, u^*_{1,2^{\hat{M}-1}-2}(k), \\ & u^*_{1,2^{\hat{M}-1}-1}(k), u^*_f(k)\Big]^T\end{aligned}$$

$\mathcal{U}(k)$ and $\mathcal{U}^*(k)$ are related through a unitary transformation:

$$\mathcal{U}^*(k) = V_u\mathcal{U}(k); \qquad V_u = \text{diag}\{V, 1\}$$

V is a $2^{\hat{M}} \times 2^{\hat{M}}$ orthogonal matrix that performs a change of basis from the pulse basis to the wavelet basis.

$\mathcal{U}_{high}(k)$ and $\mathcal{U}_{low}(k)$ can also be transformed to $\mathcal{U}^*_{high}(k)$ and $\mathcal{U}^*_{low}(k)$ via the same transformation.

Similarly, we can apply the same wavelet transformation to the first $2^{\hat{P}}$ elements of $\mathcal{Y}(k|k)$, $\mathcal{R}(k)$, $\mathcal{Y}_{high}(k)$ and $\mathcal{Y}_{low}(k)$.

$$\begin{aligned}\mathcal{Y}^*(k|k) &= W_y\mathcal{Y}(k|k) \\ \mathcal{R}^*(k) &= W_y\mathcal{R}(k) \\ \mathcal{Y}^*_{high}(k) &= W_y\mathcal{Y}_{high}(k) \\ \mathcal{Y}^*_{low}(k) &= W_y\mathcal{Y}_{low}(k)\end{aligned}$$

where W_y is an orthogonal matrix with the following structure:

$$W_y = \text{diag}\{W, 1\}$$

W is a $2^{\hat{P}} \times 2^{\hat{P}}$ orthogonal matrix representing a change of basis from the pulse basis to the wavelet basis.

The objective (9) and constraints (10)-(11) can be expressed in terms of the transformed variables as follows:

$$\min_{\mathcal{U}^*(k)} \left\|\Lambda_y W_y^T \mathcal{E}^*(k)\right\|_2^2 + \left\|\Lambda_u \; \Pi \; V_u^T \mathcal{U}^*(k)\right\|_2^2 \tag{12}$$

with constraints

$$\begin{aligned} \mathcal{E}^*(k) &= W_y \left\{\mathcal{R}(k) - \mathcal{S}^x \Delta x(k|k) - \mathcal{H}y(k|k) - \mathcal{S}^U \Pi V_u^T \mathcal{U}^*(k)\right\} & (13) \\ \mathcal{C}_u^* \mathcal{U}^*(k) &\geq \mathcal{C}(k|k) & (14) \end{aligned}$$

$\mathcal{C}_u^* = \mathcal{C}_u \Pi V_u^T$ and $\mathcal{C}(k|k)$ can be written in terms of the transformed variables as follows:

$$\mathcal{C}(k|k) = \begin{bmatrix} -V_u^T \mathcal{U}_{high}^* \\ V_u^T \mathcal{U}_{low}^* \\ -\Delta\mathcal{U}_{max} \\ -\Delta\mathcal{U}_{max} \\ W_y^T \left\{W_y \left(\mathcal{S}^x \Delta x(k|k) + \mathcal{H}_c y(k|k)\right) - \mathcal{Y}_{high}^*(k)\right\} \\ -W_y^T \left\{W_y \left(\mathcal{S}^x \Delta x(k|k) + \mathcal{H}_c y(k|k)\right) - \mathcal{Y}_{low}^*(k)\right\} \end{bmatrix} \tag{15}$$

We should note here that the QP required for the transformed MPC objective and constraints is of the same dimension as that of the original formulation. In the next section we will see how the wavelet domain formulation and the resulting time-frequency decompositions of the input and output variables can be utilized in performing basis reduction for reduced on-line computational requirements. The technique will be referred to as input *blocking* and output *condensation*.

3.2 Blocking and Condensation in Wavelet Domain

It is important from a practical stand point that, in addition to good performance, a control algorithm be computationally efficient for easy, inexpensive on-line implementation. *Blocking* and *Condensation* are the techniques that can be employed to reduce the computational requirements of the MPC algorithm. In blocking, the dimension of the input vector to be computed is reduced by constraining it to a lower dimensional subspace. This can be done by eliminating some of the insignificant linear combinations of the input moves from the computation. The "blocked" input vector can be expressed as:

$$\Delta\mathcal{U}_B(k) = \mathbf{B}_t \Delta\mathcal{U}(k) \quad \rightarrow \quad \Delta\mathcal{U}(k) = \mathbf{B}_t{}^T \Delta\mathcal{U}_B(k)$$

The blocking matrix $\mathbf{B}_t$ is a "fat" matrix derived from an orthogonal matrix by deleting the rows corresponding to the particular linear combinations of the input moves that are set to zero *a priori* and therefore omitted from the computation. Blocking strategies can range from simple elimination of the last few moves to more complex variations of intervals between successive moves.

The computational complexity can be further reduced by eliminating the insignificant linear combinations of the future output tracking error as well. This technique shall be referred to as "Output Condensation." The condensed output tracking error vector can be generated similarly as before:

$$\mathcal{E}_H(k) = \mathbf{H}_t \mathcal{E}(k) \quad \rightarrow \quad \mathcal{E}(k) = {\mathbf{H}_t}^T \mathcal{E}_H(k)$$

where the condensation matrix $\mathbf{H}_t$ is again a full-rank "fat" matrix derived from an orthogonal matrix by deleting the rows corresponding to the eliminated linear combinations.

The key to successful blocking and condensation lies in the proper choice of the blocking and condensation matrices. It is desired to reduce the complexity of the optimization without substantial performance degradation. However, a proper choice for the blocking and condensation matrices is a nontrivial task. A time-domain approach is not well suited, given the fact that most of the chemical engineering systems exhibit behaviors on multiple time scales. In the next few pages, we present a novel approach based on the wavelet domain formulation of the MPC problem. We will use the time-frequency information of the Haar decompositions of the input and output vectors to choose the blocking and condensation matrices.

In the previous section we have seen that, in the wavelet domain, the transformed input vector

$$\mathcal{U}^*(k) = \left[u^*_{\hat{M},+}, u^*_{\hat{M},0}, u^*_{\hat{M}-1,1}, u^*_{\hat{M}-1,1}, \cdots, u^*_{1,2^{\hat{M}-1}-2}, u^*_{1,2^{\hat{M}-1}-1}, u^*_f\right]^T$$

implies

$$\mathcal{U}_c(k,t) = u^*_{\hat{M},+}\phi_{\hat{M},0} + \sum_{\ell=1}^{\hat{M}} \sum_{j=0}^{2^{\hat{M}-\ell}-1} u^*_{\ell,j}\psi_{\ell,j} + u^*_f \phi_s$$

For the basis $\psi_{\ell,j}$, indices ℓ and j indicate the local position in frequency and time respectively. As we move from $\ell = 1$ to $\ell = \hat{M}$, the frequency range that the Fourier transform of $\psi_{\ell,j}$ occupies moves to increasingly lower frequencies. Hence, the wavelet decomposition provides a convenient way to choose different control move horizon lengths for different frequency ranges. In order to implement this idea of frequency-varying horizon length, for each ℓ, we can retain the first m_ℓ elements of the ℓth level wavelets and eliminate the rest. m_ℓ is chosen according to the horizon length appropriate for the frequency range corresponding to $\psi_{\ell,\cdot}$, the ℓth-level wavelets. In other words, *constrain* $\mathcal{U}^*(k)$ to be

$$\mathcal{U}^*(k) = \mathbf{B}^T \mathcal{U}_B^*(k)$$

where $\mathcal{U}_B^*(k) = \left[u^*_{\hat{M},+}, u^*_{\hat{M},0}, \left\{u^*_{\hat{M}-1,j}\right\}_{j=0,m_{\hat{M}-1}}, \ldots, \left\{u^*_{1,j}\right\}_{j=0,m_1}, u^*_f\right]^T$ and $\mathbf{B}$ is the blocking matrix derived from an identity matrix by eliminating the rows corresponding to the elements of $\mathcal{U}^*(k)$ that are to be blocked. For example, for $\hat{M} = 4$ and $m_\ell = 1$ for $\ell = 1, 2, 3, 4$,

$$\mathcal{U}^*(k) = \begin{bmatrix} u^*_{4,+}(k) \\ u^*_{4,0}(k) \\ u^*_{3,0}(k) \\ u^*_{3,1}(k) \\ u^*_{2,0}(k) \\ \vdots \\ u^*_{2,3}(k) \\ u^*_{1,0}(k) \\ \vdots \\ u^*_{1,7}(k) \\ u^*_f(k) \end{bmatrix}; \quad \mathcal{U}_B^*(k) = \begin{bmatrix} u^*_{4,+}(k) \\ u^*_{4,0}(k) \\ u^*_{3,0}(k) \\ u^*_{2,0}(k) \\ u^*_{1,0}(k) \\ u^*_f(k) \end{bmatrix}$$

$$\mathbf{B} = \begin{bmatrix} [1] & 0 & 0 & 0 & 0 & 0 \\ 0 & [1] & 0 & 0 & 0 & 0 \\ 0 & 0 & \begin{bmatrix} 1 \\ 0 \end{bmatrix} & 0 & 0 & 0 \\ & & & 0 & 0 & 0 \\ 0 & 0 & 0 & \begin{bmatrix} 1 \\ 0 \\ 0 \\ 0 \end{bmatrix} & 0 & 0 \\ & & & & 0 & 0 \\ & & & & 0 & 0 \\ & & & & 0 & 0 \\ 0 & 0 & 0 & 0 & \begin{bmatrix} 1 \\ 0 \\ \vdots \\ 0 \end{bmatrix} & 0 \\ & & & & & 0 \\ \vdots & \vdots & \vdots & \vdots & & \vdots \\ 0 & 0 & 0 & 0 & & 0 \\ 0 & 0 & 0 & 0 & 0 & 1 \end{bmatrix}^T$$

The same blocking can be applied to the constraint vectors $\mathcal{U}^*_{high}$ and $\mathcal{U}^*_{low}$ as well in order to reduce the number of inequality constraints. In almost all cases, blocking of these vectors does *not* imply removing some of the constraints since the input constraints are either constant or exhibit little variations across the control horizon.

A similar method can be applied in condensing the output vectors. The transformed future error and output vectors, $\mathcal{E}^*(k)$ and $\mathcal{Y}^*(k|k)$ are condensed as follows:

$$\begin{aligned} \mathcal{E}_H^*(k) &= \mathbf{H}\mathcal{E}^*(k) \Rightarrow \mathcal{E}^*(k) = \mathbf{H}^T\mathcal{E}_H^*(k) \\ \mathcal{Y}_H^*(k|k) &= \mathbf{H}\mathcal{Y}^*(k|k) \Rightarrow \mathcal{Y}^*(k) = \mathbf{H}^T\mathcal{Y}_H^*(k|k) \end{aligned}$$

Here $\mathbf{H}$ is the condensation matrix, a fat matrix derived from an identity matrix by eliminating the rows corresponding to the outputs to be "condensed." Again, from each ℓth-level wavelet, first p_ℓ elements are retained and the rest are eliminated. p_ℓ can be interpreted as the prediction horizon (counted in terms of the time unit of 2^ℓ) for the frequency range corresponding to $\psi_{\ell,\cdot}$, the ℓth-level wavelet. The same condensation operator can be applied to the constraint vectors $\mathcal{Y}_{high}^*(k)$ and $\mathcal{Y}_{low}^*(k)$, thus reducing the number of inequality constraints that need to be checked.

The objective (12) and constraints (13)-(14) can be rewritten in terms of the blocked and condensed vectors as follows:

$$\min_{\mathcal{U}_B^*(k)} \left\|\Lambda_y {W_y}^T\mathbf{H}^T\mathcal{E}_H^*(k)\right\|_2^2 + \left\|\Lambda_u \, \Pi {V_u}^T\mathbf{B}^T\mathcal{U}_B^*(k)\right\|_2^2 \tag{16}$$

with constraints

$$\mathcal{E}_H^*(k) = \mathbf{H}W_y \left\{\mathcal{R}(k) - \mathcal{S}^x\Delta x(k|k) - \mathcal{H}y(k|k) - \mathcal{S}^U \Pi V_u^T\mathbf{B}^T\mathcal{U}_B^*(k)\right\} \tag{17}$$

$$(\mathcal{C}_u^*)_B\,\mathcal{U}_B^*(k) \;\geq\; \mathcal{C}_H(k|k) \tag{18}$$

where

$$(\mathcal{C}_u^*)_B = \begin{bmatrix} -{V_u}^T\mathbf{B}^T \\ {V_u}^T\mathbf{B}^T \\ -\Pi {V_u}^T\mathbf{B}^T \\ \Pi {V_u}^T\mathbf{B}^T \\ -{W_y}^T\mathbf{H}^T\mathbf{H}W_y\mathcal{S}^{\mathcal{U}}\,\Pi\, V_u^T\mathbf{B}^T \\ {W_y}^T\mathbf{H}^T\mathbf{H}W_y\mathcal{S}^{\mathcal{U}}\,\Pi\, {V_u}^T\mathbf{B}^T \end{bmatrix}$$

$$\mathcal{C}_H(k|k) = \begin{bmatrix} -{V_u}^T\mathbf{B}^T\left(\mathcal{U}_{high}^*\right)_B \\ {V_u}^T\mathbf{B}^T\left(\mathcal{U}_{low}^*\right)_B \\ -\Delta\mathcal{U}_{max} \\ -\Delta\mathcal{U}_{max} \\ {W_y}^T\mathbf{H}^T\mathbf{H}W_y\left\{\mathcal{S}^x\Delta x(k|k) + \mathcal{H}_c y(k|k) - \mathcal{Y}_{high}(k)\right\} \\ -{W_y}^T\mathbf{H}^T\mathbf{H}W_y\left\{\mathcal{S}^x\Delta x(k|k) + \mathcal{H}_c y(k|k) - \mathcal{Y}_{low}(k)\right\} \end{bmatrix}$$

Note that this is a QP again, but of a reduced dimension:

$$\left(\sum_{\ell=1}^{\hat{M}} m_\ell + 2\right)$$

as opposed to $(2^{\hat{M}}+1)$. In addition, the number of inequality constraints are also reduced. It should also be noted that the inequality constraints corresponding to the repeated rows of $V_u^T \mathbf{B}^T$ and $W_y^T \mathbf{H}^T \mathbf{H} W_y$ and zero rows of $\Pi\ V_u^T \mathbf{B}^T$ are those that are redundant and are satisfied automatically.

For the particular wavelet basis considered in this application, namely the Haar wavelet, owing to the special structure of the transformation matrices V_u and W_y, the number of inequality constraints can be significantly reduced this way. In each level ℓ ($\ell = 1 \cdots, \hat{M}$), the number of non-redundant input and output magnitude constraints are equal to m_ℓ and p_ℓ respectively. For the constraints on $\Delta\mathcal{U}(k)$, only m_ℓ equations need to be considered for each level ℓ, since all other constraints correspond to the zero rows of $\Pi V_u^T \mathbf{B}^T$ and hence are automatically satisfied. In effect the transformed constraint vector has a dimension of

$$4 \times \left[\sum_{\ell=1}^{\hat{M}} m_\ell + 2\right] + 2 \times \left[\sum_{\ell=1}^{\hat{P}} p_\ell + 2\right]$$

This is considerably less compared to the dimension of the time domain constraint vector:

$$4 \times \left[(2^{\hat{M}} + 1)\right] + 2 \times \left[(2^{\hat{P}} + 1)\right]$$

Based on practical considerations, it is suggested to choose the same m_ℓ and p_ℓ for all ℓ. Using a uniform m_ℓ implies that an increasingly larger control move horizon is used as frequency decreases (equivalently, ℓ increases) since this makes the horizon length for the ℓth-level wavelet $m_\ell \times 2^\ell$. Since typical power spectrums for disturbances and setpoint changes in chemical processes are predominantly concentrated in the "low" frequency region, the length of the control move horizon needed for "high" frequency variations of the inputs is shorter than that for the low frequency variations. Similarly for p_ℓ, one can choose them to be $m_\ell + (t_s)_\ell$, the control move horizon plus the settling time appropriate for the frequency channel for the ℓth-level wavelet (counted in terms of the time unit of 2^ℓ). Often, the output vectors can be condensed substantially this way since the settling time decreases as ℓ decreases (as frequency increases). From an approximation perspective this can also be viewed as using finer resolution in the near future and a coarser approximation far into the horizon. This is appropriate from a practical stand point, since MPC is implemented in receding horizon fashion. Since the frequency localization of the wavelet bases is not perfect unlike the sinusoidal basis, the argument of frequency decoupled input/output responses cannot be justified rigorously. Nevertheless, the method should provide a good starting point for finding a proper choice of the blocking and condensation matrices.

As mentioned earlier, the extension of the above methodology to multivariable systems is straightforward. The input blocking and output condensation can be applied to each input and output separately. The technique is particularly suitable for multi-time-scale systems.

As described, the wavelet decompositions and the resulting time-frequency interpretations of the input and output signals, provide an intuitive and efficient framework for performing tasks which are more conveniently posed in frequency domain, such as basis reduction, for a time-domain technique like MPC. Next we will illustrate this technique by applying it to the following examples.

3.3 Numerical Examples:

Example 1

The following is a non-square system with two inputs and one output.

$$y(s) = \begin{bmatrix} \frac{1}{s+1} & \frac{50s-1}{(50s+1)(30s+1)} \end{bmatrix} \begin{bmatrix} u_1(s) \\ u_2(s) \end{bmatrix}$$

with the following constraints on the inputs and the output

$$\begin{aligned} -0.5 \leq u_1 \leq 0.5 \\ -1.0 \leq u_2 \leq 1.0 \\ -0.1 \leq y \leq 0.1 \end{aligned}$$

The first input acts through much faster dynamics than the second input. The second input is more expensive and is to be used only when the first input saturates. The following MPC controller was designed, for the above given regulatory requirements and operating constraints:

$$P = M = 17; \quad \Lambda_y = \mathrm{diag}\{\tilde{\Lambda}_y, \cdots, \tilde{\Lambda}_y, \tilde{\Lambda}_t\}; \quad \Lambda_u = \mathrm{diag}\{\tilde{\Lambda}_u, \cdots, \tilde{\Lambda}_u\}$$

where

$$\tilde{\Lambda}_y = \tilde{\Lambda}_t = 10^3; \quad \tilde{\Lambda}_u = \begin{bmatrix} 10^{-3} & 0 \\ 0 & 1 \end{bmatrix}$$

For a unit step disturbance entering through the first-order dynamics of $\frac{1}{50s+1}$, the process with the above controller, was observed to be unstable. The

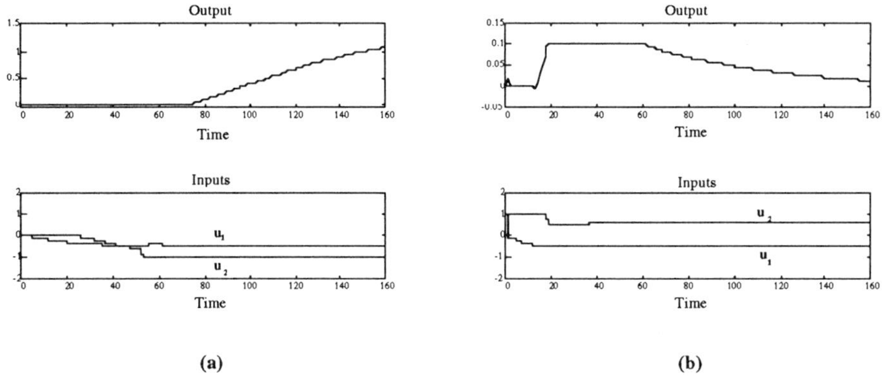

Figure 5: Example 1: Closed-Loop Responses for (a) $P = 17$ (b) $P = 129$

input/output response is shown in figure 5(a). This is because the particular choice of P, though quite significant from the computation standpoint, was not large enough to prevent the inversion of the RHP zero associated with the second input.

Here, the process settling time $t_s \approx 300$. An equivalent infinite horizon objective would require $P \geq M + t_s = 317$, with the terminal weight $\tilde{\Lambda}_t = \lim_{k\to\infty} k$. It was observed that an equally good performance can be achieved for a choice of $P = 129$ as well. Figure 5(b) shows the simulated response for the above process with $P = 129$ and the remaining settings being same as in the previous case. The simulation clearly shows the desirability of a large prediction horizon and equivalently confirms the guaranteed closed-loop stability for the infinite horizon objective (Rawlings and Muske,1991).However this leads to a large dimensional QP. In the present case, the dimension of the optimization problem is 34 and the number of constraints to be checked is 394.

Example 2

Consider the following non-square system with two outputs and one input:

$$\begin{bmatrix} y_1(s) \\ y_2(s) \end{bmatrix} = \begin{bmatrix} \frac{1}{(3s+1)(2s+1)(10s+1)} \\ \frac{100s-5}{(50s+1)(3s+1)} \end{bmatrix} u(s)$$

y_1 is the controlled variable to be kept as close to the setpoint as possible and y_2 is an associated variable that doesn't have a regulatory requirement, but must be kept between -1 and 1. The input u also has the magnitude limit of ± 1.5

The following MPC controller was designed and its performance was simu-

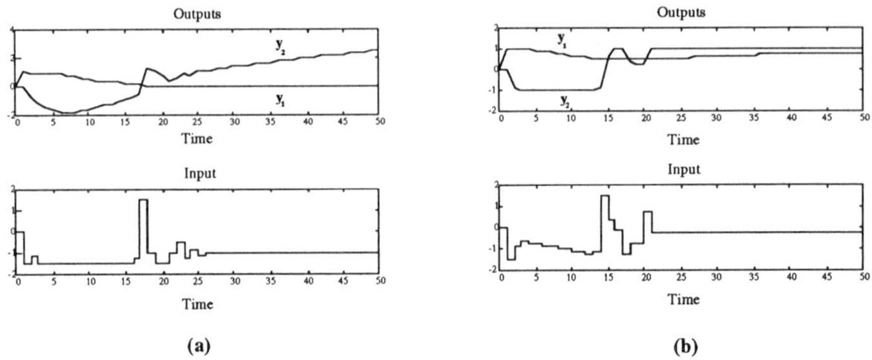

Figure 6: Example 2: Closed-Loop Responses for (a) $P = 17$ (b) $P = 129$

lated against a step disturbance entering the first output:

$$P = M = 17;\ \ \Lambda_y = \text{diag}\{\tilde{\Lambda}_y, \cdots, \tilde{\Lambda}_y, \tilde{\Lambda}_t\};\ \ \Lambda_u = \text{diag}\{\tilde{\Lambda}_u, \cdots, \tilde{\Lambda}_u\}$$

where

$$\tilde{\Lambda}_y = \tilde{\Lambda}_t = \begin{bmatrix} 1 & 0 \\ 0 & 0 \end{bmatrix};\ \ \tilde{\Lambda}_u = 10^{-6}$$

Figure 6(a) shows the simulated responses of the input and outputs. The constraint on the second output is clearly violated. This is because the particular choice of P was not large enough to predict the detrimental effects of the earlier input moves.

An equivalent infinite horizon objective would require $P = 17 + t_s$, where the process settling time $t_s \approx 300$, with $\tilde{\Lambda}_t = \lim_{k\to\infty} k \begin{bmatrix} 1 & 0 \\ 0 & 0 \end{bmatrix}$. We performed the simulations with $P = 129$ and the performance was observed to be as good as that of the infinite horizon objective. Figure 6(b) shows the simulated response with $P = 129$ and the other settings being same as in the earlier case. The simulation again confirms the desirability of a large prediction horizon. But again, this results in a large size optimization problem of dimension 17 with 584 constraint equations.

The above two examples demonstrate the importance of the choice of the prediction and control horizons in MPC. However, the infinite horizon formulation, or its equivalent for the special case of openloop stable systems, result in optimization problems of very large dimensions.

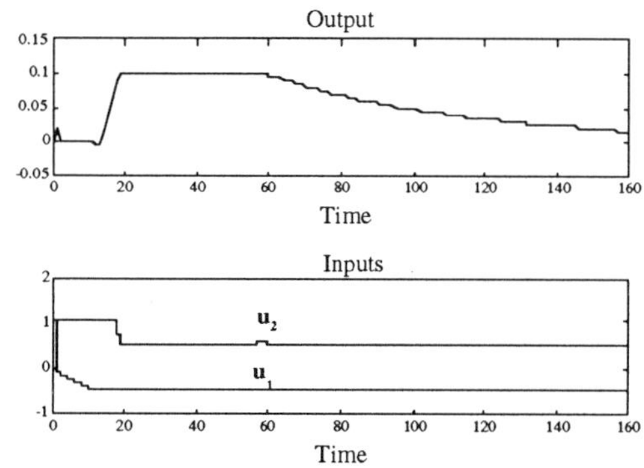

Figure 7: Example 1 : MPC with Blocking and Condensation, $P = 129$.

Wavelet Domain Solutions

Now we revisit the two examples again, for which the infinite horizon objective proved to be attractive. We can significantly reduce the computational load by applying the wavelet domain blocking and condensation techniques.

Example 1

After the Haar transformation of the problem, the blocking and condensation techniques, with $m_\ell = 1$ and $p_\ell = 1$ for all ℓ, were applied and the problem was solved for the same choice of M, $\tilde{\Lambda}_y$, $\tilde{\Lambda}_u$, and $P = 129$. The resulting optimization problem is of dimension 12. When compared to the original unblocked/uncondensed MPC which had a QP of dimension 34, significant reduction in computation is achieved by using the new techniques. The number of constraints also reduces significantly. There are only 66 non-redundant constraint equations compared to 394 for the original problem. The simulated responses are shown in Figure 7. We should note that there is no degradation in the controller performance despite the significant reduction in the computational complexity.

Example 2

Similarly for the second example, we again applied the blocking and condensation techniques, with $m_\ell = 1$ and $p_\ell = 2$ for all ℓ, after the Haar transformation. The transformed problem has a dimension of 6 compared to 17 of the original problem. The dimension of the constraint vector reduces from 584 to 88. The blocked/condensed solution is shown in Figure 8. In spite of the significant reduction in the on-line computation, the controller again shows equally good

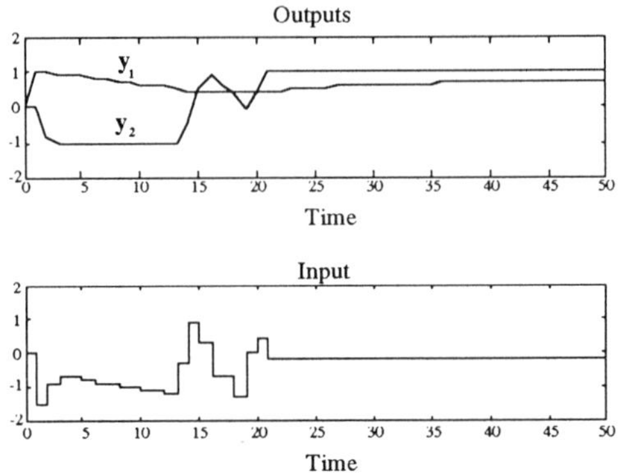

Figure 8: Example 2: MPC with Blocking and Condensation, $P = 129$.

performance and output constraint-handling.

3.4 Frequency Domain Tuning of MPC

The Wavelet Domain Transformation of the MPC problem also allows a suitable frame work for MPC tuning. Model Predictive Controllers are traditionally tuned through a number of different parameters including prediction horizon, control horizon, input/output weights, estimator parameters and in some cases artificially imposed input/output constraints. Though they provide an intuitive tuning framework for time domain performance they do not, in general, have any frequency domain interpretations. For cases where it is more appropriate to tune for a frequency domain performance function (*e.g.* H_∞-norm, desired loop-shapes *etc.*), MPC in its present time domain form is not conducive to intuitive tuning. The use of the time-domain MPC would require painstaking trial-and-error tuning. A reformulation of MPC in the wavelet domain enables more direct tuning for frequency-domain performance objectives. The wavelet domain formulation of MPC presented earlier can be extended to include wavelet domain tuning parameters, specifically the input/output weights. We can further utilize the time-frequency information of the wavelet decompositions of the input/output signals, in tuning the controller. In the transformed problem the objective and constraints are written in terms of the wavelet basis as follows:

$$\min_{\mathcal{U}^*(k)} \left\|\Lambda_y^* \mathcal{E}^*(k)\right\|_2^2 + \|\Lambda_u^* \mathcal{U}^*(k)\|_2^2 \tag{19}$$

The Λ_y^* and Λ_u^* can be viewed as projections of the time-domain weighting matrices into the wavelet domain. An equivalent time-domain formulation would be:

$$\min_{\mathcal{U}(k)} \left\|{W_y}^T \Lambda_y^* W_y \mathcal{E}(k)\right\|_2^2 + \left\|{V_u}^T \Lambda_u^* V_u \mathcal{U}(k)\right\|_2^2 \tag{20}$$

The weighting matrices in the two formulations are related as follows:

$$\begin{aligned}\Lambda_y &= {W_y}^T {\Lambda_y}^* W_y \\ \Lambda_u &= {V_u}^T {\Lambda_u}^* V_u\end{aligned}$$

In the new formulation, in effect we are considering a time-varying, frequency based weighting over the prediction and control horizons. Each component of the transformed input/output corresponds to a certain frequency *band* that the respective wavelet function occupies. By weighting each of these components appropriately, the input/output behavior can be shaped in frequency domain. In addition, the above formulation allows for the constraints specified in the wavelet (time-frequency) domain as well as the usual time domain constraints.

One important observation to be noted here is that the MPC formulation presented earlier is for a state-space model. In order to allow for frequency domain weighting, it is more convenient to use a FIR model, the states of which are the past input values. Obviously, having the past inputs as the states is very convenient, since in frequency weighting of the input profile, at any given time we should consider the entire profile that is effective in loopshaping. This "weighting horizon" corresponds to all the inputs of relevance at any given time k, which includes M values into the future as well as n values into the past. Here n corresponds to the number of impulse response coefficients in the FIR model and M, the length of the control horizon. In other words, at time k we need to weigh the past inputs as well, which are dynamically still active.

3.5 Formulations with Other Wavelet Bases

Two of the bases applied so far in MPC transformation are the wavelet *like* basis of Alpert(1992) by Elias Juarez and Kantor (1992) and the Haar transformation presented by Lee *et al* (1992). Alpert's basis is perfectly localized in time. But the functions are, to some extent, delocalized in frequency domain. We observed that the Haar basis is not perfectly localized in the frequency domain and its Fourier transform decays like $|\omega|^{-1}$ for $\omega \longrightarrow \infty$. For infinite-domain signals (signals of V_i) there are other orthogonal wavelet bases with better frequency localization than the Haar basis (Daubechies,1988). However, unlike the Haar basis, these infinite-domain wavelets do not apply to the finite-length signals, i.e., signals of V_i^ℓ, straightforwardly. For instance, the problem with Daubechies'(1988) wavelets is the wrapping around of some of the functions over the interval, because of the assumption of periodicity on finite intervals. Because of this "delocalization" in time domain, these bases are not suitable for representing finite dimensional signals.

Functions which are independent linear combinations of the pulse functions can also serve as basis for finite length piece-wise continuous signals belonging to ${V_0}^\ell$. The Haar wavelet can be viewed as a special case of above class of basis functions. Evaluation and construction of such wavelet *like* basis best suited for MPC transformation is part of our current focus of research. From a frequency domain tuning perspective, it is desirable to have a basis, with good localization and *minimal* overlap among different functions, in the frequency domain. Such a basis allows good frequency decoupling in the resulting decompositions. On a more global level, the challenge of constructing wavelets with best allowable frequency localization determined by the Heisenberg's uncertainty principle, is an active area of research for the wavelet theoreticians.

4 Conclusions

We presented the wavelet theory from the process control perspective. We gave a brief overview of the existing results and possibilities for future applications. The readers should realize that the research in this area is still in rudimentary stages. Most of the applications in process engineering are yet to be explored and the realization of the full power of the wavelet transforms will take some more time. The transformation of the conceptual possibilities presented here and elsewhere in this volume into concrete, rigorous tools can be expected in near future. Nevertheless, few of the results that already exist, provide a glimpse of the applications to come.

References

[1] Alpert, B.K. (1992). Wavelets and Other Bases for Fast Numerical Linear Algebra. In *Wavelets: A Tutorial in Theory and Applications*, Edited by Charles Chui. pp 181-216. Academic Press, San Diego, CA.

[2] Bakshi R.B. and Stephanopoulos G. (1992). Representation of Process Trends Part III: Multi-Scale Extraction of Trends from Process Data. Submitted to *Computers & Chem. Engng.*

[3] Bakshi R.B. and Stephanopoulos G. (1993). Wave-Net: A Multiresolution, Hierarchical Neural Network with Localized Learning. *AIChE Journal*, **39**, pp. 57-81.

[4] Beylkin G., Coifman R. and Rokhlin V. (1991). Fast Wavelet Transforms and Numerical Algorithms I. *Communications on Pure and Applied Mathematics*, **44**, pp.141-183.

[5] Carrier J.F. and Stephanopoulos G. (1992). Multiresolution Theory for Model Identification and Control. *Presented at the AIChE Annual Meeting*, Miami Beach, FL.

[6] Coifman R., Meyer Y. and Wickerhauser V. (1992). Wavelet Analysis and Signal Processing. In *Wavelets and their Applications*, Edited by Mary B. Ruskai, pp. 153-178. Jones and Bartlett Publishers, Boston, MA.

[7] Daubechies I. (1988). Orthonormal Bases of Compactly Supported Wavelets. *Communications on Pure and Applied Mathematics*, **41**, pp. 909-996.

[8] Daubechies I. (1992). *Ten Lectures on Wavelets.* **SIAM**, Philadelphia, PA.

[9] Elias-Juarez A. and Kantor J.C. (1992). On the Application of Wavelets to Model Predictive Control. *Presented at the American Control Conference*, Chicago, IL.

[10] Joshi, A. and Motard R.L. (1991). Linear Time-Frequency Analysis in Automated Fault Detection. *Presented at the AIChE Annual Meeting*, Los Angeles, CA.

[11] Lee J.H., Chikkula Y., Yu Z.H. and Kantor J.C. (1992a). Improving the Computational Efficiency of the Model Predictive Control Algorithm Using the Wavelet Transformation. Submitted to *International Journal of Control.*

[12] Lee J.H. and Yu Z.H. (1992b). Robust Tuning of Model Predictive Controllers. Accepted for publication in *Computers & Chem. Engng.*

[13] Lee J.H., Gelormino M.S. and Morari M. (1992c). Model Predictive Control of Multi-Rate Sampled-Data Systems: A State-Space Approach. *International Journal of Control*, **55**, pp. 153-191.

[14] Leonard J.A. and Kramer M.A. (1991). Radial Basis Function Networks for Classifying Process Faults. *IEEE Control Systems,* pp. 31-38, April, 1991.

[15] Leonard J.A. and Kramer M.A. and Ungar L.H. (1992). A Neural Network Architecture that Computes Its Own Reliability. *Computers & Chem. Engng*, **16**, 819-35.

[16] Liandrat J., Perrier V. and Tchamitchian Ph. (1992). Numerical Resolution of Nonlinear Partial Differential Equations Using the Wavelet Approach. In *Wavelets and their Applications*, Edited by Mary B. Ruskai, pp. 227-238. Jones and Bartlett Publishers, Boston, MA.

[17] Mallat S. (1989a). Multiresolution Approximation and Wavelet Orthonormal bases of $L^2(\mathbf{R})$. *Transactions of the American Mathematical Society*, **315**, pp. 69-87.

[18] Mallat S. (1989b). A Theory for Multiresolution Signal Decomposition: The Wavelet Representation. *IEEE Transactions on Pattern Analysis and Machine Intelligence,* **11** pp.674-693.

[19] Meyer Y. (1985). Principe d'incertitude, bases hilbertiennes et algèbres d' opérateurs. *Séminaire Bourbaki*, no. 662.

[20] Meyer Y. (1986). Ondelettes et functions splines. *Séminaire Equations aux Derivees Partielles*, Ecole Polytechnique, Paris, France.

[21] Nikolaou M. and Yong You (1992). Solution of Partial Differential Equations Using Wavelets. *Presented at the AIChE Annual Meeting*, Miami Beach, FL.

[22] Palavajjhala S., Motard R.L. and Babu Joseph (1993). Improving Robustness and Performance of QDMC Using Blocking and Condensing in the Wavelet Domain. Submitted to *I&EC Research.*

[23] Rawlings, J.B. and Muske K.R. (1991). The Stability of Constrained Receding Horizon Control. Submitted to *IEEE Trans. Automatic Control.*

[24] Stephanopoulos G. and Carrier J.F. (1991). Generation and Validation of Models for the Design of Process Controllers. *Presented at the AIChE Annual Meeting*, Los Angeles, CA.

[25] Strang G. (1989). Wavelets and Dilation Equations: A Brief Introduction. *SIAM Review*, **31**, pp.614-627.

Chapter 7

Use of Wavelets for Numerical Solution of Differential Equations

Michael Nikolaou†
Yong You

Department of Chemical Engineering
Texas A&M University
College Station, TX 77843-3122

Abstract

In contrast to the traditional trigonometric basis functions which have infinite support, wavelets may have compact support, thus being able to approximate a functions not by cancellation, but through placement of the right wavelets at appropriate locations. The multi-resolution analysis (MRA) properties of wavelets render them attractive candidates for functions in terms of which numerical solutions of differential equations can be represented. In particular, Mallat's MRA algorithm applied on orthonormal or semi-orthogonal wavelets can be used to generate adaptive grid methods. In this work, two classes of compactly supported wavelets, namely, Daubechies' orthonormal wavelet bases and Chui-Wang's semi-orthogonal B-spline wavelets, are used with the method of weighted residuals for numerical solution of differential equations. A variation relying on Laplace transform is also examined. Based on the MRA properties of wavelet bases, two-scale multi-level methods are proposed, which generate higher level approximate wavelet solutions from the results of lower level calculations. Adaptive grid refinement can be applied only to regions where the residual at lower level is relatively large. The proposed methods may be advantageous when solutions exhibit abrupt changes. Several numerical examples based on these methods are given. Comparisons with traditional methods for numerical solution of differential equations are made and directions for further development are given.

---

† Author to whom all correspondence should be addressed.

1. Introduction

The basic idea of wavelets, i.e. translation and dilation, can be traced back to the early 1960's (Daubechies, 1992; Chui, 1992). Enthusiasm has peaked recently, due to the development of the multiresolution analysis (MRA) algorithm based on wavelets (Mallat, 1989) and the construction of compactly supported orthonormal wavelet bases (Meyer, 1985; Daubechies, 1988).

Wavelets constitute unconditional (Riesz) bases for $L^2(\mathcal{R})$, the space of all square integrable functions on the real line. This means that a function in $L^2(\mathcal{R})$ can be decomposed and stably reconstructed in terms of wavelets. As explained in the sequel, wavelet bases can have orthonormality or semi-orthogonality, and compact support in time or frequency domain. Together with the corresponding wavelets, scaling functions generate an MRA, i.e. they form a nested set of unconditional bases for a sequence of nested subspaces of increasing resolution which tends to $L^2(\mathcal{R})$.

In contrast to the traditional trigonometric sine and cosine basis functions, which have infinite support and are not in $L^2(\mathcal{R})$, wavelet bases are $L^2(\mathcal{R})$ functions and may have compact support. Unlike trigonometric approximation, approximation with wavelet bases does not rely on cancellation. When an abrupt change, such as a shock wave or a spike, occurs in a function, only local coefficients in a wavelet approximation will be affected. No global cancellation is needed. Orthonormality of wavelet bases can simplify computations. Due to the MRA properties, wavelets are inherent multi-level bases. Mallat's decomposition and reconstruction algorithm provides a fast transition between these different levels. We will elaborate on these and other useful wavelet properties in later sections.

Due to the above properties, wavelets have been used in a variety of areas, with primary applications in signal processing, image analysis, and data compression. It is expected that wavelets can lead to efficient methods for numerical solution of differential equations (DEs). Wavelets can naturally fit in multigrid or adaptive grid methods, for which we expect the MRA algorithm to be used effectively.

Wavelets have been used for numerical solution of DEs: (i) directly as bases with Galerkin's method or other methods (Glowinski et al., 1989; Jaffard, 1992; Qian and Weiss, 1992, 1993); (ii) for algebraic manipulations to simplify the results using existing discretization methods (Beylkin, 1993); and (iii) after adaptation to specific differential operators when Galerkin's method is used (Lorentz and Madych, 1990; Liandrat et al., 1992; Jaffard, 1992; Xu and Shann, 1992; Dahlke and Weinreich, 1993; Jawerth and Sweldens, 1993).

Glowinski et al. (1990) used Daubechies' compactly supported orthonormal wavelets (Daubechies, 1988) and variational methods to solve one-dimensional linear and nonlinear ordinary differential equations (ODEs) and partial differential equations (PDEs). They also discussed use of multigrid methods for linear elliptic equations.

Lorentz and Madych (1990) discussed solution of the equation $u^{(2m)}(x) = g(x)$ with strong boundary conditions and the compatibility condition on $g(x)$. Using Chui-Wang's compactly supported semi-orthogonal spline wavelets (Chui and Wang, 1991; Chui, 1992), they derived another wavelet basis which is adapted to the DE. Using the new basis function and Galerkin's method, they generated a diagonal system of linear equations.

Qian and Weiss (1992, 1993) discussed solution of Helmholz equations in nonseparable domains. Daubechies' compactly supported wavelet bases were used, based on the capacitance matrix methods. Special treatment of boundary conditions was also used to periodize the problem.

Liandrat et al. (1992) proposed an adaptive method for solving PDEs using orthonormal wavelet bases. Special test functions were constructed from the wavelet basis and a constant coefficient elliptic differential operator of second order. A flexible adaptation algorithm was obtained.

Jaffard (1992) used Daubechies wavelets for discretization of elliptic problems by Galerkin method. A diagonal conditioning method was used to bound the conditioning number of the matrix obtained after discretization. Orthonormal wavelets were also constructed for elliptic problems with Neumann boundary conditions.

Xu and Shann (1992) used the anti-derivative of Daubechies' wavelets as trial function to solve two-point boundary value (TPBV) problems, based on the concept of *frames* and Sobolev spaces.

Beylkin (1993) discussed the use of Daubechies' wavelet bases for solving TPBV problems. A finite difference method was used for discretization. Diagonal preconditioning based on wavelets was used for better condition number of the resulting linear system. Calculation of Green's function for the periodized elliptic operators was thus simplified.

Dahlke and Weinreich (1993), using analogous ideas to those in Lorentz and Madych (1991), constructed compactly supported bi-orthogonal wavelets which are adapted to similar differential equations discussed by Lorentz and Madych (1991).

Briggs and Henson (1993) noticed the similarity between the multiresolution analysis of wavelets and the classical multigrid methods. No specific method was given.

Jawerth and Sweldens (1993) developed adapted wavelet bases on a closed subset of the real line for solution of boundary value problems. Using Galerkin's method, they obtained diagonal linear systems.

In this work, we discuss the use of wavelet bases for numerical solutions of ODEs and PDEs. Daubechies' compactly supported orthonormal bases (Daubechies, 1988, 1992) and Chui-Wang's semi-orthogonal B-spline wavelets (Chui and Wang, 1991; Chui, 1992) are used with the method of weighted residuals (MWR), specifically Galerkin's method and collocation method (Finlayson, 1972; Fletcher, 1986). Two-scale multi-level methods are developed based on the MRA properties of wavelet bases and Galerkin's method. These methods generate approximate wavelet solutions at higher resolution level from the results of lower level calculations. Adaptive grid refinement can be applied only to regions where the residual at the immediately lower level is relatively large. Thus, the properties of MRA can be used in a natural and efficient way to maintain different resolution level for different regions of the solution domain. The Laplace transform and its numerical inversion are also used to eliminate time dependency of parabolic PDEs. The Laplace transform removes the time coordinate and eases boundary condition treatment.

Several numerical examples are given, including initial value problems (IVP), TPBV problems, and nonlinear regularized Burgers' equations with different parameters. A two dimensional example is also presented. Comparisons of wavelet methods with traditional methods are also made. The general conclusion is that wavelets can be advantageous for problems for which multi-level methods are warranted.

In the following sections, wavelet theory is discussed first from a multiresolution approximation standpoint, followed by details on specific wavelet bases including Daubechies' compactly supported wavelets and Chui-Wang's B-spline wavelets. In the next section, use of wavelet bases with Galerkin's method (Wavelet-Galerkin Method) is illustrated by some numerical examples. Use of Laplace transform techniques with wavelets is discussed in the subsequent section. Two-scale methods, which make use of the multiresolution analysis properties to obtain higher level approximations from lower ones, are discussed, and numerical examples are also given. Finally, an assessment of our results is given and further developments are recommended.

2. Multiresolution Analysis

In this section, we will briefly review the basic ideas of multiresolution analysis (MRA) and approximation with (semi-)orthogonal wavelets. More details can be found in Mallat (1989), Meyer (1989), Daubechies (1988, 1992), Chui (1992), and earlier chapters in this volume. The following presentation is slightly more general than usual expositions based on orthonormal wavelets.

2.1. Scaling Functions and Multiresolution Analysis

Construction of wavelet functions can start from the building of a so-called *scaling* or *dilation* function, $\phi(x)$, and a set of related coefficients, $\{p(k)\}_{k\in\mathcal{Z}}$, which satisfy the so-called *two-scale relation* or *refinement equation*,

$$\phi(x) = \sum_k p(k)\phi(2x-k) \tag{1}$$

and some additional conditions. The scaling function $\phi(x)$ has *compact support* if and only if finitely many coefficients $p(k)$ are non-zero.

Translations of the scaling function, $\{\phi(x-k)\}$, form a Riesz or unconditional basis of a subspace $V_0 \subset L^2(\mathcal{R})$. Furthermore, through translation of ϕ by a factor of 2^n and dilation by a factor of $k \cdot 2^{-n}$, a Riesz basis, $\{\phi_{n,k}(x)\}_{k\in\mathcal{Z}}$, is obtained for the subspace $V_n \subset L^2(\mathcal{R})$, where

$$\phi_{n,k}(x) = 2^{n/2}\phi(2^n x - k) \tag{2}$$

corresponding to resolution level n.

If translations of the scaling function, $\{\phi(x)\}$, at the same resolution level n satisfy the condition,

$$\int_{-\infty}^{\infty} \phi_{n,k}(x)\phi_{n,l}(x)dx = \delta_{k,l} \qquad n,k,l \in \mathcal{Z} \tag{3}$$

then $\phi(x)$ is *orthonormal*. In this case, the best approximation of an $L^2(\mathcal{R})$ function, $f(x)$, by a function $A_n f(x)$ in the subspace V_n of $L^2(\mathcal{R})$ (i.e. at *resolution* level n) is given by the orthogonal projection of f on V_n, as follows:

$$\mathcal{A}_n f(x) = \sum_k a_{n,k}\phi_{n,k}(x) \tag{4}$$

where $a_{n,k}$ is given by the inner product of $f(x)$ and $\phi_{n,k}(x)$,

$$a_{n,k} = \langle f, \phi_{n,k} \rangle = \int_{-\infty}^{\infty} f(x)\phi_{n,k}(x)dx \tag{5}$$

The operator $\mathcal{A}_n : L^2(\mathcal{R}) \ni f(x) \rightarrow \mathcal{A}_n f(x) \in V_n$, is a linear orthogonal projection operator (Mallat, 1989) resulting in the best approximation of $f(x)$ in V_n.

Approximation of a function, $f(x)$, can be conducted at different resolution levels (corresponding to different n's), and the approximations in the subspaces, $\cdots, V_{n-1}, V_n, V_{n+1}, \ldots$, have the following relationships:

(i) $V_n \subset V_{n+1}$;
(ii) $f(x) \in V_n \Leftrightarrow f(2x) \in V_{n+1}$;
(iii) $f(x) \in V_n \Leftrightarrow f(x + 2^{-n}) \in V_n$;
(iv) $\lim_{n\to\infty} V_n = \bigcup_n V_n$ is dense in $L^2(\mathcal{R})$;
(v) $\lim_{n\to-\infty} \bigcap_n V_n = \{0\}$;
(vi) The set $\{\phi(x-k)\}_{k\in\mathcal{Z}}$ forms a Riesz or unconditional basis for V_0, i.e. there exist constants A and B, with $0 < A \leq B < \infty$, such that,

$$A \sum_{k\in\mathcal{Z}} |c_k|^2 \leq \left\| \sum_{k\in\mathcal{Z}} c_k \phi(x-k) \right\|_2^2 \leq B \sum_{k\in\mathcal{Z}} |c_k|^2$$

for any sequence $\{c_k\} \in l_2$, the space of all square summable sequences ($A = B = 1$ for o. n. basis).

In this case, $\phi(x) \in L^2$ is called a scaling function that generates a multiresolution analysis (MRA) with the above properties.

2.2. Wavelet Functions and Detail Signals

From property (i) of scaling functions ($V_n \subset V_{n+1}$) it is obvious that there is some information lost when the approximation goes from level $n+1$ to level n. To characterize the difference, also called *detail lost*, let the orthogonal complement of V_n to V_{n+1} be W_n such that, $V_n \oplus W_n = V_{n+1}$, $\forall\, n$. It turns out (Mallat, 1988; Chui, 1992) that a basis for W_n can be obtained by dilating and translating a single function, $\psi(x)$, called *basic (mother, analyzing) wavelet*, which is defined by

$$\psi(x) = \sum_k q(k)\phi(2x-k) \tag{6}$$

where $q(k)$ are a set of coefficients for the two-scale relationship of wavelet basis.

A Riesz basis for the subspace W_n is given by $\{\psi_{n,k}(x)\}_{k\in\mathcal{Z}}$, where

$$\psi_{n,k}(x) = 2^{n/2}\psi(2^n x - k) \tag{7}$$

The set $\{\psi_{n,k}(x)\}_{n,k\in\mathcal{Z}}$ is a basis for $L^2(\mathcal{R})$.

A wavelet basis is *orthonormal* if any two translated and/or dilated wavelets satisfy the orthogonality condition:

$$\int_{-\infty}^{\infty} \psi_{n,k}(x)\psi_{m,l}(x)dx = \delta_{n,m}\delta_{k,l}, \qquad n,m,l,k \in \mathcal{Z} \tag{8}$$

A wavelet basis is *semi-orthogonal* if wavelets at different resolution levels only satisfy the semi-orthogonality condition

$$\int_{-\infty}^{\infty} \psi_{n,k}(x)\psi_{m,l}(x)dx = 0, \quad n \neq m, \ n,m,l,k \in \mathcal{Z}$$

For semi-orthogonal wavelet ψ, a dual basis, $\tilde{\psi}_{n,k}$, exists (Chui, 1992), which satisfies the *bi-orthogonality* condition,

$$\int_{-\infty}^{\infty} \psi_{n,k}(x)\tilde{\psi}_{m,l}(x)dx = \delta_{n,m}\delta_{k,l} \qquad n,m,k,l \in \mathcal{Z} \tag{9}$$

and must be used in place of $\psi_{n,k}(x)$ to evaluate the approximation coefficient $b_{n,k}$ below. For an orthonormal basis, the dual is the basis itself.

The subspaces W_n have the following properties,

(i) $V_{n+1} = V_n \oplus W_n$, i.e. W_n is the orthogonal complement of V_n to V_{n+1};
(ii) W_{n_1} is orthogonal to W_{n_2} if $n_1 \neq n_2$;
(iii) $\oplus_{n=-\infty}^{\infty} W_n = L^2(\mathcal{R})$.

We can define an operator $\mathcal{D}_n : L^2(\mathcal{R}) \ni f(x) \to \mathcal{D}_n f(x) \in W_n$ similar to $\mathcal{A}_n$. Then,

$$\mathcal{D}_n f(x) = \sum_k b_{n,k}\psi_{n,k}(x) \tag{10}$$

where $\{b_{n,k}\}$ are coefficients that can be determined through an equation analogous to (5), i.e.

$$b_{n,k} = \langle f, \psi_{n,k}\rangle = \int_{-\infty}^{\infty} f(x)\psi_{n,k}(x)dx$$

Thus, for a function, $f(x)$, we have, at resolution level n,

$$\begin{aligned} \mathcal{A}_n f(x) &= \sum_k a_{n,k}\phi_{n,k}(x) \\ &= \underbrace{\sum_k a_{n-1,k}\phi_{n-1,k}(x)}_{\mathcal{A}_{n-1}f(x)\in V_{n-1}} + \underbrace{\sum_k b_{n-1,k}\psi_{n-1,k}(x)}_{\mathcal{D}_{n-1}f(x)\in W_{n-1}} \end{aligned} \tag{11}$$

$\mathcal{A}_{n-1}f(x)$ can be decomposed again and this decomposition can be conducted recursively to a desired level.

2.3. Mallat's Decomposition and Reconstruction Algorithms

Mallat (1989) developed a very efficient algorithm for multi-resolution analysis using wavelets. Using this algorithm, the relationships between the approximation coefficients at different levels, e.g. $\{a_{n,k}\}$ at level n, and $\{a_{n-1,k}\}$ and $\{b_{n-1,k}\}$ at level $n-1$, are determined through some sequences of constants which depend only on the wavelet basis being used. Such sequences will be given in the next section for Daubechies wavelets and Chui-Wang wavelets. Next we briefly present Mallat's decomposition and reconstruction algorithms:

Recall that $\{\phi_{n-1,k}(x)\}_k$ and $\{\psi_{n-1,k}(x)\}_k$ are unconditional bases for V_{n-1} and W_{n-1} respectively. Thus a scaling function $\phi_{n,k}(x) \in V_n = V_{n-1} \oplus W_{n-1}$ can be represented in terms of these bases, using coefficients $\frac{1}{2}g(l-2k)$ and $\frac{1}{2}h(l-2k)$, as

$$\psi_{n,l}(x) = \sum_k \left[\frac{1}{2}g(l-2k)\phi_{n-1,k}(x) + \frac{1}{2}h(l-2k)\psi_{n-1,k}(x)\right], \quad l \in \mathcal{Z} \tag{12}$$

where the sequences $\{\frac{1}{2}g(k)\}$ and $\{\frac{1}{2}h(k)\}$ are called *decomposition sequences* and depend only on the mother wavelet. From this equation and (11), we get the *decomposition algorithm*,

$$\begin{aligned} a_{n-1,l} &= \frac{1}{2}\sum_k g(l-2k)a_{n,k} \\ b_{n-1,l} &= \frac{1}{2}\sum_k h(l-2k)a_{n,k} \end{aligned} \tag{13}$$

Thus, if we know $\{a_{n,k}\}$, we can go from a higher level n to a lower level $n-l$, by applying the following tree structure calculation (sequence in bold is known):

$$
\begin{array}{ccccccccc}
\{\mathbf{a_{n,k}}\} & \rightarrow & \{a_{n-1,k}\} & \rightarrow & \{a_{n-2,k}\} & \rightarrow & \cdots & \rightarrow & \{a_{n-l,k}\} \\
& \searrow & & \searrow & & \searrow & & \searrow & \\
& & \{b_{n-1,k}\} & & \{b_{n-2,k}\} & & \cdots & & \{b_{n-l,k}\}
\end{array}
$$

By applying the two-scale relations of scaling and wavelet functions to the right-hand side of Eq. (11), we obtain the following *reconstruction algorithm*,

$$a_{n+1,l} = \sum_k \left[p(l-2k)a_{n,k} + q(l-2k)b_{n,k}\right] \tag{14}$$

where $\{p(k)\}$ and $\{q(k)\}$ are called *reconstruction sequences.* So, if we know $\{a_{n-l,k}\}$ and $\{b_{j,k}\}_{j=n-l}^{n-1}$, we can reconstruct the approximation at level n from the lower level $n-l$, as follows (sequences in in bold are known):

$$
\begin{array}{ccccccccc}
\{a_{n,k}\} & \leftarrow & \{a_{n-1,k}\} & \leftarrow & \cdots & \leftarrow & \{a_{n-l+1,k}\} & \leftarrow & \{\mathbf{a_{n-l,k}}\} \\
& \nwarrow & & \nwarrow & & \nwarrow & & \nwarrow & \\
& & \{\mathbf{b_{n-1,k}}\} & & \cdots & & \{\mathbf{b_{n-l+1,k}}\} & & \{\mathbf{b_{n-l,k}}\}
\end{array}
$$

The above decomposition and reconstruction algorithm yields the so-called *(fast) discrete wavelet transform.*

Reconstruction and decomposition sequences, i.e. $\{p(k)\}$, $\{q(k)\}$, $\{g(k)\}$, and $\{h(k)\}$, vary for different wavelet bases. In general, the reconstruction sequences, $\{p(k)\}$, $\{q(k)\}$, are the two-scale coefficients for scaling and wavelet functions, Eqs. (1) and (6). Decomposition sequences, $\{g(k)\}$, and $\{h(k)\}$, depend on the orthogonality (or on the dual) of the wavelet. Some details will be given in the next section.

3. Wavelet Bases

An example of orthonormal wavelet basis of $L^2(\mathcal{R})$ is the Haar basis (1910),

$$\psi_H(x) = \begin{cases} 1, & 0 \le x < \frac{1}{2} \\ -1, & \frac{1}{2} \le x < 1 \\ 0, & \text{otherwise} \end{cases} \tag{15}$$

However, approximation with the Haar basis is not very accurate and lacks smoothness (Strang, 1990). In recent years, some more powerful wavelet bases have been developed. Strömberg (1981) constructed the first wavelet in recent literature, based on splines.

Meyer's wavelets are constructed in the frequency domain by specifying desired frequency characteristics (Meyer, 1985). They have compact support in the frequency domain. Meyer's wavelets are orthonormal and are C^∞-functions, i.e., infinitely differentiable functions. However, these wavelets do not have compact support in the time domain and the two-scale coefficients do not decay very fast.

Battle-Lemarié spline wavelets (Battle, 1987; Lemarié, 1988) are constructed by orthonormalization of B-spline functions. They are orthonormal and C^k-functions, but do not have compact support, although their decay is fast.

Daubechies' wavelets (1988) have compact support and are orthonormal. These bases are C^λ-functions, i.e., Hölder continuous functions with continuity index λ. Daubechies wavelets do not have an explicit or analytic mathematical expression, but are constructed through a recursive algorithm.

Chui-Wang's B-spline wavelets (Chui and Wang, 1991, 1992a, 1992b, 1993; Chui, 1992a) have compact support and are in C^k. They are semi-orthogonal. Both Chui-Wang wavelets and their duals have analytic expressions. However, the dual wavelets do not have compact support.

Biorthogonal spline wavelets have also been developed (Cohen, 1992; Cohen et al, 1992). These bases do not have orthogonality or semi-orthogonality properties. The bases and their duals satisfy the biorthogonality condition given in Eq. (9). These wavelets have compact support and analytic expressions. The dual wavelets also have compact support but do not have analytic expressions.

More wavelet bases have been developed (see Jawerth and Sweldens, 1993, for a review) to suit specific requirements. Here we will give more information about Daubechies wavelets and Chui-Wang wavelets which we will use in the sequel.

3.1. Daubechies' Wavelet Bases

The construction of Daubechies wavelets starts from finding a sequence with finite non-zero terms, $p(k)$, $k = 0, \cdots, 2N - 1$, for the two-scale relation of the scaling function (Eq. 1), where N is an index of the wavelet functions. N determines the number of non-zero coefficients, the support, and the regularity of the wavelet functions. $N = 1$ leads to the Haar wavelet given by Eq. (15). Daubechies (1988, 1992) gave the numerical values of the reconstruction and decomposition sequences for different N (not unique). For low order bases, the reconstruction sequences have explicit expressions. For example, the non-zero two-scale coefficients $\{p(k)\}$ for the scaling function ϕ are,

$$N=2 \qquad \begin{cases} p(0)=(1+\sqrt{3})/4, & p(1)=(3+\sqrt{3})/4 \\ p(2)=(3-\sqrt{3})/4, & p(3)=(1-\sqrt{3})/4 \end{cases}$$

$$N=3 \qquad \begin{cases} p(0)=(1+\sqrt{10}+\sqrt{5+2\sqrt{10}})/16 \\ p(1)=(5+\sqrt{10}+3\sqrt{5+2\sqrt{10}})/16 \\ p(2)=(10-2\sqrt{10}+2\sqrt{5+2\sqrt{10}})/16 \\ p(3)=(10-2\sqrt{10}-2\sqrt{5+2\sqrt{10}})/16 \\ p(4)=(5+\sqrt{10}-3\sqrt{5+2\sqrt{10}})/16 \\ p(5)=(1+\sqrt{10}-\sqrt{5+2\sqrt{10}})/16 \end{cases}$$

Since Daubechies wavelets are orthonormal, the non-zero two-scale coefficients $\{p(k)\}$ for the wavelet function ψ (Eq. 9) are given by,

$$q(k)=(-1)^k p(1-k), \qquad k=2-2N, \ \cdots, \ 1 \tag{16}$$

Note that these two sequences, $\{p(k)\}$ and $\{q(k)\}$, have only a finite number of non-zero terms. The decomposition sequences are given by,

$$\begin{aligned} g(k) &= p(-k), \quad k=0, \ \cdots, \ 2N-1 \\ h(k) &= q(-k), \quad k=2-2N, \ \cdots, \ 1 \end{aligned} \tag{17}$$

The support for the scaling and the wavelet functions are

$$\begin{aligned} \text{supp } \phi_{n,k} &= [2^{-n}k, 2^{-n}(k+2N-1)] \\ \text{supp } \psi_{n,k} &= [2^{-n}(k+1-N), 2^{-n}(k+N)] \end{aligned}$$

These bases also have the continuity property, $\phi_{n,k}(x)$, $\psi_{n,k}(x) \in C^{\lambda_N}$, i.e they belong to the space of Hölder continuous functions with index λ_N, which satisfy

$$\left|f^{(k)}(x) - f^{(k)}(x+h)\right| \le K\left|h\right|^{\alpha}$$

where $k=[\lambda_N]$; $0 \le \alpha = \lambda_N - k < 1$; and K a positive constant. The value of λ_N is determined by the wavelet order N. The differentiability of these wavelet functions increases with the index N. When $N=2$, the Daubechies wavelet is continuous everywhere, left-differentiable at dyadic points (i.e. at $x = k \cdot 2^{-l}$), and nowhere right-differentiable at dyadic points (Pollen, 1992). Daubechies and Lagarias (1992) estimated the continuity of the bases as, $\lambda_2 \approx 0.55$, $\lambda_3 \approx 1.088$, $\lambda_4 \approx 1.617$. For larger N, $\lambda_N \approx 0.3485N$.

Daubechies wavelet bases also have up to N zero moments, i.e.

$$\mathcal{M}_{\psi} = \int_{-\infty}^{\infty} x^k \psi(x)dx = 0, \qquad k=0, \ \cdots \ , \ N-1 \tag{18}$$

which means that the scaling bases can represent polynomials of degrees up to $N-1$ exactly.

Figs. 1 and 2 present Daubechies' scaling and wavelet functions of order $N=3$, and $N=4$, respectively. From these figures, it can be seen that the regularity of the basis functions increases with the order N. However, the support increases as well.

3.1.1. Computational Issues

Daubechies scaling functions and wavelets do not have analytic expressions, but are implicitly determined from the two-scale coefficients $\{p(k)\}$, and $\{q(k)\}$, and Eqs. (1) and (6) (Daubechies, 1988, 1992, 1993). The sequences $\{p(k)\}$ and $\{q(k)\}$ have only finitely many non-zero terms. Although this fact allows the computation of approximate values of Daubechies bases, exact values of these bases at dyadic points (i.e. $x = k \cdot 2^{-n}$) can be calculated recursively starting from the values at integer points (Strang, 1989; Alpert, 1992) as follows:

Let x take the integer values $1,\ 2,\ \cdots,\ 2N-2$, in Eq. (1). That yields a set of $2N-2$ linear equations with $2N-2$ unknowns $\phi(1),\ \phi(2),\ \cdots,\ \phi(2N-2)$,

$$\phi(j) = \sum_{k=1}^{2N-2} p(k)\phi(2j-k), \quad j = 1,\ 2,\ \cdots,\ 2N-2$$

However, these equations are homogeneous. After replacing any of these equations by

$$\sum_{k=1}^{2N-2} \phi(k) = 1$$

we can obtain the value of the scaling function at the integer points. Then, by using Eq. (1) again, we can calculate the scaling function values at half integer points (these values depend on the values at integer points only). This method can recursively generate the scaling function values at all desired dyadic points.

Scaling functions can also be approximately represented by the *graphical display* method given by Daubechies (1988).

Recently, some efficient algorithms involving calculations with Daubechies wavelets have been developed. Latto et al. (1992) developed an exact method for evaluating the integral which often appears in Galerkin methods for PDEs,

$$\mathcal{W}(l) = \int_{-\infty}^{\infty} \phi^{(n_1)}(x)\phi^{(n_2)}(x-l)dx \tag{19}$$

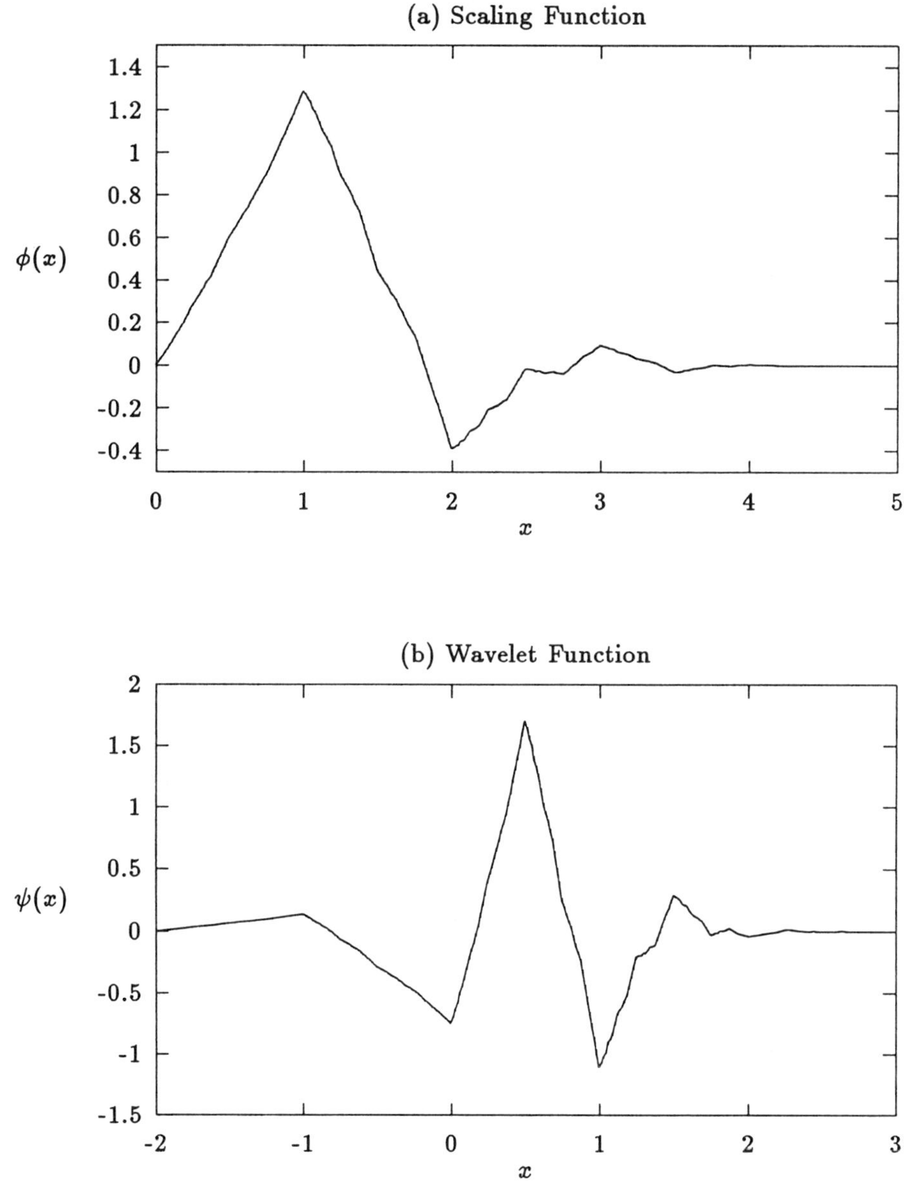

Figure 1. Daubechies' Scaling/Wavelet Basis: $N = 3$.

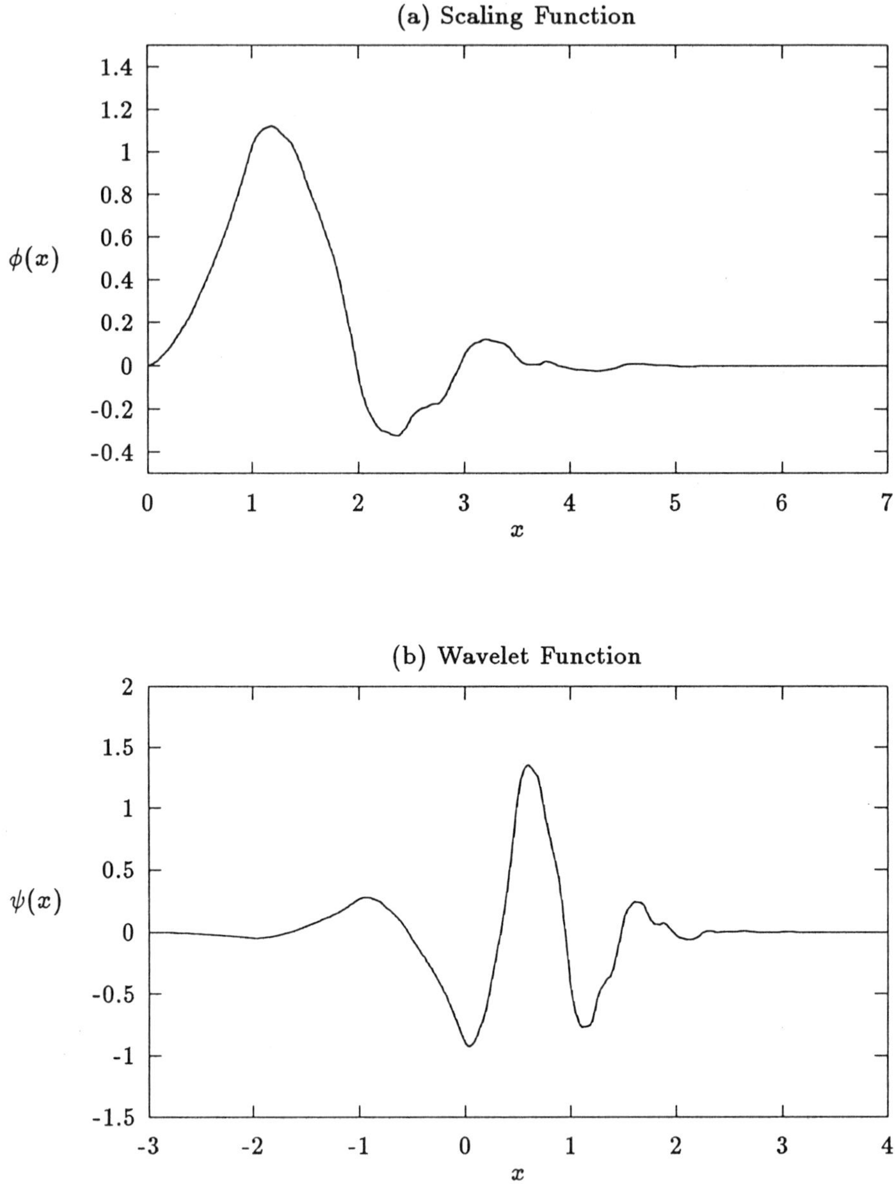

Figure 2. Daubechies' Scaling/Wavelet Basis: $N = 4$.

where $\mathcal{W}(l)$ are called *connection coefficients*, and n_1 and n_2 are the orders of derivatives of the corresponding basis. Their method is based on the two-scale relationship of the basis functions as well as the moment properties of Daubechies wavelets. Using the two-scale relationship of the scaling functions, i.e. Eq. (1), the above equation (19) can be written as,

$$\mathcal{W}(l) = 2^{(n_1+n_2)} \int_{-\infty}^{\infty} \sum_j p(j)\phi^{(n_1)}(2x-j) \sum_k p(k)\phi^{(n_2)}(2x-2l-k)dx$$

By changing variables and indices, we have,

$$\mathcal{W}(l) = 2^{(n_1+n_2-1)} \sum_j p(j) \sum_k p(j+k-2l)\mathcal{W}(k)$$

The above equation generates a set of homogeneous linear equations of $\mathcal{W}(l)$. We can get an additional nonhomogeneous equation as follows.

$$\mathcal{M}_\phi^r(l) = \int_{-\infty}^{\infty} x^r \phi(x-l)dx, \qquad r = 1,\ 2,\ \cdots$$

Due to the vanishing moment property of the wavelet bases (Eq. 18), polynomials of degree up to $N-1$ can be represented by scaling functions only (wavelet coefficients are zero). Thus, we have,

$$x^r = \sum_k c_k \phi(x-k), \qquad r \le N-1$$

where

$$c_k = \mathcal{M}_\phi^r(k) = \int_{-\infty}^{\infty} x^r \phi(x-k)dx, \qquad r = 1,\ 2,\ \cdots$$

is the moment of the scaling function. Using an appropriate r, say $r = n_2$, differentiating the above equation n_2 times, and then multiplying it with $\phi^{(n_1)}(x)$ and integrating both sides on the real line, we can obtain a nonhomogeneous equation of $\mathcal{W}(j)$ with the scaling moments as coefficients. Combination of this equation with the above homogeneous equations results in a linear system of equations in $\mathcal{W}(l)$. Calculation of the scaling moments also uses the two-scale relation of scaling functions (see Latto et al., 1992, for more details). The computation of integrals of the type in Eq. (19) has also been studied by Dahmen and Micchelli (1993).

More expressions (e.g. integrals including wavelet bases) can be computed by the same method. Such integrals appear very frequently when using the wavelet-Galerkin method. Their values can be calculated exactly in advance to save computation time. Note that integrals involving bases at higher resolution level can be reduced to the form of Eq. (19) by variable transformation. Thus, evaluation of such integrals at different resolution levels requires computation at one level only, and subsequent scaling.

Expansions of smooth functions in terms of Daubechies' scaling function bases for approximation purposes have also been studied. Some quadrature formulas have been developed (Beylkin et al, 1991; Sweldens and Piessens, 1992a, 1992b, 1993). Using these formulas, the scaling function coefficients for approximation of a smooth function $f(x)$, Eq. (4), can be determined by,

$$a_{n,k} = \int_{-\infty}^{\infty} f(x)\phi_{n,k}(x)dx = \sum_{k=0}^{r} d_k f(x_k) \tag{20}$$

where, x_k and d_k are fixed values for the quadrature formula (independent of the function $f(x)$); and r is the number of points used in the formula. Thus, to evaluate the scaling function coefficients for approximation of a smooth function, we only need to evaluate the function itself at a few points.

3.1.2. Variations

Daubechies developed some variations of the original basis (Daubechies, 1988) to have better symmetry, regularity, moments of scaling functions, and frequency properties (Daubechies, 1992, 1993; Cohen and Daubechies, 1993). Better symmetry is obtained by selecting the linear phase solution for the coefficients $\{p(k)\}$ involved in the two-scale equation for the scaling function (Eq. 1). Figure 3 shows the basis functions of $N = 4$. More regularity was obtained with larger support. Bases with vanishing scaling function moments, called *Coiflets*, have also been constructed. In this case, inner products of the scaling functions with other functions are easier to evaluate (Beylkin et al., 1991; Daubechies, 1992). Wavelet bases with better frequency properties have also been constructed (Cohen and Daubechies, 1993). They are special cases of wavelet packets (Coifman and Meyer, 1990).

3.2. Chui-Wang Wavelet Bases

Chui and Wang (1991, 1992a, 1992b, 1993; Chui, 1992) developed semi-orthogonal wavelets based on B-spline functions, which are defined by the recursive convolution formula,

$$N_m(x) = \int_{-\infty}^{\infty} N_{m-1}(x-t)N_1(t)dt = \int_0^1 N_{m-1}(x-t)dt \tag{21}$$

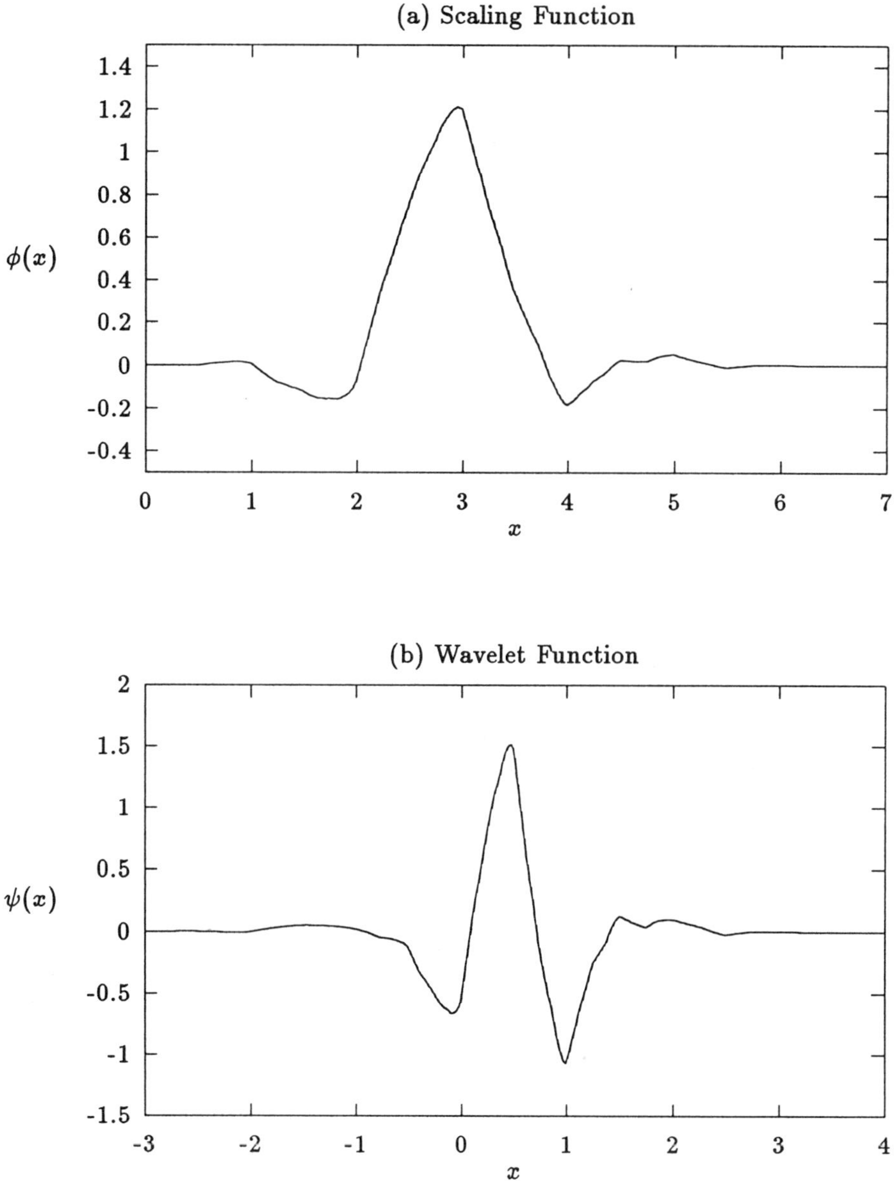

Figure 3. Daubechies Scaling/Wavelet Basis with Least Asymmetry: $N = 4$.

starting from the box function,

$$N_1(x) = \begin{cases} 1, & \text{if } 0 \leq x < 1 \\ 0, & \text{otherwise} \end{cases} \tag{22}$$

It can be shown that $N_m(x)$ is a scaling function that generates an MRA. The two-scale relation for this scaling function is,

$$\phi(x) = N_m(x) = \sum_{k=0}^{m} 2^{-m+1} \binom{m}{k} \phi(2x - k) = \sum_{k=0}^{m} p(k)\phi(2x - k) \tag{23}$$

The wavelet basis is given by

$$\psi(x) = 2^{-m+1} \sum_{k=0}^{2m-2} (-1)^k N_{2m}(k+1) N_{2m}^{(m)}(2x - k) \tag{24}$$

where

$$N_{2m}^{(m)}(x) = \sum_{k=0}^{m} (-1)^k \binom{m}{k} \phi(x - k) \tag{25}$$

Chui-Wang wavelets are semi-orthogonal. The basis and its dual satisfy the bi-orthogonality condition given in Eq. (8). Chui-Wang wavelets are compactly supported, smooth (in C^k), and symmetric. Their drawback is that the dual wavelet and scaling functions do not have compact supports, although they have good decay far from their centers. The reconstruction sequences are thus infinite, but have fast (exponential) decay. The support for the scaling and the wavelet functions are

$$\text{supp } \phi_{n,k} = [2^{-n}k, 2^{-n}(k+m)]$$
$$\text{supp } \psi_{n,k} = [2^{-n}k, 2^{-n}(k+2m-1)]$$

One the the major advantages of Chui-Wang bases is that there exist analytic expressions for scaling functions, wavelet functions and their duals (piece-wise polynomials) (Chui, 1992). Theory on B-splines is well developed (de Boor, 1978; Schoenberg, 1973). Fig. 4 gives the scaling and wavelet bases of order $m = 4$.

For MRA, the reconstruction sequences are the coefficients in the two-scale relations of scaling functions (Eq. 23) and wavelet functions (Eqs. 24, 25), i.e.,

$$\begin{aligned} p(k) &= 2^{-m+1} \binom{m}{k}, \quad k = 0, \cdots, m \\ q(k) &= (-1)^k 2^{-m+1} \sum_{l=0}^{m} \binom{m}{l} N_{2m}(k+1-l), \quad k = 0, \cdots, 3m-2 \end{aligned} \tag{26}$$

(a) Scaling Function

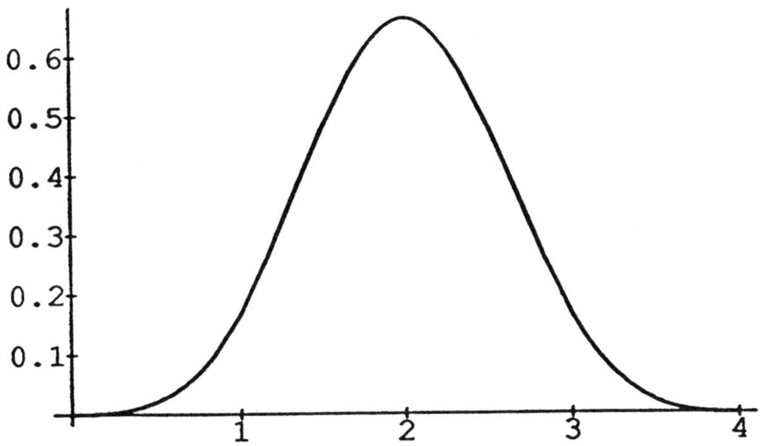

(b) Wavelet Function

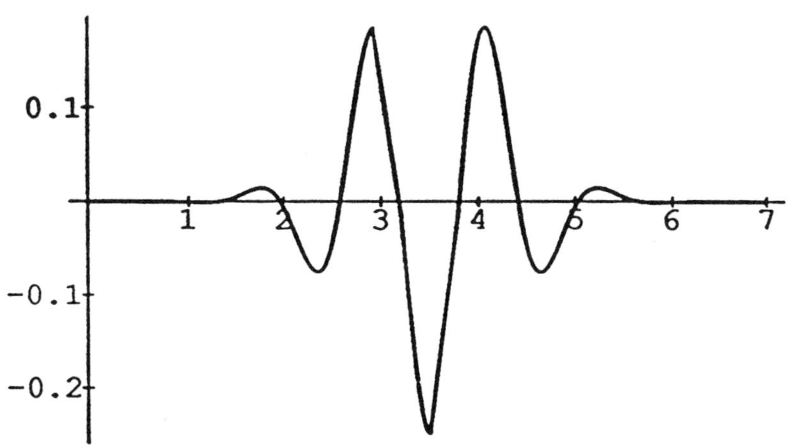

Figure 4. Chui-Wang Scaling/Wavelet Functions: $m = 4$.

for the Chui-Wang basis of mth order. Only a finite number of terms are non-zero. However, the decomposition sequences have infinite non-zero terms, which, however, have fast (exponential) decay. For linear B-wavelets,

$$\frac{1}{2}g(k) = \begin{cases} \frac{1}{2}g(-1) = (1+\sqrt{3})/4 \\ \frac{1}{2}g(2n-1) = \frac{1}{2}g(-2n-1) = \sqrt{3}/4\ (\sqrt{3}-2)^n,\ n = 1,2,\cdots \\ \frac{1}{2}g(2n) = \frac{1}{2}g(-2n-2) = (3-\sqrt{3})/4\ (\sqrt{3}-2)^n,\ n = 0,1,\cdots \end{cases}$$

$$\frac{1}{2}h(k) = \begin{cases} \frac{1}{2}h(-2) = \sqrt{3}/2 \\ \frac{1}{2}h(2n-3) = \frac{1}{2}h(-2n-1) = (3+\sqrt{3})/4\ (\sqrt{3}-2)^n,\ n = 1,2,\cdots \\ \frac{1}{2}h(2n-2) = \frac{1}{2}h(-2n-2) = \sqrt{3}/2\ (\sqrt{3}-2)^n,\ n = 1,2,\cdots \end{cases}$$

The numerical values of these sequences for higher order Chui-Wang wavelets can be found in Chui (1992).

3.3. Multidimensional Wavelet Bases

Multi-dimensional wavelet bases can be constructed from one-dimensional ones by tensor product (Mallat, 1989). For example, two dimensional wavelet bases can be defined from one-dimensional basis as follows. For scaling functions, we have:

$$\Phi_{n,k_1,k_2}(x,y) = \phi \otimes \phi(x,y) = \phi_{n,k_1}(x)\phi_{n,k_2}(y)$$

For wavelets, there are three different sets:

$$\Psi^{(1)}_{n,k1,k2}(x,y) = \phi \otimes \psi(x,y) = \phi_{n,k_1}(x)\psi_{n,k_2}(y)$$

$$\Psi^{(2)}_{n,k1,k2}(x,y) = \psi \otimes \phi(x,y) = \psi_{n,k_1}(x)\phi_{n,k_2}(y)$$

$$\Psi^{(3)}_{n,k1,k2}(x,y) = \psi \otimes \psi(x,y) = \psi_{n,k_1}(x)\psi_{n,k_2}(y)$$

Two-dimensional Daubechies' scaling and wavelet functions of order $N = 4$ as defined above shown given in Fig. 5.

Nonseparable multivariate wavelet bases have also been developed recently (Riemenschneider and Shen, 1992; Stöckler, 1992; Cohen and Schlenker, 1993; de Boor, DeVore, and Ron, 1993).

4. Wavelet-Galerkin Method

In this section, we will discuss use of wavelet bases with Galerkin's method. Then several numerical examples will be given to illustrate the wavelet-Galerkin method.

(a) $\Phi\ (x, y)$

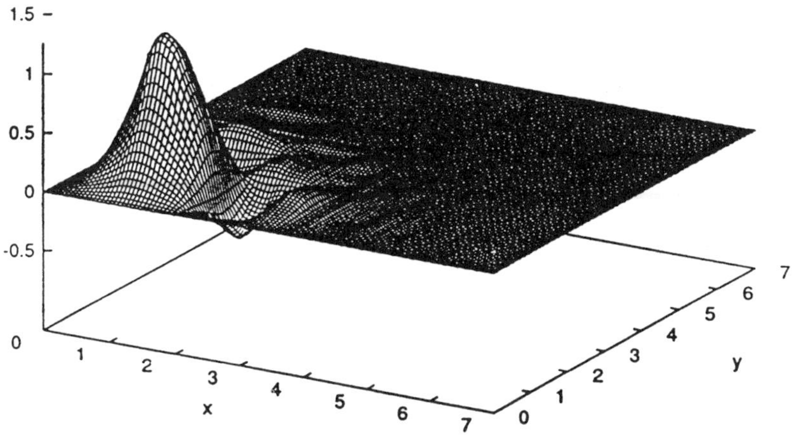

(b) $\Psi^{(1)}\ (x, y)$

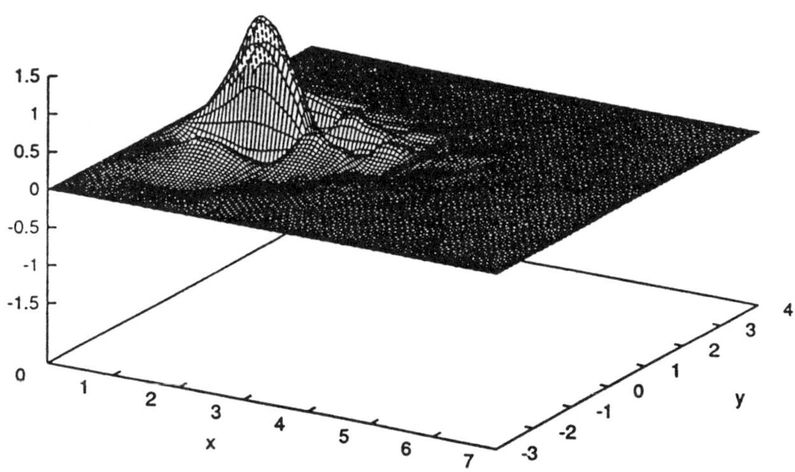

Figure 5. Two-Dimensional Daubechies Basis: $N = 4$.

(c) $\Psi^{(2)}\ (x, y)$

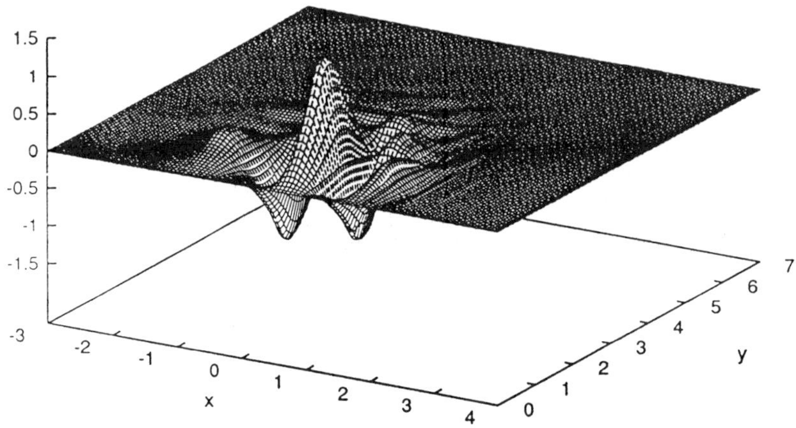

(d) $\Psi^{(3)}\ (x, y)$

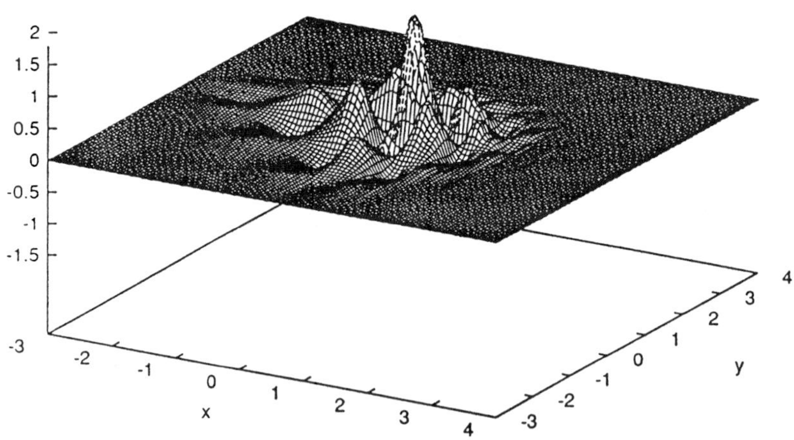

Figure 5. Continued.

4.1. The Method of Weighted Residuals and Galerkin's Method

In the method of weighted residuals (MWR), the solution of a differential equation, $u(x,t)$, is approximated by $\tilde{u}(x,t)$ represented by a finite series of functions, $\phi_i(x)$, as follows:

$$u(x,t) \approx \tilde{u}(x,t) = \sum_{i=1}^{p} a_i(t)\phi_i(x) \tag{27}$$

where $\{\phi_i(x)\}$ are called basis or trial functions and $\{a_i(t)\}$ are coefficients (possibly time dependent) to be determined to satisfy a given DE; p is the number of basis functions. In general, the approximate solution, $\tilde{u}(x,t)$, can not satisfy the original equation exactly and substitution of the approximate solution into the original DE results in a residual. Using Eq. (27) in the differential equation

$$\mathcal{L}u(x,t) - f(x,t) = 0, \quad x \in \Omega,\ t > 0 \tag{28}$$

we obtain

$$\mathcal{L}\tilde{u}(x,t) - f(x,t) = R(x,t) \tag{29}$$

where $\mathcal{L}$ is a differential operator, and $R(x,t)$ is a residual function. Thus, the coefficients $\{a_i(t)\}$ in Eq. (27) should be determined such that the residual is minimized in some sense.

The method of weighted residuals minimizes $R(x,t)$ by forcing it to zero in the domain Ω using *weighting (test) functions*, $w_j(x)$, such that, for every weighting function,

$$\int_{\Omega} R(x,t)w_j(x)dx = 0, \qquad \text{for } j = 1, \cdots, q$$

where q is the number of weighting functions to be determined by the type of boundary conditions and the number of basis functions, p. This discretization, in general, leads to a linear or nonlinear system of equations, from which the coefficients $\{a_i(t)\}$ in the approximate solution can be determined.

Choosing different weighting functions results in different schemes. The mostly used methods are the Galerkin, collocation, subdomain, and least squares methods (Finlayson, 1972; Fletcher, 1984; Lapidus and Pinder, 1982; Celia and Gray, 1992).

In Galerkin's method, the weighting functions are chosen to be the basis functions, i.e., $w_j(x) = \phi_j(x)$. Thus we have,

$$\int_{\Omega} R(x,t)\phi_j(x)dx = \int_{\Omega} [\mathcal{L}\tilde{u}(x,t) - f(x,t)]\phi_j(x)dx = 0, \quad j = 1, \cdots, q$$

Basis functions used in the MWR are usually linear or piece-wise linear, quadratic polynomials, cubic polynomials, Lagrange polynomials, Hermite polynomials, Chebyshev polynomials, etc. (Fletcher, 1984; Celia and Gray, 1992; Lapidus and Pinder, 1982).

In this work, we will use scaling functions and wavelet functions as the Galerkin bases. As mentioned earlier, wavelets may have orthonormality or semi-orthogonality, compact support, have multi-level structure, and are unconditional bases for L^2 spaces, Sobolev spaces, etc. (Daubechies, 1988, 1992); thus they can provide accurate approximation of solutions of DEs (Glowinski et al., 1990). Our target, in the sequel, is to finally show how wavelets can be used in adaptive grid methods.

Using scaling functions as bases, an approximate solution at resolution level n can be expressed as,

$$u(x,t) \approx \mathcal{A}_n u(x,t) = \sum_k a_{n,k}(t)\phi_{n,k}(x)$$

Then the residual is,

$$R_n(x,t) = (\mathcal{L}\mathcal{A}_n u)(x,t) - f(x,t)$$

By applying the standard Galerkin method with weighting functions, $\{\phi_{n,j}(x)\}$, we get,

$$\int_\Omega [\mathcal{L}(\mathcal{A}_n u) - f]\,\phi_{n,j}\,dx = 0$$

This is the general equation when scaling functions are used as bases with Galerkin's method. Boundary conditions may impose some constraints on the solution and result in additional equations for the unknown coefficients $\{a_{n,k}\}$ (Gottlieb and Orszag, 1977).

Several numerical examples will be given in the sequel to illustrate the details of how to use this method.

4.2. Initial Value Problems

In this section, we will give some numerical examples of using the wavelet-Galerkin method. Daubechies and Chui-Wang wavelets will be used.

Consider an IVP in the general form

$$\begin{aligned} &\frac{du}{dx} + p(x)u = q(x), \qquad 0 < x \le 1 \\ &u(0) = u_0 \end{aligned} \tag{30}$$

Looking for approximate solution of the problem at resolution level n, we assume

$$u(x) \approx \mathcal{A}_n u(x) = \sum_k a_{n,k}\phi_{n,k}(x) \tag{31}$$

where $\{a_{n,k}\}$ are unknown coefficients; $\{\phi_{n,k}\}$ are Daubechies' scaling functions with index N. Recall that Daubechies scaling functions have compact support, given by,

$$\text{supp } \phi_{n,k} = \left[k \cdot 2^{-n},\ (2N-1+k)2^{-n}\right]$$

Thus, for the domain [0, 1], the only non-zero terms (k) in the summation in eqn. (31) range from $2-2N$ to 2^n-1. This gives a total of $p = 2^n + 2N - 2$ unknown coefficients in the approximate solution.

Using this approximate solution, the initial condition becomes,

$$\mathcal{A}_n u(0) = \sum_{k=2-2N}^{2^n-1} a_{n,k}\phi_{n,k}(0) = u_0 \tag{32}$$

Using Galerkin's method, with $\phi_{n,j}$ as test functions, gives,

$$\int_0^1 [\frac{d\mathcal{A}_n u}{dx} + p(x)\mathcal{A}_n u]\,\phi_{n,j}(x)dx = \int_0^1 q(x)\phi_{n,j}(x)dx$$

where j ranges from $3-2N$, to 2^n-1, because we already have one equation in (32). We can rewrite this equation in matrix form as,

$$\mathbf{K}\mathbf{a_n} = \mathbf{f} \tag{33}$$

where $\mathbf{a_n}$ is the vector containing the unknown coefficients:

$$\mathbf{a_n} = \left[a_{n,2-2N},\ a_{n,3-2N},\ \cdots,\ a_{n,2^n-1}\right]^T \tag{34}$$

and the elements of the vector $\mathbf{f}$ and matrix $\mathbf{K}$ are

$$f_j = \int_0^1 q(x)\phi_{n,j}(x)dx$$

$$K_{j,k} = \int_0^1 \left[\phi_{n,j}(x)\phi'_{n,k}(x) + p(x)\phi_{n,j}\phi_{n,k}\right]dx$$

for $j, k = 3-2N,\ \cdots,\ 2^n-1$. From eqn. (32) we have,

$$K_{2-2N,k} = \phi_{n,k}(0), \qquad f_{2-2N} = u_0$$

Thus we have a linear system of p equations of the p unknown coefficients, namely, $\{a_{n,k}\}$. We can obtain the coefficients of the approximate solution, $\mathcal{A}_n u(x)$, by solving this linear system.

4.2.1. Example 1. An Initial Value Problem (IVP).

As a numerical example of an initial value problem, Eq. (30) is solved with $p(x) = q(x) = 1$, $u_0 = 2$. Daubechies orthonormal basis of order three ($N = 3$) is used in the approximate solution shown in Fig. 6. The approximate solution is obtained at resolution level $n = 0$ and the total number of coefficients is 5. Fig. 7 gives the corresponding error distribution of the approximate solution Fig. 8 shows the error distribution using $N = 3$, $n = 2$, and $p = 8$. Fig. 9 presents an error measure versus number of unknowns (i.e. grid points) for order (N) 2, 3, and 4 bases. The error measure used here is the standard l_2-norm of the errors at 21 evenly distributed points on [0, 1]. The approximation levels range from 0 to 5 (the corresponding number of unknowns are from 7 to 38). Note that for fixed N, refinement of the grid (i.e. increasing the number of coefficients) initially improves the accuracy of the solution but quickly brings the error to a plateau.

Chui-Wang's cardinal B-spline bases are also used for this problem. In this case, the range of k in Eq. (31) is $[1-m,\ 2^n-1]$, with the total number of unknowns at level n being $p = 2^n + m - 1$. Fig. 10 gives the error distribution with $m = 4$, $n = 2$, and $p = 7$. The same error measure versus number of coefficients for $m = 2$ and $m = 4$ is given in Fig. 11.

It should be remarked that in using Chui-Wang B-spline functions of order $m = 2$ at level 0, the basis is, up to a constant, exactly the same as the piece-wise linear Lagrange polynomials defined on the discretization points, 0, 0.5, and 1 (Celia and Gray, 1992).

4.3. Two Point Boundary Value (TPBV) Problems

We consider the problem

$$\begin{aligned} &\frac{d^2u}{dx^2} + f(x)\frac{du}{dx} + p(x)u = q(x), \qquad 0 < x < 1 \\ &u(0) = u_0, \qquad u(1) = u_1 \end{aligned} \tag{35}$$

Again, we assume a scaling basis approximation as in Eq. (31). Applying Galerkin's method we have,

$$\int_0^1 \phi_{n,j}\left[\frac{d^2\mathcal{A}_n u}{dx^2} + f(x)\frac{d\mathcal{A}_n u}{dx} + p(x)\mathcal{A}_n u - q(x)\right]dx = 0$$

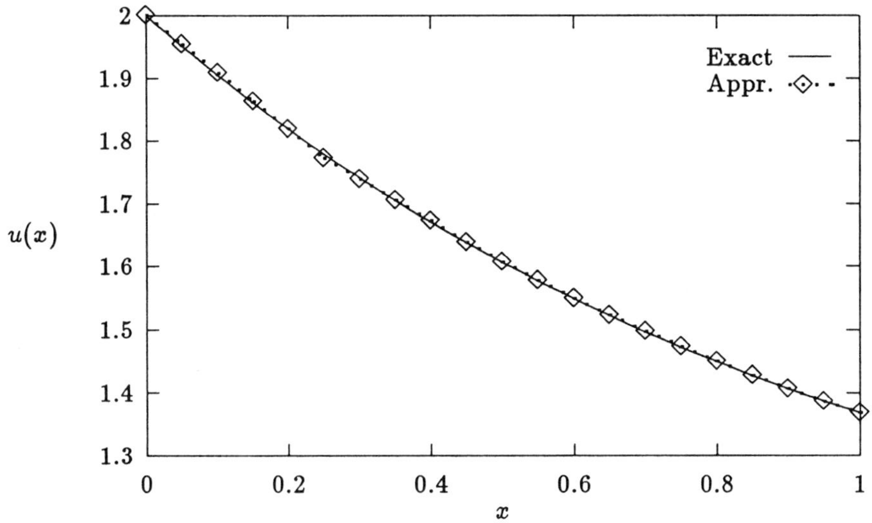

Figure 6. Initial Value Problem: $N = 3,\ n = 0,\ p = 5.$

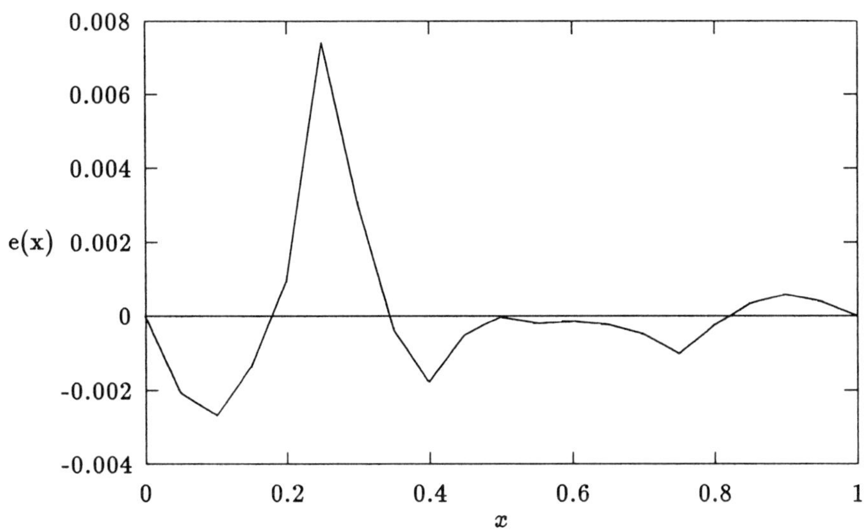

Figure 7. Error Distribution: $N = 3,\ n = 0,\ p = 5.$

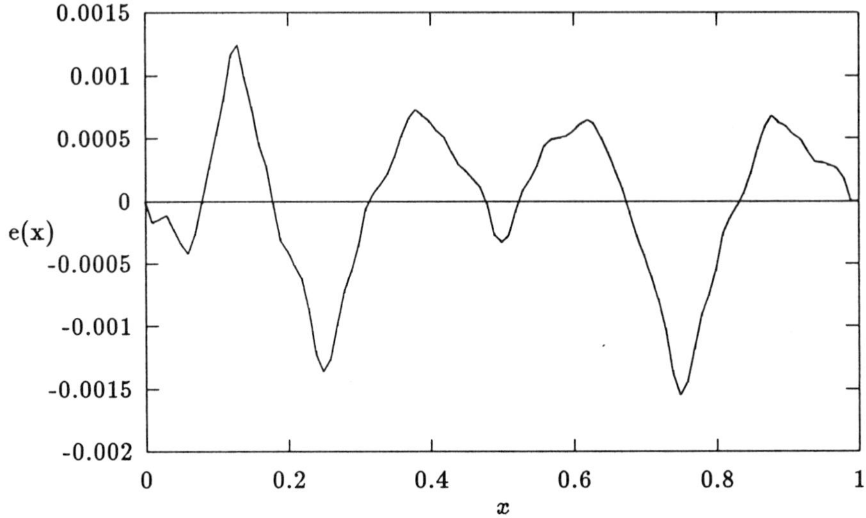

Figure 8. Error Distribution: $N = 3,\ n = 2,\ p = 10$.

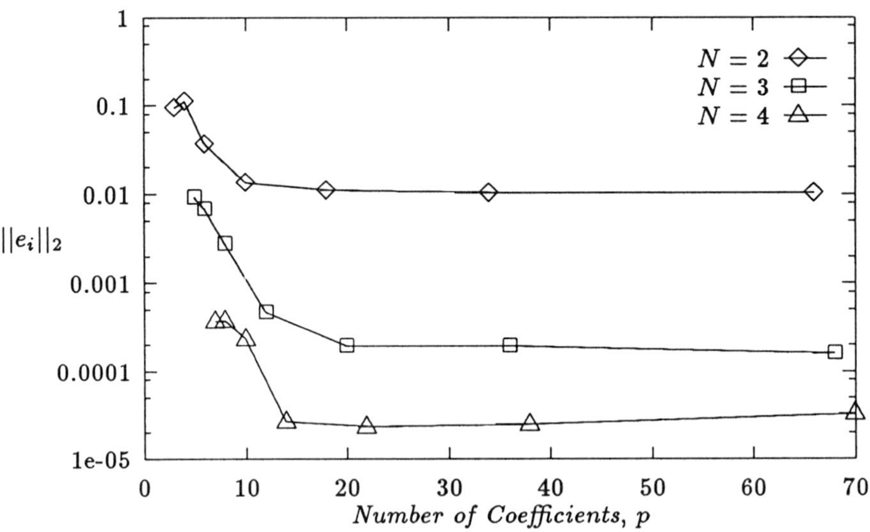

Figure 9. Number of Coefficients vs. Error Measure.

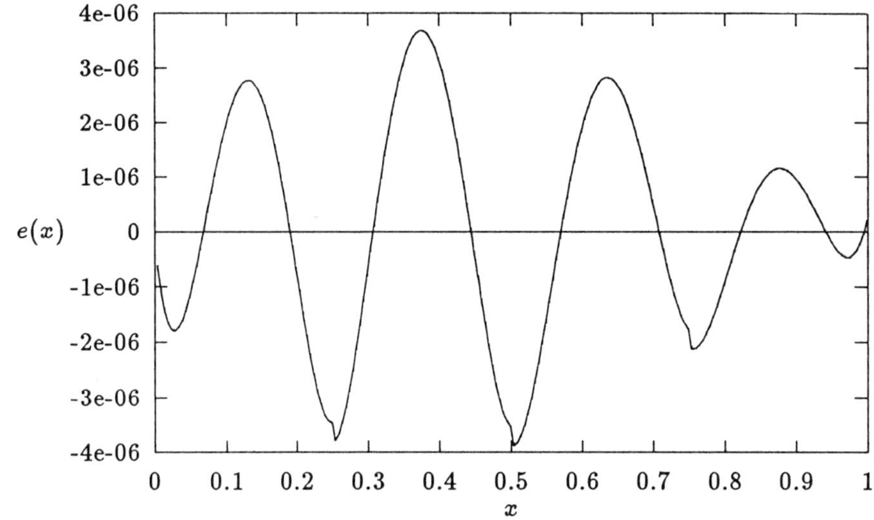

Figure 10. Error Distribution: $m = 4$, $n = 2$, $p = 7$.

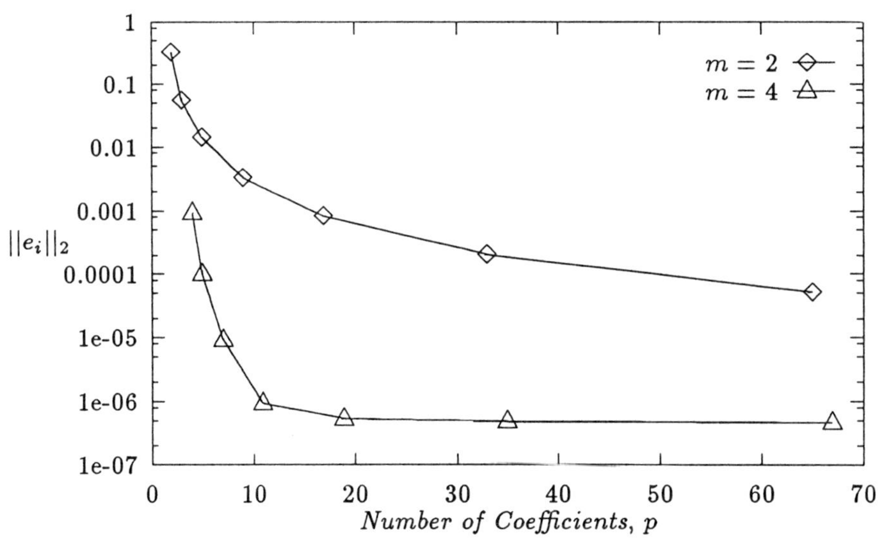

Figure 11. Number of Coefficients vs. Error Measure.

Using integration by parts, we can obtain the following linear equation,

$$\mathbf{K a_n} = \mathbf{f} \tag{36}$$

where $\mathbf{a_n}$ is the vector of unknown coefficients given in (34). The vector $\mathbf{f} = \{f_j\}$ and the matrix $\mathbf{K} = \{K_{j,k}\}$ are defined by,

$$f_j = \int_0^1 q(x)\phi_{n,j}(x)dx$$

$$K_{j,k} = \int_0^1 \left[-\phi'_{n,j}\phi'_{n,k} + f(x)\phi_{n,j}\phi'_{n,k} + p(x)\phi_{n,j}\phi_{n,k}\right]dx + \phi_{n,j}\phi'_{n,k}\Big|_0^1$$

for $3 - 2N \leq j, k \leq 2^n - 2$. Boundary conditions give two additional equations for $\{a_{n,k}\}$,

$$K_{2-2N,k} = \phi_{n,k}(0), \qquad f_{2-2N} = u_0$$

$$K_{2^n-1,k} = \phi_{n,k}(1), \qquad f_{2^n-1} = u_1$$

for $k = 2 - 2N$ to $2^n - 1$. Thus, we end up with a linear system of p equations with p unknowns. The coefficients can be obtained by solving this system, thus the approximate solution, $\mathcal{A}_n u$, of the original TPBV problem.

4.3.1. Example 2. A TPBV Problem.

Heat conduction at steady state is described by the following TPBV problem,

$$\begin{aligned} \frac{d^2T}{dx^2} &= U(T - T_a), \qquad 0 < x < 1 \\ T_a &= \begin{cases} 25, & \text{if } 0 \leq x \leq 0.45; \\ 300, & \text{if } 0.45 < x < 0.55; \\ 25, & \text{if } 0.55 \leq x \leq 1; \end{cases} \\ T(0) &= 100, \qquad T(1) = 0 \end{aligned} \tag{37}$$

Note the profile of T_a. The results for $N = 4$; $n = 3$, $n = 4$, and $n = 5$ are given in Figs. 12, 13, and 14 respectively. The corresponding numbers of coefficients are $p = 14$ for $n = 3$, $p = 22$ for $n = 4$, and $p = 38$ for $n = 5$. Fig. 15 shows error measure versus number of unknows. Here the error measure is the l_2-norm of the errors at 101 evenly distributed points on the domain.

Figs. 16 and 17 give the results when the Chui-Wang B-spline basis of order four ($m = 4$) is used. Fig. 16 gives the results at level $n = 4$, $p = 19$. In Fig. 17, the values of the same error measure used with Daubechies wavelets are given for different number of coefficients.

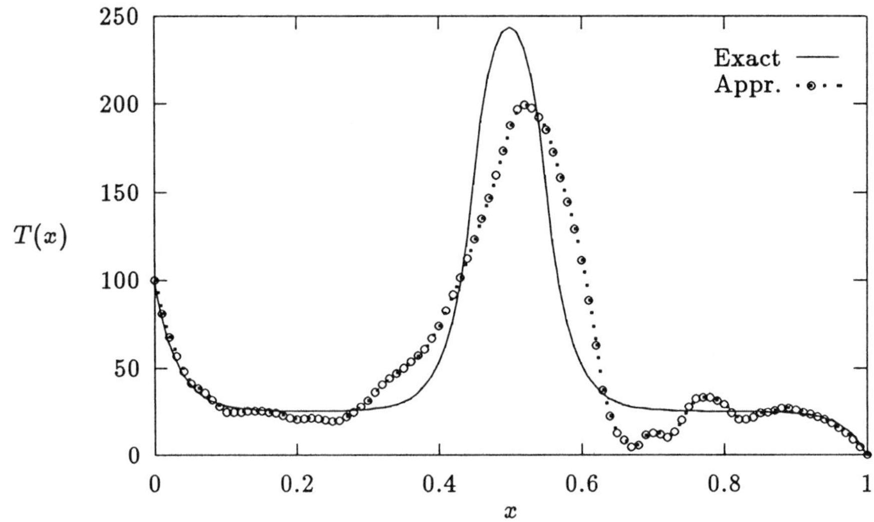

Figure 12. TPBV Problem: $N = 4,\ n = 3,\ p = 14.$

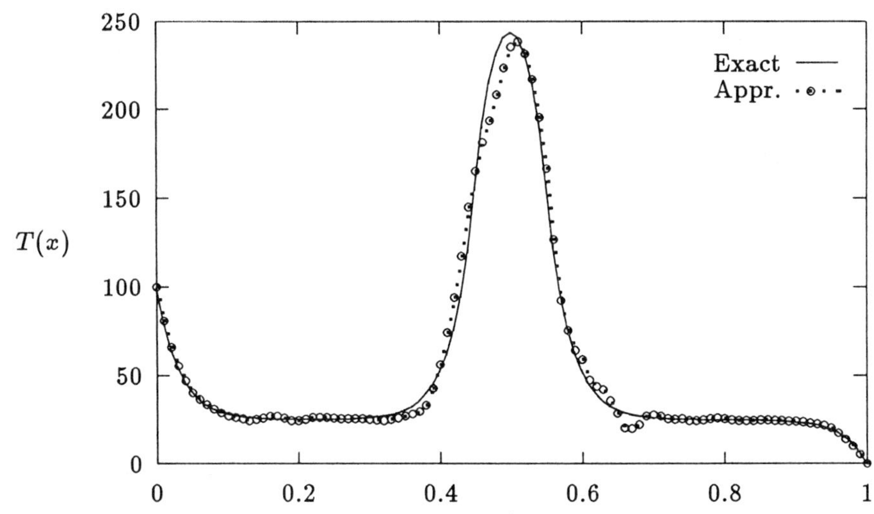

Figure 13. TPBV Problem: $N = 4,\ n = 4,\ p = 22.$

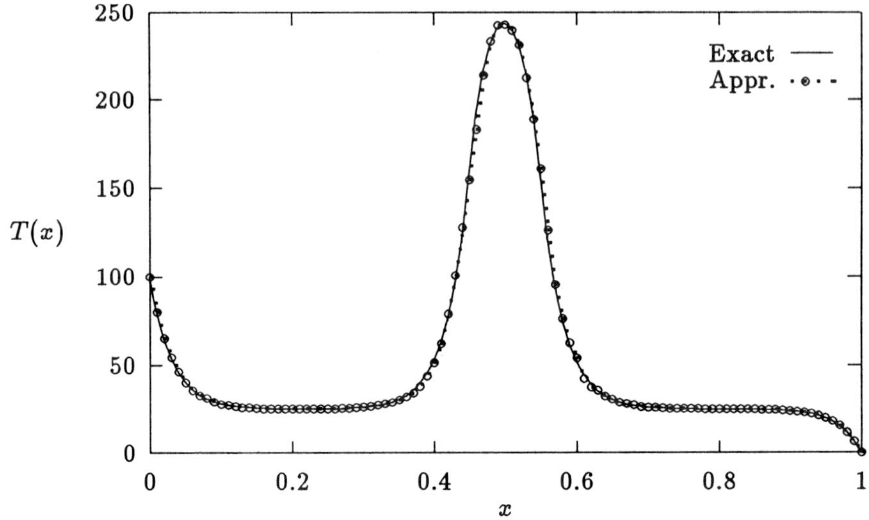

Figure 14. TPBV Problem: $N = 4$, $n = 5$, $p = 38$.

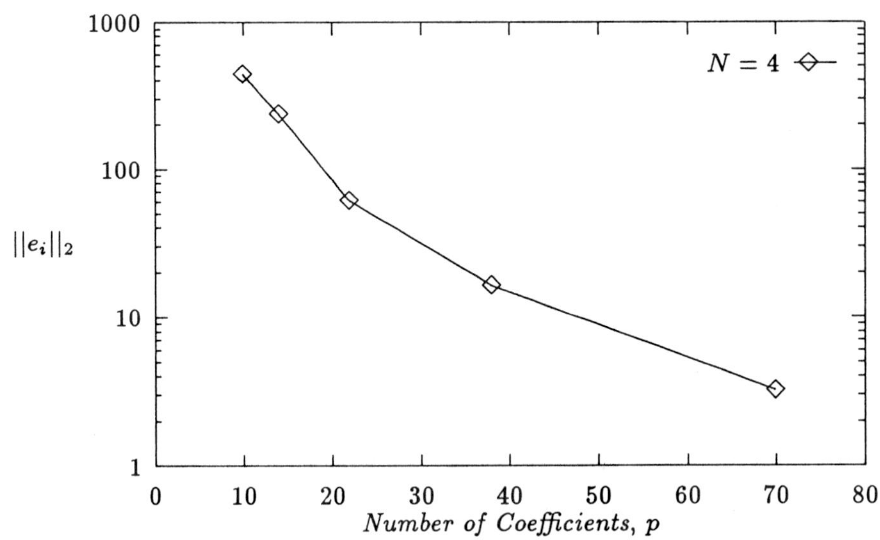

Figure 15. Number of Coefficients vs. Error Measure.

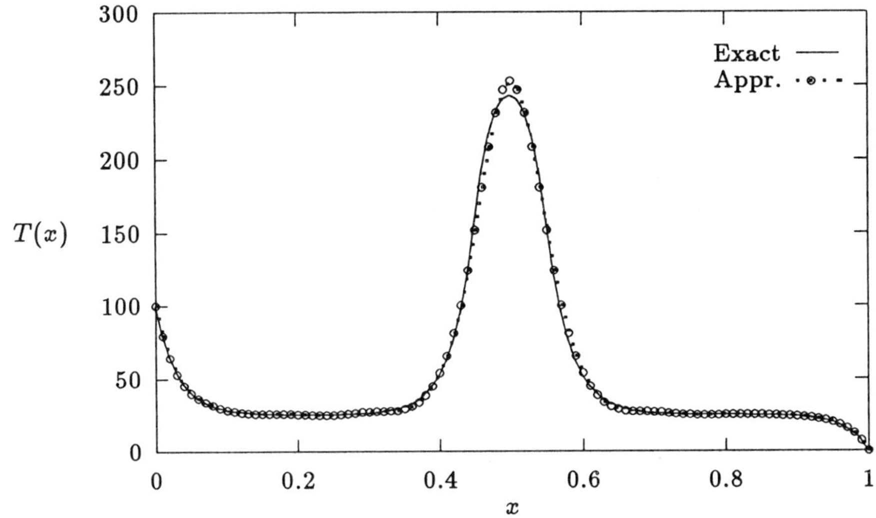

Figure 16. Error Distribution: $m = 4$, $n = 4$, $p = 19$.

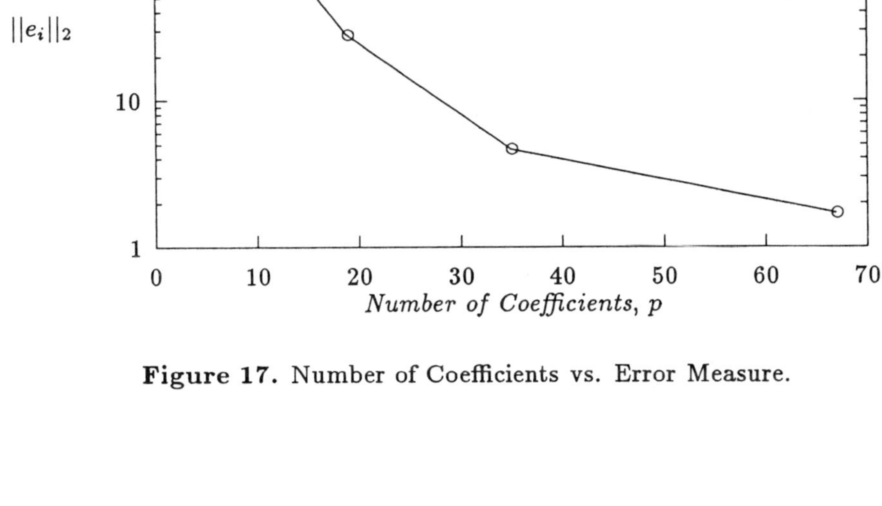

Figure 17. Number of Coefficients vs. Error Measure.

4.4. Burgers' Equation

The general form of Burgers' equation is given by,

$$u_t + uu_x - \nu u_{xx} = 0 \tag{38}$$

where ν is a viscosity-like parameter which can also be considered as the reciprocal of the Reynolds number in the dimensionless case; u_t is a time-varying term indicating no steady-state nature; uu_x is a nonlinear convective term; and νu_{xx} is a dissipative term. Depending on what terms it contains and values of the Reynolds number, Burgers' equation may exhibit different behaviors. Thus, Burgers' equation is considered as a typical problem for testing numerical methods.

4.4.1. Example 3. Burgers' Equations

Consider the following Burgers' equation (Fletcher, 1982),

$$u_t + uu_x - \frac{1}{Re} u_{xx} = 0, \qquad -1 < x < 1 \tag{39}$$

with the boundary conditions

$$u(-1,t) = 1, \qquad u(1,t) = 0 \tag{40}$$

and initial condition

$$u(x,0) = \begin{cases} 1, & \text{if } -1 \le x \le 0 \\ 0, & \text{if } 0 < x \le 1 \end{cases} \tag{41}$$

To solve this problem, we first transform the equation and the boundary and initial conditions to the domain [0, 1]. Let

$$u(x,t) \approx \mathcal{A}_n u(x,t) = \sum_{k=2-2N}^{2^n-1} a_{n,k}(t)\phi_{n,k}(x) \tag{42}$$

Then Galerkin's method gives,

$$\int_0^1 \phi_{n,j}(x)\left[\frac{\partial \mathcal{A}_n u}{\partial t} + \frac{1}{2}\mathcal{A}_n u \frac{\partial \mathcal{A}_n u}{\partial x} - \frac{1}{4Re}\frac{\partial^2 \mathcal{A}_n u}{\partial x^2}\right] dx = 0$$

Assume the domain of this equation can be extended to infinity. Apply integration by parts and use boundary conditions, to get

$$\frac{d\mathbf{a_n}(t)}{dt} = [\mathbf{M}(t) + \mathbf{K}]\mathbf{a_n}(t) \tag{43}$$

where $\mathbf{a_n}(t)$ can be regarded as a vector defined by,

$$\mathbf{a_n}(t) = [a_{n,2-2N}(t),\ a_{n,3-2N}(t),\ \ldots,\ a_{n,2^n-1}(t)]^T,$$

and the matrices $\mathbf{M}$ $=\{M_{j,k}\}$ and $\mathbf{K}$ $=\{M_{j,k}\}$ are given by,

$$M_{j,k}(t) = -\frac{1}{2}\sum_{m=2-2N}^{2^n-1} a_{n,m}(t)\int \phi_{n,j}\phi_{n,k}\phi'_{n,m}dx$$

$$K_{j,k} = -\frac{1}{4Re}\int \phi'_{n,j}(x)\phi'_{n,k}(x)dx$$

for $j,k = 2-2N,\ 2^n-1$. This is an ordinary differential equation for the unknown coefficients $\{a_{n,k}(t)\}$. The initial values for the coefficients are given by,

$$a_{n,k}(0) = \langle u(x,0), \phi_{n,k}(x)\rangle = \int_0^1 u(x,0)\phi_{n,k}(x)dx$$

where $k \in [2-2N,\ 2^n-1]$. The boundary conditions,

$$\begin{aligned} \mathcal{A}_n u_x(0,t) &= \sum_{k=2-2N}^{2^n-1} a_{n,k}(t)\phi_{n,k}(0) = 1 \\ \mathcal{A}_n u_x(1,t) &= \sum_{k=2-2N}^{2^n-1} a_{n,k}(t)\phi_{n,k}(1) = 0 \end{aligned} \tag{44}$$

are used during the solution of Eq. (43) to adjust the coefficients to satisfy the boundary conditions.

With different Reynolds numbers, the solution of Burgers' equation will have different shapes. Fig. 18 gives the results for $Re = 1$ at different time points. (In the Figures for Burgers equation, the exact solution is given only for the latest time step.) The parameters $N = 3$, $n = 3$, and $p = 12$ are used in this simulation.

Fig. 19 gives the approximate solution when $Re = 10$. The values $N = 3$, $n = 3$, $p = 12$ are used.

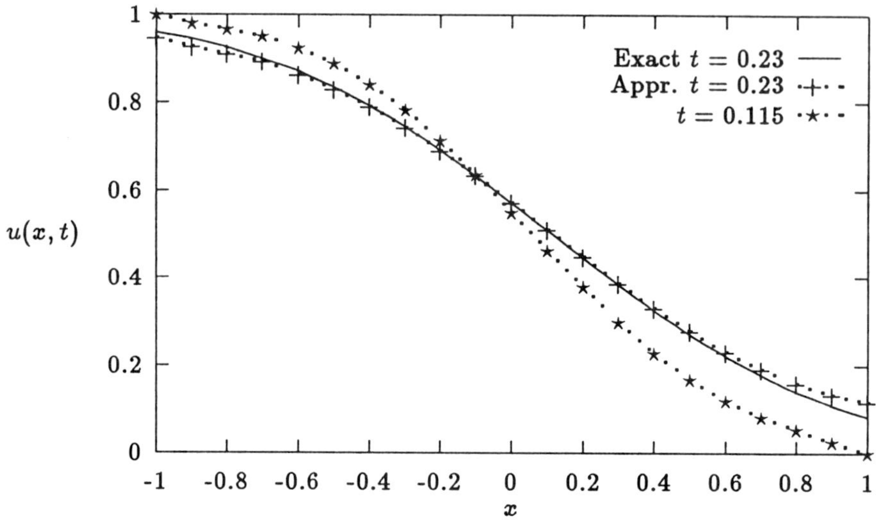

Figure 18. Burgers' Equation $(Re = 1): \ N = 4, \ n = 3, \ p = 14.$

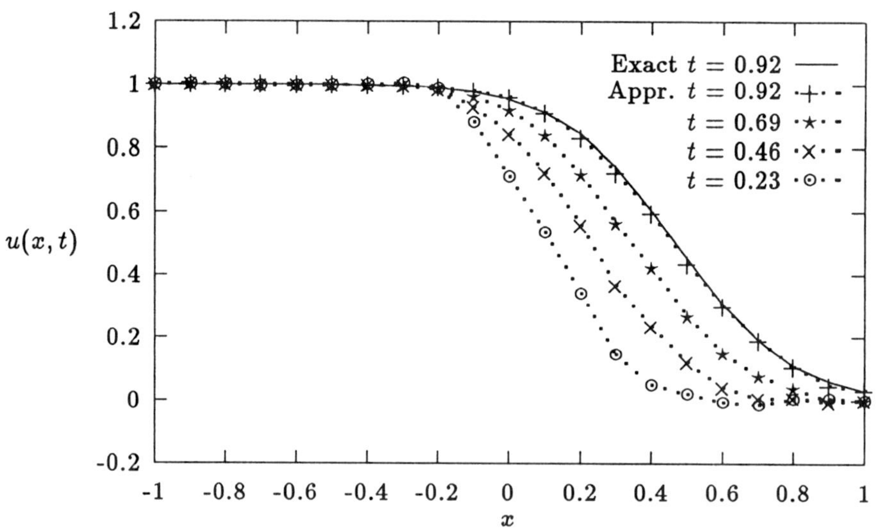

Figure 19. Burgers' Equation $(Re = 10): \ N = 4, \ n = 3, \ p = 14.$

Results for $Re = 100$ are given in Figs. 20 and 21, where the values $N = 4$, $n = 4$, $p = 22$ and $N = 4$, $n = 5$, $p = 38$ are used, respectively. We can see that the oscillation around the wave front is reduced.

4.4.2. Comparison with Chebyshev Polynomials as Basis Functions

Fletcher (1982) solved Eqs. (39-41) with Chebyshev polynomials as the basis functions. Chebyshev polynomials (Dahlquist and Bjorck, 1974) are defined by the recursive formula on the domain $x \in [-1,\ 1]$,

$$\begin{cases} T_0(x) = 1 \\ T_1(x) = x \\ T_{n+1}(x) = 2xT_n(x) - T_{n-1}(x), \qquad \text{for } n \geq 1 \end{cases} \tag{45}$$

It is well shown that Chebyshev polynomials results in more accurate results than polynomial trial functions (Fletcher, 1982).

Simulation results with wavelets of different orders at different levels are given in Table 1, along with the results given by Fletcher (1982). The error measure used in the table is the l_2-norm of the errors at 21 evenly distributed points on the domain.

Table 1. *Error Measure of the Wavelet Approximate Solutions of Burgers' Equation Using Daubechies' Bases.*

Basis Order N	Approx. Level n	No. of params p	$Re = 1$, $t = 0.23$	$Re = 10$, $t = 0.92$	$Re = 100$, $t = 0.92$
3	2	8	0.0687	0.1935	1.0246
	3	12	0.0798	0.0594	0.5833
	4	20	1.1257	0.0440	0.1404
4	2	10	0.0350	0.2198	0.9830
	3	14	0.0580	0.0385	0.6366
	4	22	0.0851	0.0341	0.1720
5	2	12	0.0325	0.1558	0.9573
	3	16	0.0473	0.0346	0.6420
	4	24	0.0786	0.0281	0.2909
Chebyshev Poly.†		9	0.0022	0.0083	0.1049

† From Fletcher (1982).

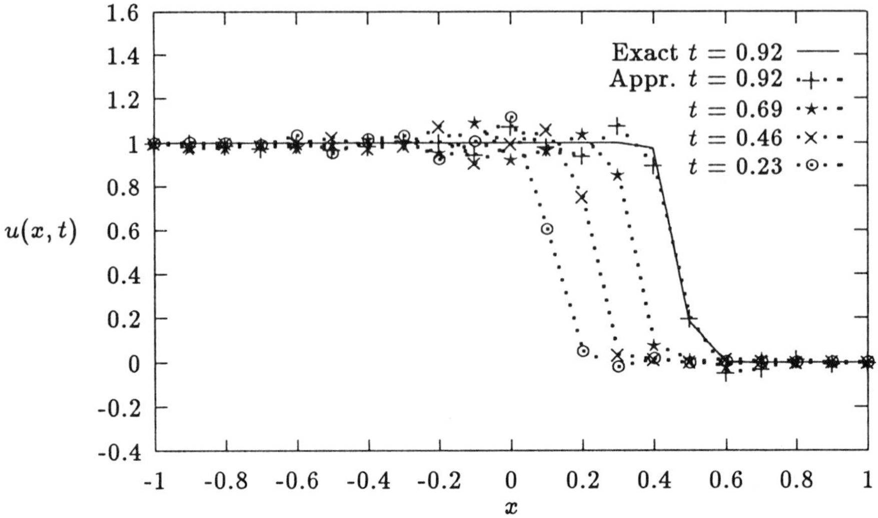

Figure 20. Burgers' Equation ($Re = 100$) : $N = 4,\ n = 4,\ p = 22.$

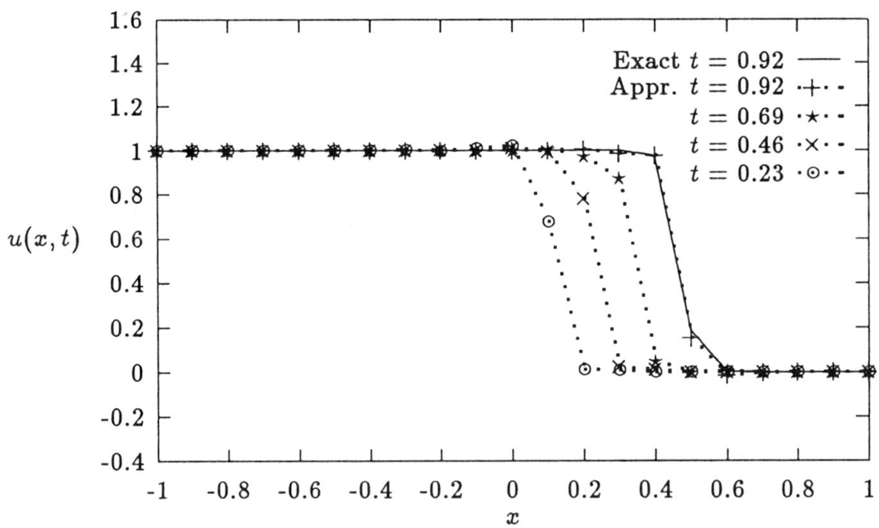

Figure 21. Burgers' Equation ($Re = 100$) : $N = 4,\ n = 5,\ p = 38.$

Table 2. *Error Measure of the Wavelet Approximate Solutions of Burgers' Equation Using Chui-Wang Bases.*

Basis Order m	Approx. Level n	No. of params p	$Re = 1$ $t = 0.23$	$Re = 10$ $t = 0.92$	$Re = 100$ $t = 0.92$
	2	7	0.1761	0.2108	0.6735
4	3	11	0.1201	0.0624	0.4977
	4	19	0.1151	0.0459	0.1476
Chebyshev Poly.†		9	0.0022	0.0083	0.1049

† From Fletcher (1982).

The results using B-spline bases of order $m = 4$ are given in Table 2. Again, errors may due to boundary conditions.

The results using wavelets bases are not as accurate as those with Chebyshev polynomials. For small Reynolds number ($Re = 1$), the boundary conditions are no longer accurate, resulting in larger errors.

4.5. A Two-Dimensional Problem

Consider the following two-dimensional differential equation with boundary conditions:

$$\frac{\partial^2 u}{\partial x^2} + \frac{\partial^2 u}{\partial y^2} = f(x,y), \qquad (x,y) \in (0,\ 1) \times (0,\ 1)$$

$$u(0,y) = 1, \qquad u(1,y) = y$$

$$u(x,1) = 1, \qquad u(x,0) = 1 - x$$

Using Daubechies' basis, we assume:

$$u(x,y) \approx \sum_{k_1,k_2=2-2N}^{2^n-1} a_{n,k_1,k_2} \Phi_{n,k_1,k_2}(x,y)$$

Using the wavelet-Galerkin method, we have:

$$\sum_{k_1,k_2} a_{n,k_1,k_2} \int_0^1 \phi_{n,j_2}(y)\phi_{n,k_2}(y)dy \left[\phi_{n,j_1}\phi'_{n,k_1}\Big|_0^1 - \int_0^1 \phi'_{n,j_1}(x)\phi'_{n,k_1}(x)dx \right]$$

$$+ \sum_{k_1,k_2} a_{n,k_1,k_2} \int_0^1 \phi_{n,j_1}(x)\phi_{n,k_1}(x)dx \left[\phi_{n,j_2}\phi'_{n,k_2}\Big|_0^1 - \int_0^1 \phi'_{n,j_2}(y)\phi'_{n,k_2}(y)dx \right]$$

$$= \int_0^1 \int_0^1 f(x,y)\phi_{n,j_1}(x)\phi_{n,j_2}(y)dxdy$$

The boundary conditions become,

$$u(0,y) = 1 = \sum_{k_1,k_2} a_{n,k_1,k_2}\phi_{n,k1}(0)\phi_{n,k2}(y)$$

$$u(1,y) = y = \sum_{k_1,k_2} a_{n,k_1,k_2}\phi_{n,k1}(1)\phi_{n,k2}(y)$$

$$u(x,0) = 1-x = \sum_{k_1,k_2} a_{n,k_1,k_2}\phi_{n,k1}(x)\phi_{n,k2}(0)$$

$$u(x,1) = 1 = \sum_{k_1,k_2} a_{n,k_1,k_2}\phi_{n,k1}(x)\phi_{n,k2}(1)$$

Using a grid of $p = 2^n + 2N - 2$ evenly distributed collocation points, we can obtain a set of linear equations from each of the above boundary conditions. They will replace some of the equations obtained by WGM. Thus, we can solve for the unknown coefficients $\{a_{n,k_1,k_2}\}$.

Fig. 22 gives the exact solution when $f(x,y) = -\pi^2 \sin(\pi x)\sin(\pi y)$. Approximate wavelet solution using Daubechies' basis of order $N = 4$ at level $n = 0$ is given in Fig. 23. Figs. 24 and 25 show the error distribution with $n = 0$ and $n = 1$ respectively.

5. Two-Scale Methods

Consider again the differential equation,

$$\mathcal{L}u(x,t) = f(x,t), \qquad x \in \Omega,\ t > 0 \tag{46}$$

where $\mathcal{L}$ is a differential operator. Using the wavelet-Galerkin method, we have,

$$\int_\Omega [\mathcal{L}(\mathcal{A}_n u) - f]\,\phi_{n,j}\,dx = 0 \tag{47}$$

By solving this system for the coefficients $\{a_{n,k}\}$, we obtain an approximate solution $\mathcal{A}_n u$ at the resolution level n. However, if we find the residual $\mathcal{L}(\mathcal{A}_n u) - f$ to be unsatisfactory at this resolution level, we need to go up to the resolution level $n+1$. Using the standard Galerkin method with the scaling functions $\{\phi_{n+1,j}(x)\}$ as basis and test functions, we have,

$$\int_\Omega [\mathcal{L}(\mathcal{A}_{n+1} u) - f]\,\phi_{n+1,j}\,dx = 0 \tag{48}$$

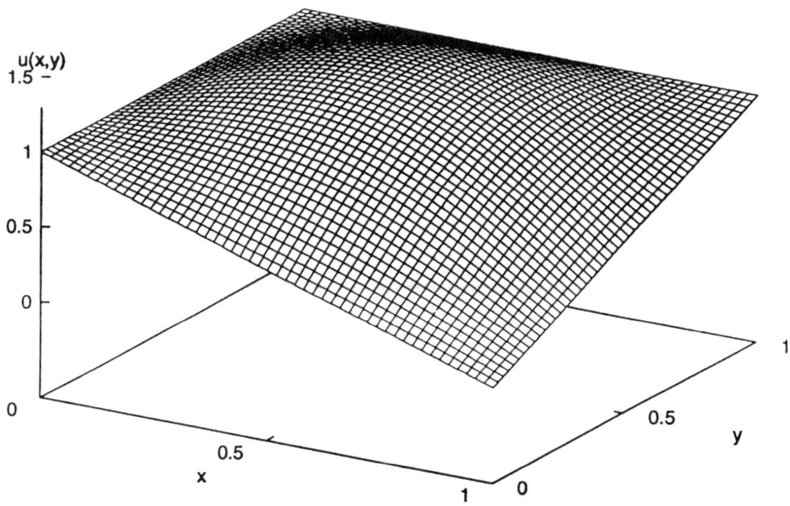

Figure 22. Exact Solution of the Two-Dimensional Problem.

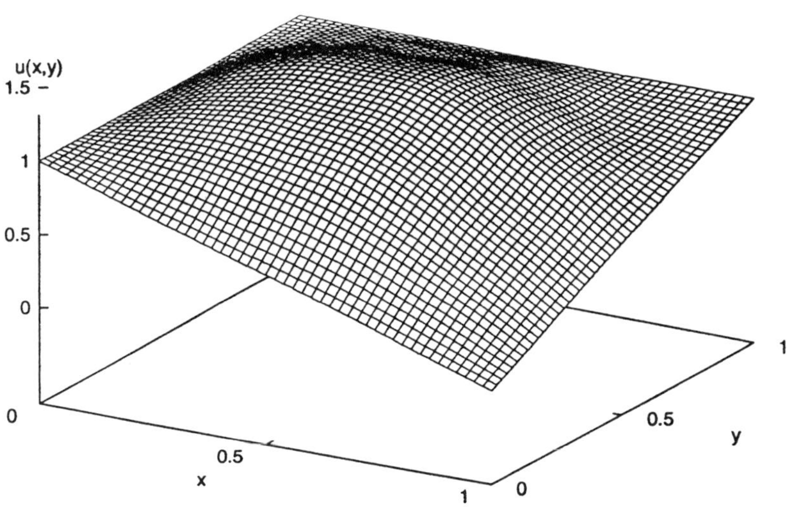

Figure 23. Approximate Solution: $N = 4,\ n = 0,\ p = 7 \times 7$.

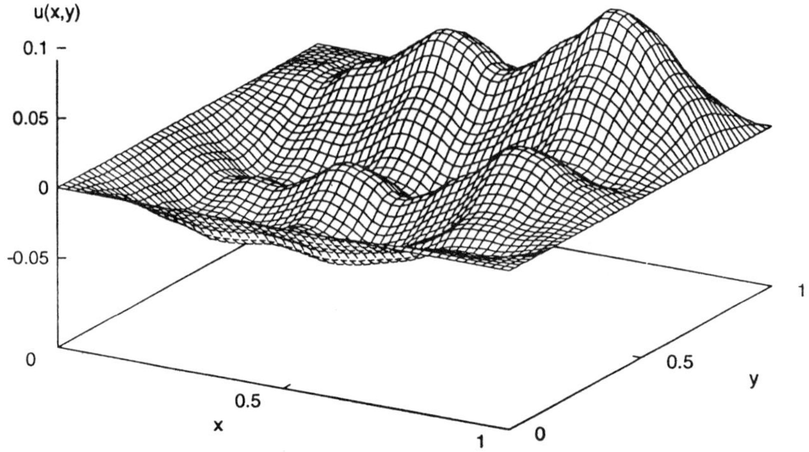

Figure 24. Error Distribution: $N = 4,\ n = 0,\ p = 7 \times 7.$

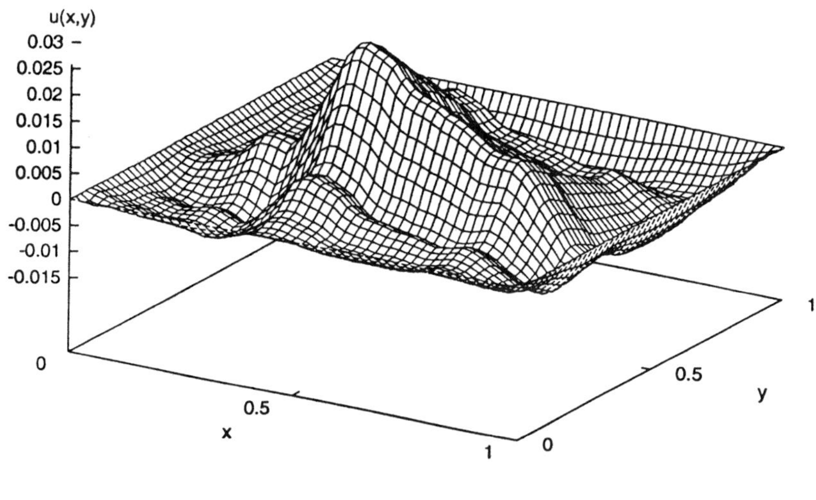

Figure 25. Error Distribution: $N = 4,\ n = 1,\ p = 8 \times 8.$

This gives a set of equations of approximately twice as many coefficients as in (47), when n is large. Can we make more of the results at the lower resolution level n? In the following, we will discuss possible methods to obtain higher level approximate solutions based on the results of the lower resolution levels. Then we will give numerical examples.

5.1. Two-Scale Methods

A direct way is to use the MRA property of the basis. Using Eq. (11) in Eq. (48) we obtain,

$$\int_{\Omega} [\mathcal{L}(\mathcal{A}_n u + \mathcal{D}_n u) - f]\phi_{n+1,j}\, dx = 0 \tag{49}$$

With the coefficients $a_{n,k}$ at level n known (i.e. $\mathcal{A}_n u$ known), we have, through (49), a system of equations for the coefficients $\{b_{n,k}\}$ of the difference signal at resolution level n. By solving this system, of approximately half dimensionality of Eq. (48) (if the resolution level is high) we can obtain these coefficients. Then, together with the scaling function coefficients, $\{a_{n,k}\}$, which we already have, we can use the reconstruction algorithm (Eq. 11 or Eq. 14), to find an approximate solution at the higher resolution level $n+1$.

To generate a second alternative, note that the test functions in Eq. (49) are at the higher level $n+1$. If we apply the decomposition relationship given in Eq. (12), we can also express the test functions $\phi_{n+1,j}$ in terms of the bases $\phi_{n,k}$ and $\psi_{n,k}$ at the lower level n, to get,

$$\sum_l \int_{\Omega} [\mathcal{L}(\mathcal{A}_n u + \mathcal{D}_n u) - f]\,[g(j-2l)\phi_{n,l} + h(j-2l)\psi_{n,l}]\, dx = 0 \tag{50}$$

Again, the coefficients $b_{n,k}$ can be solved for. In particular, when $\mathcal{L}$ is linear, the above equation (50), conbined with Eq. (47) yields

$$\sum_l \int_{\Omega} [\mathcal{L}\mathcal{D}_n u g(j-2l)\phi_{n,l} + (\mathcal{L}\mathcal{A}_n u - f + \mathcal{L}\mathcal{D}_n u)h(j-2l)\psi_{n,l}]dx = 0$$

which generates a system of linear equations in $\{b_{n,k}\}$.

An alternative view of this problem using MRA properties is as follows. Consider again the residual at level $n+1$,

$$R_{n+1} = \mathcal{L}\mathcal{A}_{n+1}u - f = (\mathcal{L}\mathcal{A}_n u - f) + \mathcal{L}\mathcal{D}_n u = R_n + \mathcal{L}\mathcal{D}_n u \tag{51}$$

Assume that the approximate solution at level n is known. Then, what we need is to find the coefficients $\{b_{n,k}\}$ of the difference signal, $\mathcal{D}_n u$, such that the residual R_{n+1} is minimized. Using Galerkin's method again with the test functions, $\psi_{n,j}$, we have

$$\int_\Omega \psi_{n,j} \mathcal{L}\mathcal{D}_n u dx = \int_\Omega \psi_{n,j}[f - \mathcal{L}\mathcal{A}_n u]dx \tag{52}$$

Using the idea discussed above, i.e. Eqs. (49), (50), and (52), to obtain improved approximate solution at higher level, we may only need to solve another set of equations of the same order at lower level. Another advantage is that, if we find the residual to be high in only a small subinterval of the domain (e.g. corresponding to a solution with abrupt changes such as a shock wave) we can use wavelet bases for improved accuracy in such an interval only, and assume zero coefficients elsewhere. Thus, only few coefficients are needed.

To summarize this two-scale adaptive method, the following procedure is given:

(0) Approximate the solution at level n by solving equation (47) for $\{a_{n,k}\}$.

(i) Check the residual $\mathcal{L}\mathcal{D}_n u - f$ *of the approximate solution according to a criterion. If the residual on the whole domain (or a subdomain) is not acceptable, go to step (ii); otherwise, stop.*

(ii) Based on the approximate solution at level n, i.e. the coefficients $\{a_{n,k}\}$, *solve for* $\{b_{n,k}\}$ *for the whole domain (or a subdomain), using Eq. (49), (50), or (52).*

(iii) Get an improved approximate solution at level $n+1$, *by using Eq. (11). Go to Step (i).*

In step (ii), the error message is given by the residual. No information on the exact solution is needed. Theoretically, this method can be applied recursively, until satisfactory approximation is achieved. However, it is recommended to use as a second-step method, i.e. start with a reasonably high level and try to obtain an approximate solution. If that solution is not good enough, then we can refine it using the above method. In this case, the *extra* work we need, if n is large, is the solution of a system of approximately $\Omega \cdot 2^n$ equations (where Ω is the domain of interest) instead of $\Omega \cdot 2^{n+1}$ equations that would be involved in direct computation of the solution at level $n+1$.

The advantage of this method is also more obvious if the solution is smooth in some subdomain but has abrupt changes in other subdomain(s). In this case, we only need lower resolution approximations in the smooth subdomains, while using higher resolution to capture the abrupt changes. This adaptive method does not require any priori knowledge of the exact solution, but can capture an abrupt change automatically through the residual.

We may also want to approximate the residuals in Eqs. (49), (50), and (52) using standard interpolation methods to carry out part of the calculation. This approach may save some computation time, especially when the operator $\mathcal{L}$ is complicated. This two-scale adaptive method may also be combined with multigrid methods for improved accuracy.

The above discussion is for a general differential operator. We now take a closer look at the method for linear differential equations.

5.2. An Analysis of Linear Differential Equations

Assume $\mathcal{L}$ is a linear operator, and let the domain Ω be [0,1]. Then Eq. (47) yields the following linear equation for $\{a_{n,k}\}$,

$$\mathbf{K}\mathbf{a_n} = \mathbf{f} \tag{53}$$

where, $\mathbf{a_n}$ is given by Eq. (34), $\mathbf{K}$ and $\mathbf{f}$ by,

$$K_{j,k} = \int_0^1 \phi_{n,j}(\mathcal{L}\phi_{n,k})dx$$

$$f_j = \int_0^1 f\phi_{n,j}dx$$

for $j,\ k = 2-2N,\ \cdots,\ 2^n-1$. Eq. (53) is a linear system of 2^n+2N-2 equations. By solving this system, we obtain an approximate solution at resolution level n.

If the approximation result at level n is not good enough, we do not need to solve a linear system of $2^{n+1} + 2N - 2$ equations to find an approximate solution at resolution level $n + 1$. Instead, we can apply Eq. (52) to get

$$\sum_{k=1-N}^{2^n+N-2} b_{n,k} \int_0^1 \psi_{n,j}\mathcal{L}\psi_{n,k}dx = \int_0^1 \psi_{n,j}f dx - \sum_{k=2-2N}^{2^n-1} a_{n,m} \int_0^1 \psi_{n,j}\mathcal{L}\phi_{n,m}dx \tag{54}$$

for $j = 1 - N,\ \cdots,\ 2^n + N - 2$. After solving the above system for $\{b_{n,k}\}$, we can get the approximation at higher level $n + 1$ by means of Eq. (11).

The savings from this two-scale approach can be assessed as follows: When direct Gaussian elimination method is used for the solution of a $p \times p$ linear system, then an order of $O(p^3)$ operations are required. If an approximation solution at level n is not satisfactory, and an approximate solution at level $n+1$ is sought through direct solution of a linear system of $2^{n+1}+2N-2$ equations, then $O((2^{n+1}+2N-2)^3)$ operations are required. However, if we use the results at level n and Eq. (54), the additional computation requires only $O\big((2^n+2N-2)^3\big)$ oprerations, which are about 1/8 of those required by the direct Gaussian elimination at level $n+1$, when n is relatively large.

The number of non-zero integrals needed for the direct method and the adaptive two-scale method are comparable, when Eqs. (49) or (52) is used.

By the reconstruction algorithm given in Eq. (14), we can obtain the higher level scaling function coefficients $\{a_{n+1,k}\}$, from the lower level scaling and wavelet coefficients, $\{a_{n,k}\}$ and $\{b_{n,k}\}$ respectively. Certainly, this refinement of the initial coarse approximate solution at resolution level n can be extended to several higher resolution levels. The approach becomes particularly attractive when refinement is effected on in areas of the domain where the residual is high.

5.2.1. A Numerical Example

To illustrate the method outlined in previous section, we use the heat equation at steady state (Eq. 37) as an example. To ensure the original boundary conditions, we impose the following boundary condition on the difference signal,

$$\mathcal{D}_n u(0) = \sum_{k=1-N}^{2^n+N-2} b_{n,k}\psi_{n,k}(0) = 0$$

$$\mathcal{D}_n u(1) = \sum_{k=1-N}^{2^n+N-2} b_{n,k}\psi_{n,k}(1) = 0$$

The adaptation method, given by Eq. (52) and (54), is used. Using Daubechies basis of order $N = 4$, we look for a solution at level $n = 3$ by solving (53). The residual is plotted in Fig. 26 (solid line). Use of the adaptation method given by Eq. (54) to solve for the difference signal in the whole domain, yields the result also given in Fig. 26 (dashed line). We can see that the difference signal captures the residual fairly well. The approximate solution at higher resolution level is given in Fig. 27, along with the analytical solution.

As mentioned above, this adaptation method can also be applied to refine the solution only in subdomains of the whole solution domain where the residual

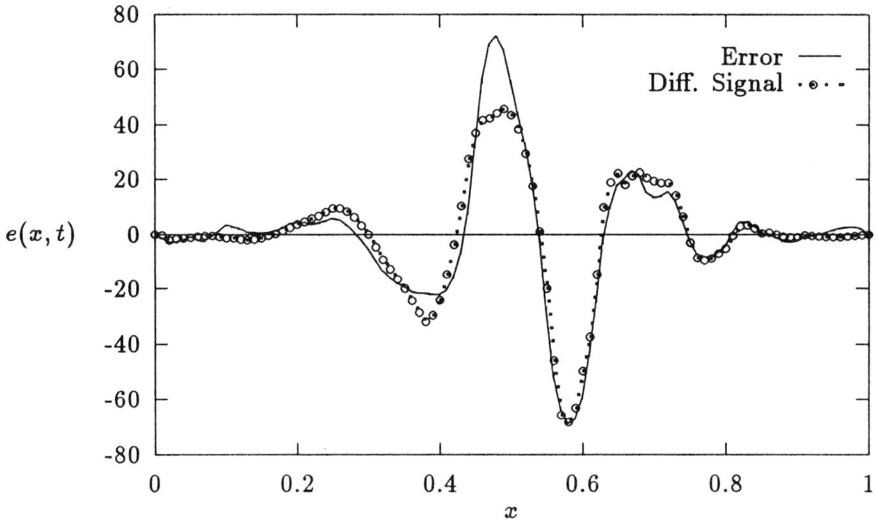

Figure 26. Error and the Difference Signal: $N = 4, \quad p = 14$.

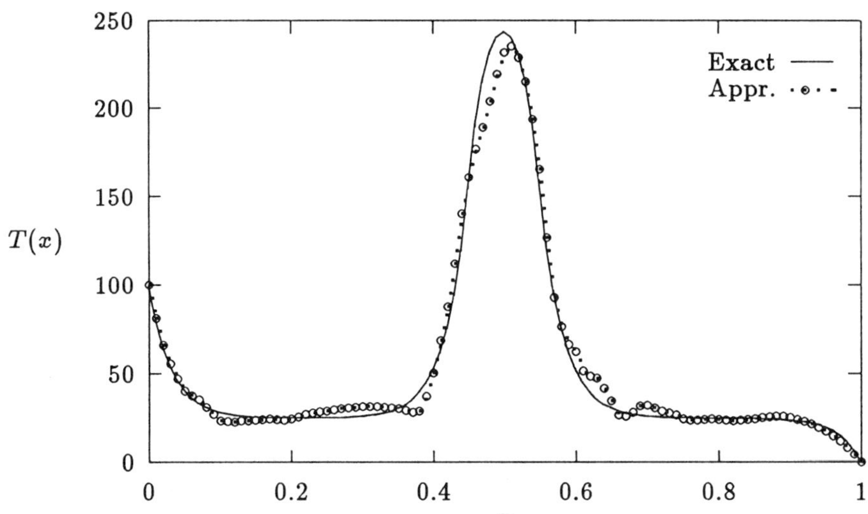

Figure 27. Reconstructed Solution at Level $n = 4$.

is large. To show this, we calculated the difference signal in the region $[\frac{1}{4}, \frac{3}{4}]$ only. The difference signal is displayed in Fig. 28, along with the residual. Fig. 29 shows the approximation result at higher level. Compared to the result in Fig. 27, the approximate solution in Fig. 29 is almost the same, but the computational cost is lower.

6. The Laplace Transform Wavelet Methods

In this section, we discuss the combination of wavelets with the Laplace transform and its numerical inversion for numerical solution of PDEs. The basic idea is to first use the Laplace transform to eliminate time dependence in the original equation; then solve the resulting time-independent differential equation in the Laplace domain; finally use an algorithm for the numerical inversion of the solution in Laplace domain to obtain the solution for the original equation in time domain.

Stehfest (1970a, 1970b) developed an algorithm for numerical inversion of Laplace transform. For a Laplace domain function, G(s), the inverse Laplace transform is given by:

$$g(t) = \frac{\ln 2}{t} \sum_{M=1}^{M_S} K_M G(s_M) \tag{55}$$

where, s is the Laplace variable; and

$$s_M = \frac{\ln 2}{t} M, \quad M = 1, \cdots, M_S \tag{56}$$

$$K_M = (-1)^{M_S/2+M} \sum_{l=[\frac{M+1}{2}]}^{\min\{M, \frac{M_S}{2}\}} \frac{l^{\frac{M_S}{2}}(2l)!}{(\frac{M_S}{2} - l)! l! (l-1)! (M-l)! (2l-M)!} \tag{57}$$

M_S is given by Stehfest (1970a) as 10 for single precision and 18 for double precision. For a fixed M_S, K_M are just a set of constants and only need to be evaluated once.

Moridis et al. (Moridis and Reddell, 1990, 1991; Moridis and Kansa, 1993) used this algorithm with finite difference, finite element, and other methods, to solve groundwater flow equations numerically.

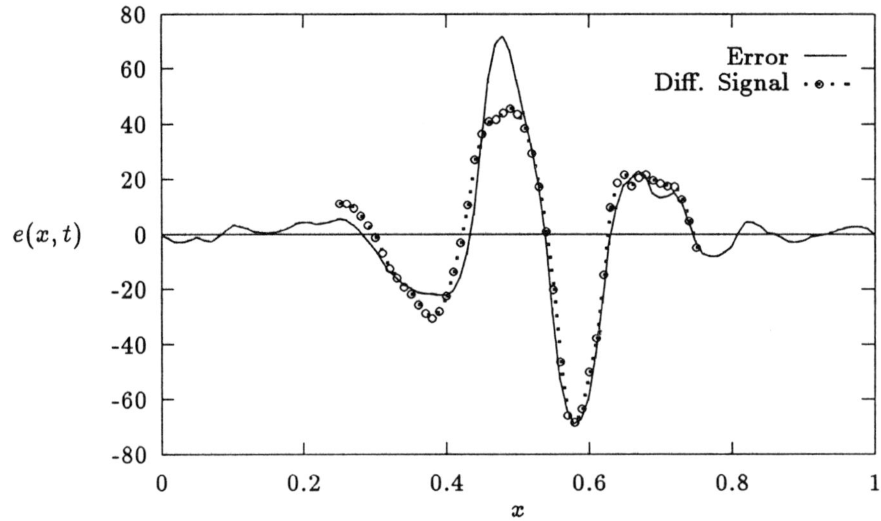

Figure 28. Error and the Difference Signal on $[\frac{1}{4}, \frac{3}{4}]$: $N = 4$.

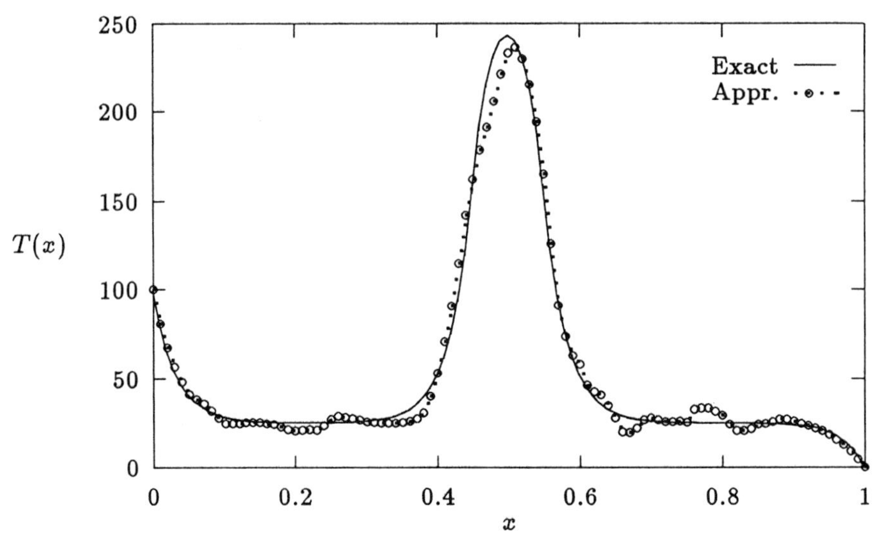

Figure 29. Reconstructed Solution at Level $n = 4$.

6.1. Laplace Transform Wavelet-Galerkin Method

Consider the parabolic equation given by:

$$\frac{\partial u}{\partial t} = \frac{\partial^2 u}{\partial x^2}, \qquad x \in (0,1) \tag{58}$$

with the initial and boundary conditions:

$$\begin{aligned} u(x,0) &= u_0(x) \\ u(0,t) &= u^0(t) \\ u(1,t) &= u^1(t) \end{aligned} \tag{59}$$

Taking the Laplace transform of the original equation, we obtain,

$$sU(x,s) - u^0(x) = U_{xx}(x,s) \tag{60}$$

which is an ODE. Using Chui-Wang B-spline basis, we can write an approximate solution at level n as:

$$U(x,s) \approx \sum_k a_k \phi_{n,k}(x) \tag{61}$$

Substitution of the approximate solution results in the following residual:

$$R_n(x,s) = s\sum_k a_k \phi_{n,k}(x) - \sum_k a_k \phi''_{n,k}(x) - u^0(x) \tag{62}$$

Applying Galerkin's method to Eq. (62), we obtain:

$$\sum_k \left[s\int_0^1 \phi_{n,j}\phi_{n,k}dx + \int_0^1 \phi'_{n,j}\phi'_{n,k}dx - \phi_{n,j}\phi'_{n,k}\Big|_0^1 \right] = \int_0^1 u(x,0)\phi_{n,j}(x)dx \tag{63}$$

This is a linear system for the coefficients $\{a_{n,k}\}$. Boundary conditions in Laplace domain are two linear equations of the same coefficients,

$$\begin{aligned} U(0,s) &= \sum_k a_k \phi_{n,k}(0) \\ U(1,s) &= \sum_k a_k \phi_{n,k}(1) \end{aligned} \tag{64}$$

Combining these equations, for any given s, we can obtain a set of coefficients and the corresponding approximate solution at the desired locations. Thus, using the numerical inversion algorithm given above, we can obtain an approximate solution at any time point.

6.2. Laplace Transform Wavelet-Collocation Method

Using the collocation method for Eq. (62), we get:

$$\sum_k a_k \phi''_{n,k}(x_j) - s \sum_k a_k \phi_{n,k}(x_j) = u_0(x_j) \tag{65}$$

where x_j's are collocation points. When we have p unknown coefficients and two boundary conditions (see Eq. 64), the total number of collocation points is $p-2$. Thus, we have a linear system of p equations with p unknowns.

The collocation points are selected to be evenly distributed in the domain of interest. For Eq. (58),

$$x_j = j * h, \qquad j = 0, \cdots, p-3 \tag{66}$$

where $h = 1/(p-3)$.

6.3. Numerical Examples

6.3.1. Example 1.

As a numerical example, we solve Eq. (58) with the initial and boundary conditions:

$$u(x,0) = u_0(x) = \sin(\pi x)$$
$$u(0,t) = u^0(t) = 0$$
$$u(1,t) = u^1(t) = 0$$

Using Galerkin's method with Chui-Wang B-spline basis of order $m = 4$ at level $n = 4$, we get the wavelet solution given in Fig. 30(a) at time $t = 0.1$, along with the exact solution. The parameter $M_S = 12$ is used for the numerical inversion of the Laplace domain solution. There are 19 coefficients in the approximate solution.

For comparison, Fig. 30(a) also gives the finite difference (FD) solution at the same time point, which was obtained by the standard explicit finite difference scheme for parabolic equations. The FD solution was evaluated at 20 grid points in x. For stability, 80 time steps are needed to reach $t = 0.1$, with the maximum time step size of 1/800.

Fig. 30(b) gives the corresponding error distributions for the Laplace transform wavelet-Galerkin solution and the FD solution. FD method gives better results.

Using the collocation method as in Eq. (65) with the same basis as above, i.e. $m = 4$ at level $n = 4$, we obtain the wavelet-collocation solution at $t = 0.1$ shown in Fig. 31(a), along with the exact solution and FD solution. Again $M_S = 12$ is used.

Fig. 31(b) shows the error distributions for the Laplace transform wavelet collocation solution and the FD solution. Laplace transform wavelet collocation solution gives better results.

To show the effects of M_s on the accuracy of the wavelet solutions, Fig. 32 gives the error measure versus M_s. The error measure used here is the l_2-norm of the error at 21 evenly distributed points on [0, 1].

6.3.2. Example 2.

Now we consider a simplified equation of fluid flow through porous media:

$$K\frac{\partial u}{\partial t} = S_0\frac{\partial^2 u}{\partial x^2}, \quad x \in (0, 1000) \tag{58}$$

with the initial and boundary conditions:

$$\begin{aligned} u(x,0) &= u_0(x) = H_1 \\ u(0,t) &= u^0(t) = H_1 \\ u(1,t) &= u^1(t) = H_2 \end{aligned}$$

where $K = 30$ is the hydraulic conductivity; $S_0 = 0.006$ is the specific aquifer storativity; $H_1 = 60$ is initial/high piezometric head; and $H_2 = 40$ is the low head.

Using Galerkin's method with Chui-Wang basis of order $m = 4$ at level $n = 3$, we obtain the approximate solution given in Fig. 33, along with the corresponding error distribution. The parameter $M_S = 8$ is used.

Fig. 34 gives the wavelet approximate solution and its error distribution when the collocation method is used with the same basis as above. Again $M_S = 8$ is used for the inverse Laplace transform.

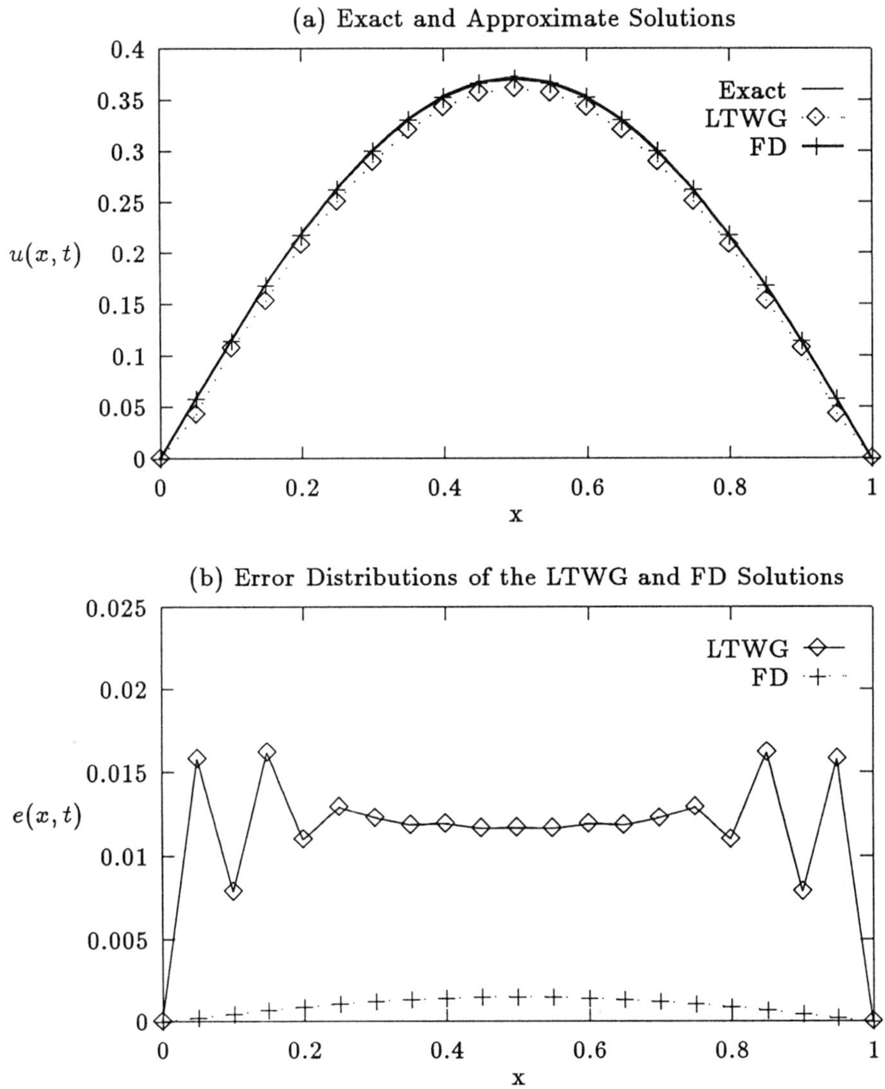

Figure 30. Laplace Transform Wavelet-Galerkin Solution at $t = 0.1$: $m = 4,\ n = 4,\ p = 19;\ M_S = 12.$

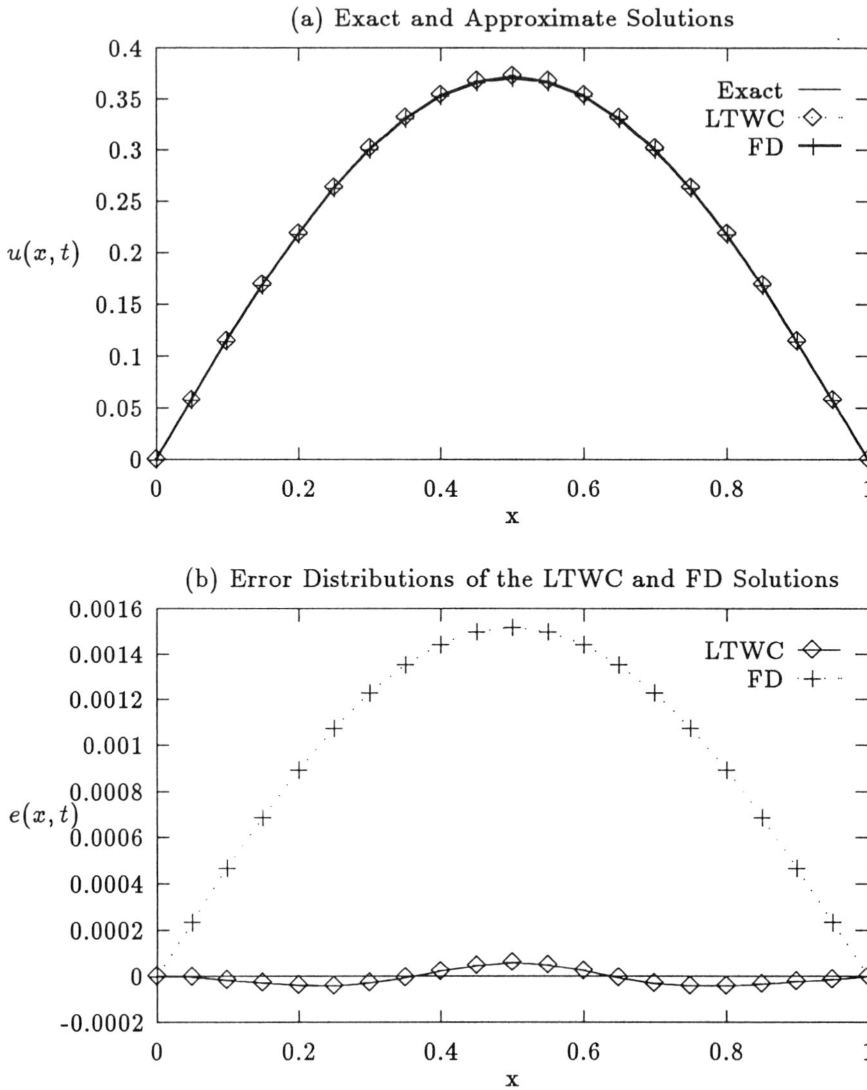

Figure 31. Laplace Transform Wavelet-Collocation Solution at $t = 0.1$: $m = 4,\ n = 4,\ p = 19;\ M_S = 12$.

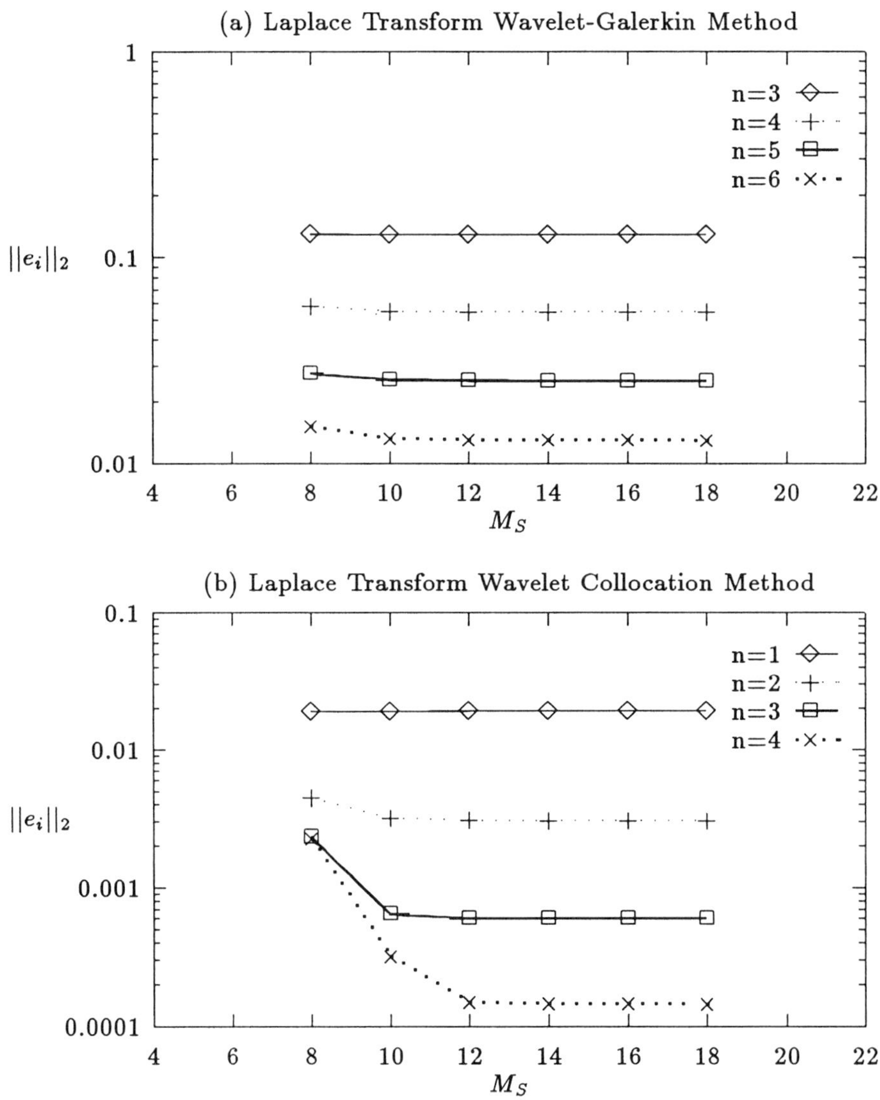

Figure 32. M_S versus Error Measure at Different Levels: $t = 0.1$, $m = 4$.

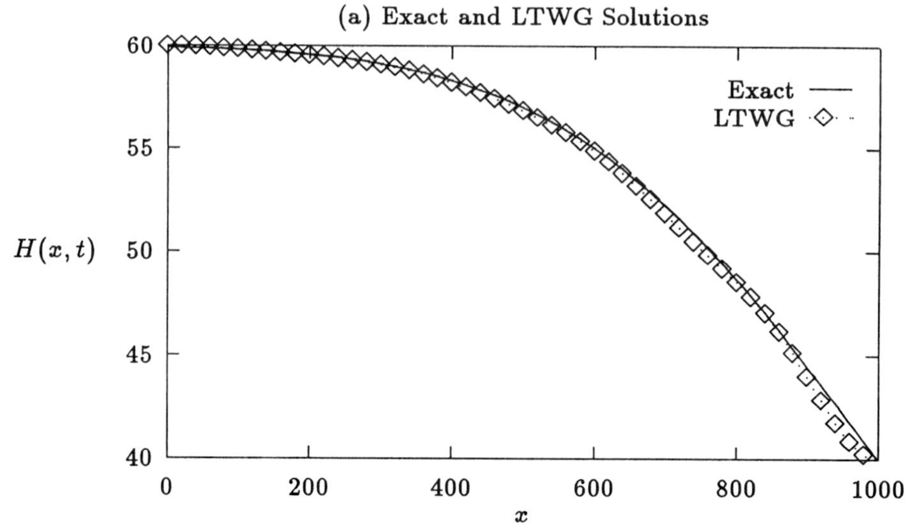

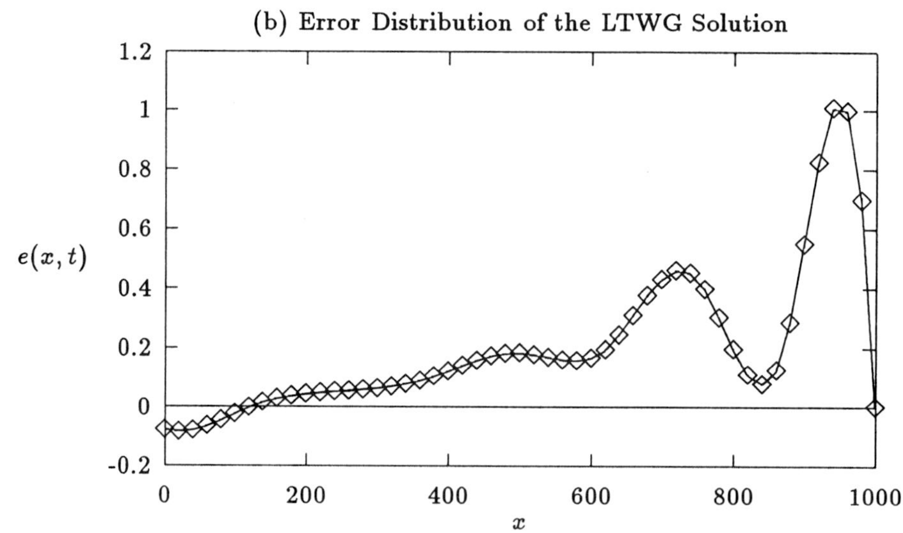

Figure 33. LTWG Solution at $t = 12:\ m = 4,\ n = 3,\ p = 11,\ M_S = 8.$

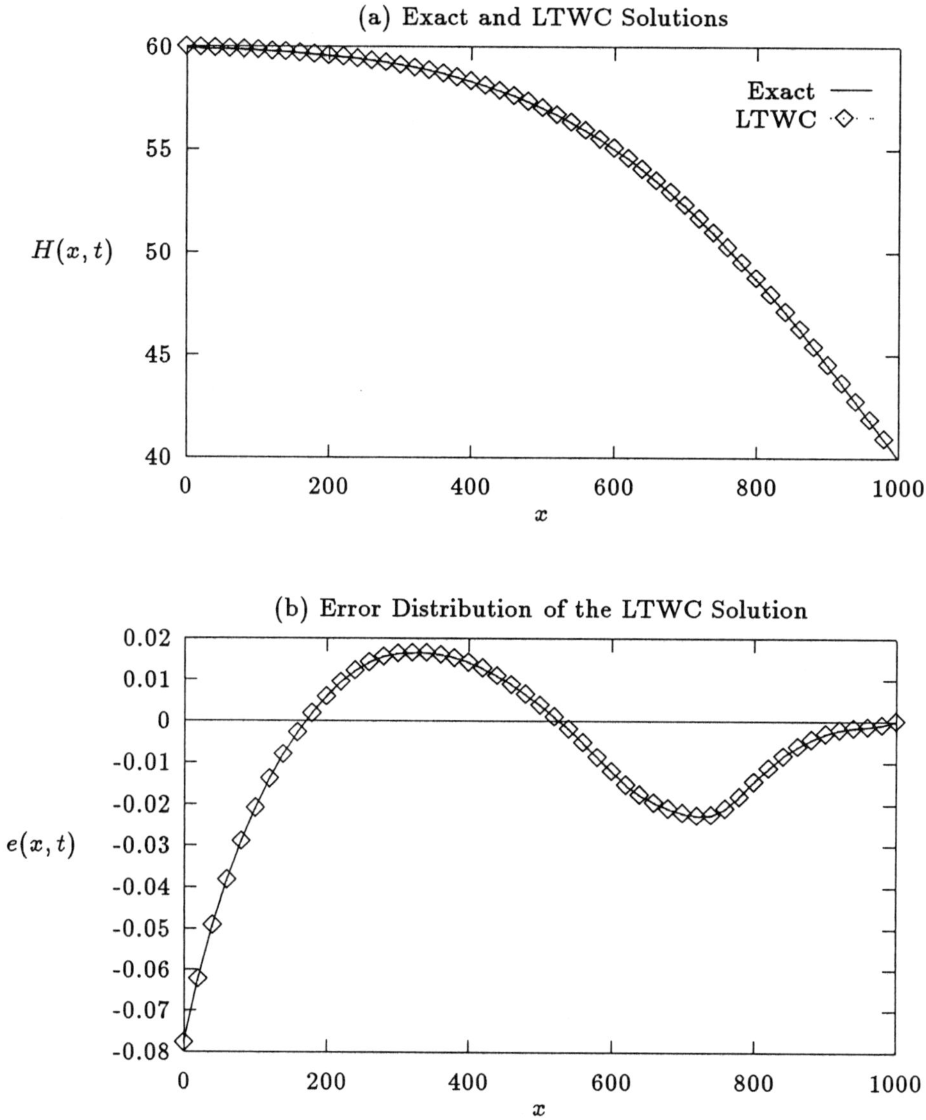

Figure 34. LTWC Solution at $t = 12:\ m = 4,\ n = 3,\ p = 11,\ M_S = 8.$

7. Discussion

Moridis and Nikolaou (1993) showed how to use the Laplace-transform-wavelet methods for problems involving fluid flow in porous media. Wavelet methods have also been used for studying of turbulence flows (Farge, 1992; Weiss, 1992).

The wavelet bases used in this work, i.e. Daubechies' orthonormal wavelets and Chui-Wang's semi-orthogonal wavelets, constitute unconditional bases on the entire real line, instead of on a finite interval. When such bases are used for problems on a finite interval of the real line, difficulties may be encountered with boundary conditions. In most cases, that we have shown in the previous sections, restriction of the bases to the solution interval can solve this problem. When such a direct method in not applicable, general methods to circumvent the difficulties are either to extend the problem to the whole real line directly or to periodize the problem, depending on the nature of a specific problem. Recently, wavelet bases on a finite interval have also been developed (Meyer, 1991; Chui and Quak, 1992; Andersson et al, 1993). This might give more flexibility for dealing with boundary conditions. With such bases, more accurate approximate solutions may be expected.

Wavelet bases that are adapted to certain differential operators have also been developed (see, for example, Lorentz and Madych, 1990; Jaffard, 1992; Liandrat et al, 1992; Dahlke and Weinreich, 1993). The methods proposed therein, however, are limited to very specific operators.

8. Conclusions

Wavelets have several potential advantages for numerical solution of differential equations. Orthogonality, compact support, and multi-level structure of the basis, among others, are the main reasons for this optimism.

In this work, an approach that takes advantage of the multiresolution analysis properties of wavelet bases was studied. This approach can generate approximate wavelet solutions at higher resolution levels based on lower level results. Wavelet-Galerkin methods were illustrated on IVP, TPBV problems and Burgers' equations, using Daubechies' orthonormal wavelets as well as Chui-Wang's semi-orthogonal B-spline wavelets. Wavelets were also compared with Chebyshev polynomials as the basis functions for solutions of Burgers' equations. Methods involving wavelets and the Laplace transform and its numerical inversion were also studied.

Our simulations show that B-spline wavelet bases give better approximate solutions of differential equations than Daubechies wavelet bases. This is in agreement with a recent research result by Sweldens and Jawerth (1993), who showed that, "for a fixed degree of polynomial approximation, Daubechies' orthogonal wavelets need in general one level more than spline wavelets to obtain an approximation with a prescribed accuracy."

Comparisons of the collocation method and Galerkin's method show that the former gives better approximation than the latter, for parabolic equations. Laplace-transform-wavelet methods were compared with the explicit finite difference scheme. Laplace-transform-wavelet-collocation resulted in more accurate results than the explicit finite difference method.

As a general remark, we would like to point out that this work is only at its beginning, and a lot of additional research is needed to fine-tune the proposed approach for higher computational savings to be realized. We do not expect wavelet-based methods to be superior to every other method for numerical solution of DEs. However, for DEs with solutions that exhibit local irregularities, wavelets have a distinct potential that deserves to be fully explored.

Acknowledgement

Financial support from the Texas Advanced Research Project (Grant N0. 999903-195) and from the Center for Energy and Mineral Resources at Texas A&M University are gratefully acknowledged. The authors would also like to thank Dr. George Moridis for providing them with an early manuscript of his work on the Laplace transform methods.

Notation

Abbreviations

DE = differential equation
IVP = initial value problem
ODE = ordinary differential equation
LTWC = Laplace-transform-wavelet-collocation
LTWG = Laplace-transform-wavelet-Galerkin
MRA = multiresolution analysis
MWR = the method of weighted residuals
PDE = partial differential equation
TPBV = two-point boundary value
WGM = wavelet-Galerkin method

Roman Letters

$A(t)$ = vector containing $a_{n,k}(t)$
$\mathcal{A}_n$ = scaling function based operator
a_i = coefficient
$\mathbf{a_n}$ = vector containing $a_{n,k}$
$a_{n,k}$ = scaling function coefficient
$a_{n,k}(t)$ = time-dependent coefficient
$\mathbf{b_n}$ = vector containing $b_{n,k}$
$b_{n,k}$ = wavelet function coefficient
C^∞ = the space of all infinitely differentiable functions
C^k = the space of all k-times continuously differentiable functions
C^λ = the space of all Hölder continuous function with index λ
$\mathcal{D}_n$ = wavelet based operator
$\mathbf{f}$ = vector, elements defined in the text
$f(x)$ = function
$g(k)$ = coefficient
H = piezometric head
$h(k)$ = coefficient
K = constant
K = hydraulic conductivity
$\mathbf{K}$ = matrix, elements defined in the text
$\mathcal{L}$ = differential operator
$L_m(x)$ = fundamental cardinal spline function
$L_2(\mathcal{R})$ = the space of all square integrable functions on the real line
$\mathbf{M}$ = matrix, elements defined in the text
M_S = parameter for inverse Laplace transform
m = order of Chui-Wang basis
$\mathcal{M}_\psi$ = moments of wavelets
$\mathbf{N}$ = matrix, elements defined in the text
N = order of Daubechies' wavelet basis
$N_m(x)$ = cardinal B-spline function
n = resolution level
$\mathbf{P}$ = matrix, elements defined in the text
p = number of coefficients/unknowns
$p(k)$ = two-scale coefficient of scaling function
$p(x)$ = function
$\mathbf{Q}$ = matrix, elements defined in the text
q = number of weighting functions
$q(k)$ = two-scale coefficient for wavelet functions
$\mathcal{R}$ = the real line
$R(x)$ = residual function
$R_n(x)$ = residual function
Re = Renold's number

S_0 = specific aquifer storativity
s = Laplace variable
T_n = Chebyshev polynomial
$u(x)$ = function
$\tilde{u}(x)$ = approximation of $u(x)$
V_n = a subspace of $L^2(\mathcal{R})$
$\mathcal{W}$ = connection coefficient
W_n = the orthogonal compliment subspace of V_n to $L^2(\mathcal{R})$
w_j = weighting function
$\mathcal{Z}$ = the set of all integers

Greek Letters

$\delta(x)$ = Dirac delta function
$\delta_{m,n}$ = Dirac delta function
ν = viscosity parameter
ϕ = scaling function
ϕ_j = basis/trial function
$\phi_{n,k}$ = $2^{n/2}\phi(2^n x - k)$
$\tilde{\phi}_{n,k}$ = dual scaling function
ψ = basic/mother/analyzing wavelet function
$\tilde{\psi}$ = dual wavelet
$\psi_{n,k}$ = $2^{n/2}\psi(2^n x - k)$
Ω = domain of interest

References

Alpert, B., "Wavelets and Other Bases for Fast Numerical Algebra," *Wavelets: A Tutorial in Theory and Applications,* C. K. Chui, ed., 181-216, Academic Press, San Diego, CA (1992).

Andersson, L., N. Hall, B. Jawerth, and G. Peters, "Wavelets on Closed Subsets of the Real Line," to appear in *Topics in the Theory and Applications of Wavelets,* L. L. Schumaker and G. Webb, eds., Academic Press, Boston, MA (1993).

Auscher, P., "Wavelets with Boundary Conditions on the Interval," *Wavelets: A Tutorial in Theory and Applications,* C. K. Chui, ed., 217-236, Academic Press, San Diego, CA (1992).

Battle, G., "A Block Spin Construction of Ondelettes. Part I: Lemarié Functions," *Commun. Math. Phys.,* **110**, 601-615 (1987).

Beylkin, G., "On Wavelet-Based Algorithms for Solving Differential Equations," *Wavelets: Mathematics and Applications,* J. J. Benedetto and M. Frazier, eds., Chap. 12, CRC Press, Boca Raton, FL(1993).

Beylkin, G., R. Coifman, and V. Rokhlin, "Fast Wavelet Transforms and Numerical Algorithms I," *Comm. Pure Appl. Math.,* **44**(2), 141-183 (1991).

de Boor, C., *A Practical Guide to Splines,* Springer-Verlag, New York, NY (1978).

de Boor, C., R. A. DeVore, and A. Ron, "On the Construction of Multivariate (Pre)wavelets," *Constr. Approx.,* **9**(2-3), 123-166 (1993).

Briggs, W. L. and V. E. Henson, "Wavelets and Multigrid," *SIAM J. Sci. Comput.,* **14**(2), 506-510 (1993).

Celia, M .A. and W. G. Gray, *Numerical Methods for Differential Equations,* Prentice Hall, Englewood Cliffs, NJ (1992).

Chui, C. K., *An Introduction to Wavelets,* Academic Press, Boston, MA (1992).

Chui, C. K. and E. Quak, "Wavelets on a Bounded Interval," *Numerical Methods of Approximation Theory,* D. Braess and L. L. Schumaker, eds., Vol. 9, 53-75, Birkhäuser Verlag, Basel (1992).

Chui, C. K. and J.-Z. Wang, "A Cardinal Spline Approach to Wavelets," *Proc. Amer. Math. Soc.,* **113**(3), 785-793 (1991).

Chui, C. K. and J.-Z. Wang, "A General Framework of Compactly Supported Splines and Wavelets," *J. Appr. Theory,* **71**(3), 263-304 (1992a).

Chui, C. K. and J.-Z. Wang, "On Compactly Supported Spline Wavelets and A Duality Principle," *Trans. Amer. Math. Soc.,* **330**(2), 903-915 (1992b).

Chui, C. K. and J.-Z. Wang, "An Analysis of Cardinal Spline-Wavelets," *J. Appr. Theory,* **72**(1), 54-68 (1993).

Cohen, A., "Biorthogonal Wavelets," *Wavelets: A Tutorial in Theory and Applications,* C. K. Chui, ed., 123-152, Academic Press, San Diego, CA (1992).

Cohen, A. and I. Daubechies, "Orthonormal Bases of Compactly Supported Wavelets III. Better Frequency Resolution," *SIAM J. Math. Anal.,* **24**(2), 520-527 (1993).

Cohen, A., I. Daubechies, and J. C. Feauveau, "Biorthogonal Bases of Compactly Supported Wavelets," *Comm. Pure Appl. Math.*, **XLV**(6), 485-560 (1992).

Cohen, A. and J.-M. Schlenker, "Compactly Supported Bidimensional Wavelet Bases with Hexagonal Symmetry," *Constr. Approx.*, **9**(2-3), 209-236 (1993).

Coifman, R. R. and Y. Meyer, "Orthonormal Wavelet Packet Bases," preprint, Yale University, New Haven, CT (1990).

Dahlquist, G. and A. Bjorck, *Numerical Methods*, Prentice-Hall, Englewood Cliffs, NJ (1974).

Dahlke, S. and I. Weinreich, "Wavelet-Galerkin Methods: An Adapted Biorthogonal Wavelet Basis," *Constr. Approx.*, **9**(2-3), 237-262 (1993).

Dahmen, W. and C. A. Micchelli, "Using the Refinement Equation for Evaluating Integrals of Wavelets," *SIAM J. Math. Anal.*, **30**(2), 507-537 (1993).

Daubechies, I., "Orthonormal Bases of Compactly Supported Wavelets," *Comm. Pure Appl. Math.*, **41**(7), 909-996 (1988).

Daubechies, I., "Orthonormal Bases of Compactly Supported Wavelets II. Variations on a Theme," *SIAM J. Math. Anal.*, **24**(2), 499-519 (1993).

Daubechies, I., *Ten Lectures on Wavelets*, CBMS-NFS Series in Applied Mathematics, SIAM, Philadelphia, PA (1992).

Daubechies, I. and J. C. Lagarias, "Two-Scale Difference Equations II. Local Regularity, Infinite Products of Matrices and Fractals," *SIAM J. Math Anal.*, **23**(4), 1031-1079 (1992).

Farge, M., "Wavelet Transforms and Their Applications to Turbulence," *Annu. Rev. Fluid Mech.*, **24**, 395-457 (1992).

Finlayson, B. A., *The Method of Weighted Residuals and Variational Principles with Application in Fluid Mechanics, Heat and Mass Transfer*, Academic Press, New York, NY (1972).

Fletcher, C. A. J., "Burgers' Equation: A Model for All Reasons," *Numerical Solutions of Partial Differential Equations*, J. Noye, ed., North-Holland, New York, NY (1982).

Fletcher, C. A. J., *Computational Galerkin Methods,* Springer-Verlag, New York, NY (1984).

Glowinski, R., W. Lawton, M. Ravachol, and E. Tenenbaum, "Wavelet Solutions of Linear and Nonlinear Elliptic, Parabolic, and Hyperbolic Problems in One Space Dimension," *Computing Methods in Applied Sciences and Engineering*, Chap. 4, R. Glowinski and A. Lichnewsky, eds., 55-120, SIAM, Philadelphia, PA (1990).

Gottlieb, D. and S. A. Orszag, *Numerical Analysis of Spectral Methods: Theory and Applications,* SIAM, Philadelphia, PA (1977).

Haar, A., "Zur Theorie der Orthogonalen Funkktionen-Systeme," *Math. Ann.*, **69**, 331-371 (1910).

Jaffard, S., "Wavelet Methods for Fast Resolution of Elliptic Problems," *SIAM J. Numer. Anal.,* **29**(4), 965-986 (1992).

Jawerth, B. and W. Sweldens, "An Overview of Wavelet Based Multiresolution Analysis," submitted to *SIAM Review* (1993).

Jawerth, B. and W. Sweldens, "Wavelet Multiresolution Analysis Adapted for the Fast Solution of Boundary Value Ordinary Differential Equations," to appear in *Proceedings of the Six Copper Maintain Multigrid Conference* (April 1993).

Lapidus, L. and G. F. Pinder, *Numerical Solutions of Partial Differential Equations in Science and Engineering,* John Wiley & Sons, New York, NY (1982).

Latto, A., H. L. Resnikoff, and E. Tenenbaum, "The Evaluation of Connection Coefficients of Compactly Supported Wavelets," *Proc. French-USA Workshop on Wavelets and Turbulence,* Y. Maday, ed., Springer-Verlag, New York, NY (1992).

Lemarié, P. G., "Ondelettes à Localisation Exponentielles," *J. de Math. Pures et Appl.,* **67**, 227-236 (1988).

Liandrat, J., V. Perrier, Ph. Tchamitchian, "Numerical Resolution of Nonlinear Partial Differential Equations Using the Wavelet Approach," *Wavelets and Applications,* M. B. Ruskai, G. Beylkin, R. Coifman, I. Daubechies, S. Mallat, Y. Meyer, and L. Raphael, eds., 227-238, Jones and Bartlett, Boston, MA (1992).

Lorentz, R.A. and W. Madych, "Spline Wavelets for Ordinary Differential Equations," preprint, GMD, St. Augustin (1990).

Mallat, S., "A Theory of Multiresolution Signal Decomposition: The Wavelet Representation," *IEEE Trans. Pattern Anal. Machine Intell.*, **11**(7), 674-693 (1989).

Meyer, Y., *Ondelettes et Fonctions Splines,* Séminaire EDP, École Polytechnique, Paris (1986).

Meyer, Y., "Ondelettes sur l'intervalle," *Rev. Mat. Iberoamericana,* **7**, 115-143, 1991).

Meyer, Y., "Wavelets and Operators," *Analysis at Urbana I: Analysis in Function Spaces,* E. Berkson, T. Peck, J. Uhl, ed., 256-365, Cambridge University Press, Cambridge (1989).

Moridis, G. J. and M. Nikolaou, "The Laplace Transform Wavelet (LTW) Method for the Solution of the Equations of Flow and Transport Through Porous Media," *EOS Trans. AGU,* **74**(43), 307 (1993).

Moridis, G. J. and D. L. Reddell, "The Laplace Transform Finite Difference (LTFD) Numerical Method for Groundwater Simulations," *EOS Trans. of AGU*, **71**(17), 123-129 (1990).

Moridis, G. J. and D. L. Reddell, "The Laplace Transform Finite Element (LTFE) Numerical Method for the Solution of Groundwater Equation Groundwater Simulations," *EOS Trans. of AGU*, **72**(17), 123-129 (1991).

Moridis, G. J. and E. Kansa, "The Method of Laplace Transform Multiquadrics (LTMQ) for the Solution of the Groundwater Flow Equation," *Proc. 7th IMACS Int. Conf. on Computer Methods for PDEs*, New Brunswick, NJ (June, 1993)

Qian, S. and J. Weiss, 'Wavelets and the Numerical Solution of Boundary Value Problems," *Appl. Math. Lett.,* **6**(1), 47-52 (1993).

Qian, S. and J. Weiss, "Wavelets and the Numerical Solution of Partial Differential Equations," submitted to *J. Comp. Phys.* (1992).

Riemenschneider, S. D. and Z. Shen, "Wavelets and Pre-Wavelets in Low Dimensions," *J. Approx. Theory*, **71**(1), 18-38 (1992).

Schoenberg, I. J., *Cardinal Spline Interpolation,* CBMS-NSF Series in Applied Mathematics, SIAM, Philadelphia, PA (1973).

Stehfest, H., "Numerical Inversion of Laplace Transforms," *Comm. of the ACM*, **17**(1), 47-49 (1970a).

Stehfest, H., "Numerical Inversion of Laplace Transforms," *Comm. of the ACM*, **17**(10), 56 (1970b).

Strang, G., "Wavelets and Dilation Equations: A Brief Introduction," *SIAM Review,* **31**(4), 614-627 (1990).

Stöckler, J., "Multivariate Wavelets," *Wavelets: A Tutorial in Theory and Applications,* C. K. Chui, ed., 325-355, Academic Press, San Diego, CA (1992).

Strömberg, J. O., "A Modified Franklin System and Higher-Order Spline Systems on $\mathcal{R}^n$ as Unconditional Basis for Hardy Spaces," *Conference on Harmonic Analysis in Honor of Antoni Zygmund,* Vol. II, W. Bechener, A. P. Calderón, R. Fefferman, and P. W. Jones, eds., 466-494, Wadsworth International, Belmont, CA (1981).

Sweldens, W. and R. Piessens, "Asymptotic Error Expansion for Wavelet Approximations of Smooth Functions," preprint (1992).

Sweldens, W. and R. Piessens, "Asymptotic Error Expansion for Wavelet Approximations of Smooth Functions II: Generalizations," preprint (1993).

Sweldens, W. and R. Piessens, "Quadratic Formula for the Calculation of the Wavelet Decomposition," preprint (1992).

Weiss, J., "Wavelets and the Study of Two Dimensional Turbulence," *Proc. French-USA Workshop on Wavelets and Turbulence,* Y. Maday, ed., Springer-Verlag, New York, NY (1992).

Xu, J.-C. and W.-C. Shann, "Galerkin-Wavelet Methods for Two-Point Boundary Value Problems," *Numer. Math.,* **63**(1), 123-142 (1992).

Chapter 8

A Parallel Two-Dimensional Wavelet Packet Transform and Some Applications in Computing and Compression Analysis

Eric Goirand[a12]
Mladen Victor Wickerhauser[a3]
Marie Farge[b2]

[a]Washington Univeristy
Department of Mathematics
1 Brookings Drive, Campus Box 1146
Saint Louis, MO 63130, USA

[b]LMD-CNRS
Ecole Normale Supérieure
24 rue Lhomond
75231 Paris Cedex 05 FRANCE

Abstract

In a first part, we study a parallel algorithm (written for a MIMD machine) to compute the two-dimensional wavelet packet transform. Then, we apply it to compute the multiplication of a matrix by a vector in parallel. In a last part, we will use it to analyze and compress a two-dimensional turbulent field.

[1]Supported in part by ONERA
[2]Supported in part by NATO contract CRG-93456
[3]Supported in part by AFOSR and NSF

8.1. INTRODUCTION

We will present an implementation on a MIMD computer of the two-dimensional periodic wavelet packet transform, the two-dimensional "best-basis" choice algorithm, the non-standard matrix multiplication algorithm [1] and some analysis and compression techniques for studying two-dimensional turbulent fields. Our implementation also works for the wavelet transform numerical algorithms of Beylkin, Coifman and Rokhlin [2]. It is one way to obtain a fast functional calculus for certain classes of linear operators ; those operators, which sparsify in the wavelet basis or the "best-basis" of the wavelet packets, can be applied to vectors in a lower order of complexity. The purpose of parallelizing the transform is to distribute the cost of the initial sparsification "investment" over a large number of machines. This one-time cost is $O(N^2 \log(N))$ with a non negligible constant; we envision applications in which $N=10^6$, for example evolutions of 2-dimensional fluid velocity fields on 1000x1000 point grids.

We will compute matrix coefficients for operators with respect to an orthonormal basis of separable wavelet packets, using the 2-dimensional version of the fast wavelet packet transform [3], [4]. The main idea is to lift an NxN matrix, which maps R^N to R^N , into the space of maps from $R^{\{N \cdot \log(N)\}}$ to $R^{\{N \cdot \log(N)\}}$. Any of a large number of NxN coefficient subsets of this lifting can be used to represent the operator, so we may pick the subset in which the matrix is most sparse. The choice is made with the "best-basis" algorithm and is itself a fast algorithm. The number of basis subsets of this type grows like 4^N for an NxN matrix, but computing all of the coefficients with respect to all of the basis functions requires just $O(N^2 \log(N))$ operations. The algorithm for choosing the "best-basis" in which the operator appears most sparse is described in [5]; it has a complexity of $O(N^2)$. We have reason to believe that the equations of motion for large physical systems sparsify by nearly an order of magnitude in this collection of bases. We thus obtain lower complexity matrix application and matrix multiplication algorithms from the new representation. Of course the method is not perfectly general : the speedup depends very much upon the problem. However, others have shown that for broad classes of operators we can expect an order of magnitude asymptotic complexity reduction for matrix multiplication.

As an application, we will use the wavelet packet representation of a field to extract parts of it, either by keeping only its significative coefficients or by selecting a particular zone. In this new representation of the field, we can then perform some well-known analyzes , to get some local information. As an example of this method, we will in the last part of this chapter analyze and compress a two-dimensional turbulent field.

8.2. PARALLEL WAVELET PACKET DECOMPOSITION

Suppose we have a two-dimensional periodic signal S = $(s_{i,j})$ with $0 \le i < 2^{Xmax}$ and $0 \le j < 2^{Ymax}$, and suppose nmax = min(Xmax,Ymax). We want to calculate all the wavelet packets ${}^{f_j}C^j$ or rather the sets of wavelet packet coefficients ${}^{f_j}C^j_{k_j,l_j}$ defined as follows :

- The indices are in the range :

$$\left.\begin{array}{l} 0 \le j \le nmax \\ 0 \le k_j < 2^{Xmax-j} \\ 0 \le l_j < 2^{Ymax-j} \\ 0 \le f_j < 4^j \end{array}\right\} \quad with\ j, k_j, l_j\ and\ f_j \in \mathrm{N}$$

- The initial packet is :

$${}^0C^0_{k,l} = s_{k,l}\ with\ 0 \le k < 2^{Xmax}, 0 \le l < 2^{Ymax}$$

- Then every other packet is defined recursively by the formula :

$${}^{f_{j+1}=4.f_j+d}C^{j+1}_{k_{j+1},l_{j+1}} = \sum_{n_j,m_j} F(d,f_j)_{n_j-2k_{j+1},m_j-2l_{j+1}} * {}^{f_j}C^j_{n_j,m_j} \quad \text{with}$$

  d = 0, 1, 2 or 3. We shall say that the packet ${}^{f_j}C^j$ is the father of the four packets ${}^{4 \cdot f_j+d}C^{j+1}$ or that they are its four children.

- From a one-dimensional wavelet Ψ (defined by its filter G) and its smoothing function Φ (defined by its filter H) we obtain four two-dimensional filters by tensor products :

$$\begin{cases} Filter(0)_{n,m} = H_n \cdot H_m \\ Filter(1)_{n,m} = H_n \cdot G_m \\ Filter(2)_{n,m} = G_n \cdot H_m \\ Filter(3)_{n,m} = G_n \cdot G_m \end{cases}$$

  In order to keep the frequencies (along the x- and y-axis) increasing, we then define $F(d, f_j)_{n,m} = Filter(gray\_code2d(d, f_j))_{n,m}$ where :

$$gray\_code2d(d, f_j) = 2 \cdot a + b$$

$$a = \begin{cases} d \bmod 2 & if\ f_j\ is\ even \\ 1-(d \bmod 2) & if\ f_j\ is\ odd \end{cases}$$

$$b = \begin{cases} \left[\frac{d}{2}\right] & \text{if } \left[\frac{f_j}{2}\right] \text{ is even} \\ 1 - \left[\frac{d}{2}\right] & \text{if } \left[\frac{f_j}{2}\right] \text{ is odd} \end{cases}$$

The symbol [x] is the integer part of x.

- At a scale j+1, there are four times as many wavelet packets as there are at scale j, and each wavelet packet has four times fewer coefficients than those at scale j.
- From the first remark, with an initial signal of 2^{Xmax} by 2^{Ymax} points, we deduce that :
  . At scale j, there are 4^j wavelet packets, each with 2^{Xmax-j} by 2^{Ymax-j} coefficients, so we still have 2^{Xmax} x 2^{Ymax} coefficients.
  . We can then use the following representation of the two-dimensional wavelet packet decomposition.

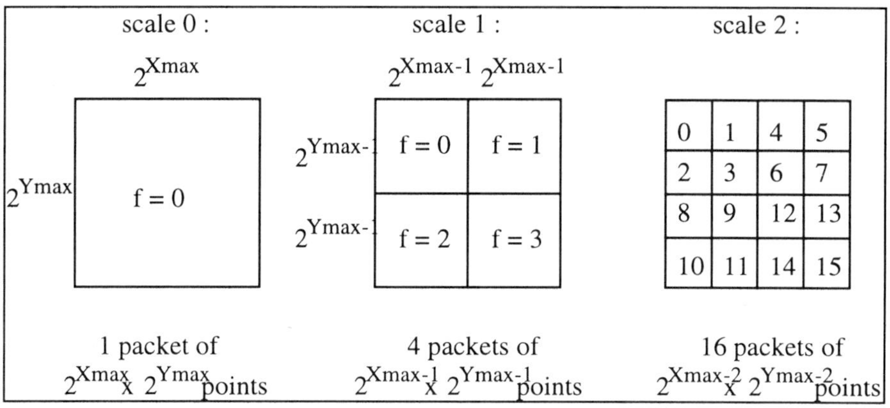

Figure 8.2.1 : Data arrangement for a 2D wavelet packet decomposition

  . If Xmax=Ymax=nmax, then at scale nmax, we have 4^{nmax} wavelet packets of 1 coefficient each.

To do the decomposition in parallel, we will separate each scale (represented by a field of 2^{Xmax} by 2^{Ymax} points) into 4^{jp} ($0 \leq jp < nmax$) parts of equal size. Each process always stores the same part of the field so we can identify the process name by the name of the packet it contains at scale jp. For example, dividing the wavelet packet onto 4 processes (jp = 1) from scale 0 to 2 gives the following :

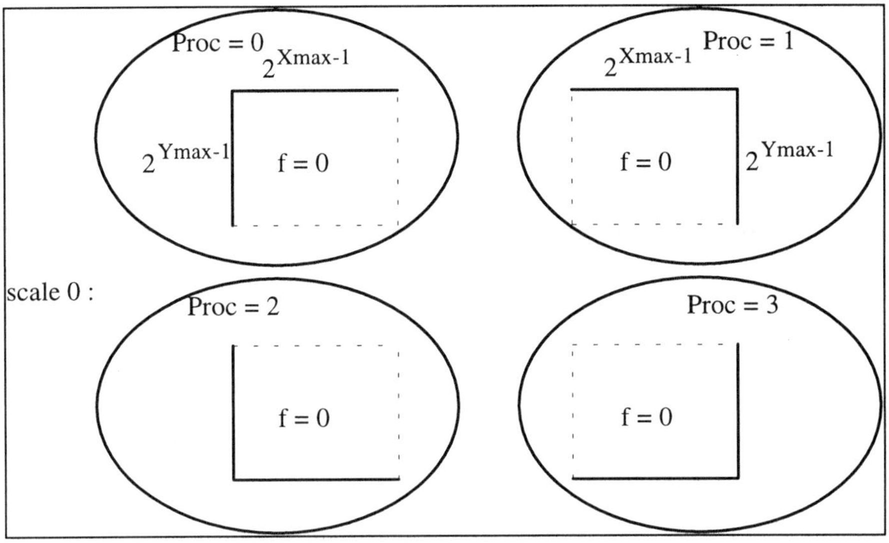

Figure 8.2.2 :
Repartition of the wavelet packets of scale 0 onto 4 processes

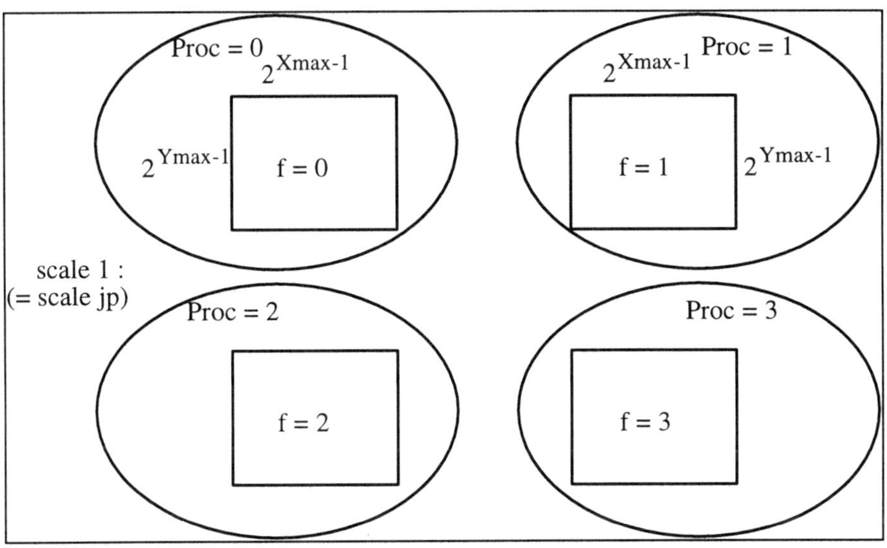

Figure 8.2.3 :
Repartition of the wavelet packets of scale 1 onto 4 processes

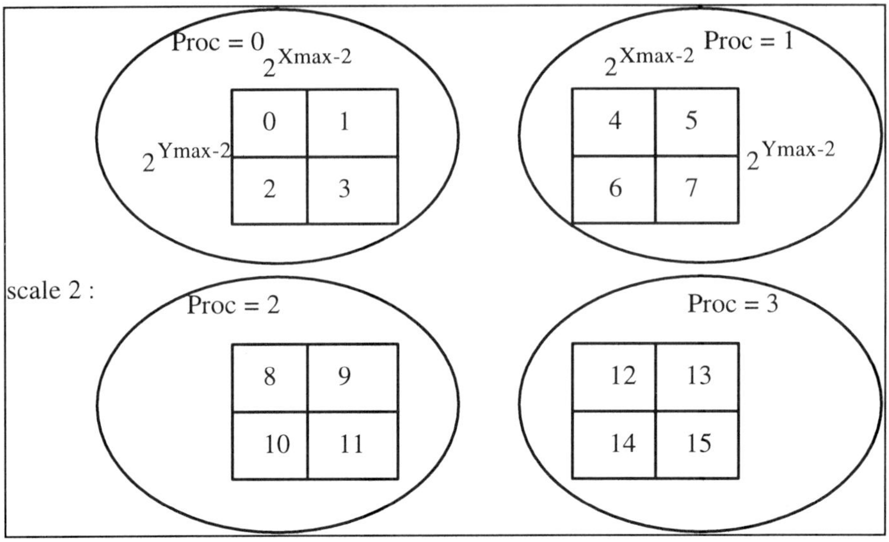

Figure 8.2.4 :
Repartition of the wavelet packets of scale 2 onto 4 processes

From the decomposition formula, one sees how the packets at scale j+1 are obtained from those at scale j. To calculate the decomposition in parallel, after cutting up the field as shown above, we will use the formula in two different ways depending upon whether j is smaller or greater than jp.

First, let us consider the case when j < jp.
A packet at scale j is shared by 4^{jp-j} processes. In the initial step, those processes will exchange their data in order to have the whole shared packet on each of them. Then, in the second step, they will calculate their own part of the packet they share at scale j+1. This part will be the entire child packet when scale j+1 equals jp.

From the decomposition formula we obtain that at scale j, the frequency f of the packet stored on process p is given by $f = \left[\frac{p}{4^{jp-j}}\right]$ where [x] is the integer part of x. Reciprocally, a packet at scale j and frequency f is shared by the processes $p = 4^{jp-j} \cdot f$ to $p = 4^{jp-j} \cdot (f+1) - 1$. We shall say that the first one is the base process and all processes sharing the same packet are companions. The base process is the one with the upper left-hand corner of the wavelet packet array.

In the same way, at scale j, the position (base_x, base_y) of the upper left corner of the packet stored on process p is given by :

$$- \; base\_x = cx\left(p,\ jp,\ 2^{Xmax}\right) - cx\left(\left[\frac{p}{4^{jp-j}}\right],\ j,\ 2^{Xmax}\right) \text{ where}$$

$$cx(p, jp, 2^{Xmax}) = \begin{cases} 0 & if\ jp=0 \\ \sum_{m=1}^{jp} 2^{Xmax-m} . \left\{\left[\frac{p}{4^{jp-m}}\right] \bmod 2\right\} & otherwise \end{cases}$$

$$- \; base\_y = cy\left(p,\ jp,\ 2^{Ymax}\right) - cy\left(\left[\frac{p}{4^{jp-j}}\right],\ j,\ 2^{Ymax}\right) \text{ where}$$

$$cy(p, jp, 2^{Ymax}) = \begin{cases} 0 & if\ jp=0 \\ \sum_{m=1}^{jp} 2^{Ymax-m} . \left\{\left[\frac{\left[\frac{p}{4^{jp-m}}\right]}{2}\right] \bmod 2\right\} & otherwise \end{cases}$$

The width and height of the part of the packet stored on a process at scale j is $2^{Xmax-jp}$ by $2^{Ymax-jp}$.

Example : Suppose we have 16 processes (jp=2) and we try to calculate which processes share the packet at scale j=1 and frequency f=2.

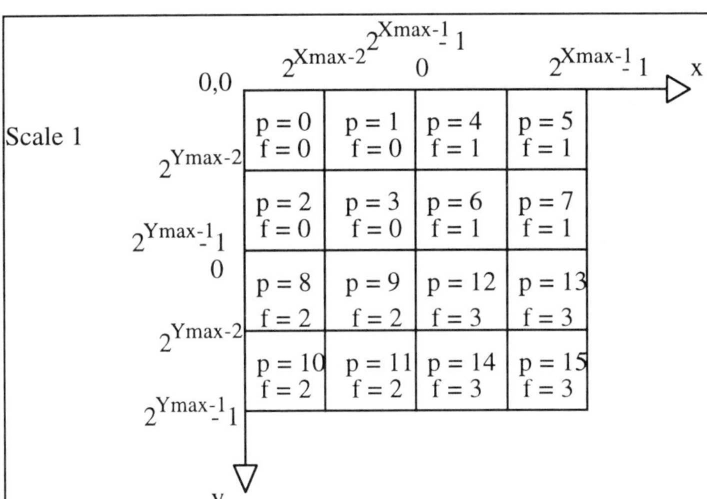

Figure 8.2.5 :
Example of the repartition of wavelet packet coefficients at scale 1 with 16 processes.

From these hypotheses, $f=\left[\frac{p}{4}\right]=2$, so we deduce from the former formulas that the processes sharing ${}^{2}C^{1}$ are numbered *8* to *11*. Each of them has wh = $2^{\text{Xmax-jp}}$ by ht = $2^{\text{Ymax-jp}}$ data points arranged as follows :

Process *8* contains the local packet ${}^{2}lC^{1}_{0..wh-1,\,0..ht-1} = {}^{2}C^{1}_{0..wh,0..ht-1}$

Process *9* contains the local packet ${}^{2}lC^{1}_{0..wh-1,\,0..ht-1} = {}^{2}C^{1}_{wh..2wh-1,0..ht-1}$

Process *10* contains the local packet ${}^{2}lC^{1}_{0..wh-1,\,0..ht-1} = {}^{2}C^{1}_{0..wh,ht..2ht-1}$

Process *11* contains the local packet ${}^{2}lC^{1}_{0..wh-1,\,0..ht-1} = {}^{2}C^{1}_{wh..2wh-1,ht..2ht-1}$

We will do the transfers as follows, looping in the index jj :

- We start with jj equals jp-1 and define a pivot process named depl, where $depl=4\cdot\left[\frac{p}{4^{jp-jj}}\right]$. In the initial step, processes p verifying $\left[\frac{p}{4^{jp-jj}}\right]=depl+i$ (A) exchange their data with processes $\left[\frac{p}{4^{jp-jj}}\right]=depl+1+i$ (B) where i equals 0 or 2. At the end of this step, the processes (A) and (B) share the same data. In the second step, processes verifying $\left[\frac{p}{4^{jp-jj}}\right]=depl+i$ (C) and processes verifying $\left[\frac{p}{4^{jp-jj}}\right]=depl+2+i$ (D) with i equals 0 or 1 will exchange the data they got from step one. At the end of these two steps, the four processes share the same data.
- We decrement jj by one and repeat until jj equals j.
- At this point, each process sharing a packet contains all the coefficients of that packet.

For definiteness, let us consider an example :

Suppose we have four processes (jp = 1) and we want to calculate the packets at scale 1 from the packet at scale 0. We have j=0, jj = jp-1=0. In this example, there will be only one loop over jj. In the first step process 0 exchanges its data with process 1 (while process 2 exchanges its data with process 3). Then in the second step process 0 exchanges its new set of data (its initial data plus process 1 initial data) with process 2 new set of data (while process 1 does the same with process 3). We exit the loop on jj, and every processes now carries the packet ${}^{0}C^{0}$.

The pseudo-code representation of the algorithms described above is given in the annex 1 "Decomposition".

To calculate the transfer complexity, we will assume that the cost of a data transfer is independent of the size of the data set. This is approximately true for a large size data set. Thus, the transfer complexity will only show an order of the number of transfers without any specification of the sizes of the data sets.

Counting passes through the jj loop, we see that the transfer complexity for the function *Data transfers* is then 4*(jp-j).

In this previous algorithm (when scale $j < jp$), the function *Data transfers* is done jp times so we deduce that the total transfer complexity is of order O(2*jp*(jp+1)). The decomposition complexity is of order $O(jp*2^{Xmax}*2^{Ymax}/4^{jp})$, where the constant depends on the length of the filters. In simple terms, if $NP=4^{jp}$ is the number of processes and if we have $N=4^{nmax}$ points (Xmax=Ymax=nmax), then the transfer complexity is $O(2*[\log_4(NP)]^2)$ and the decomposition complexity is $O(P*\log_4(NP)*N/NP)$.

Second, let us consider the case when $j \geq jp$.
Each process has 4^{j-jp} packets, and from each of these, we need to calculate their four children. In this case, there is no parallelism as the children also belong to the process. We will then use the sequential algorithm on the subset of packets each process carries at scale j.

For $j \geq jp$, there is no transfer complexity. The decomposition complexity is of order $O((nmax-jp)*2^{Xmax}*2^{Ymax}/4^{jp})$. If we consider $NP = 4^{jp}$ processes and $N = 4^{nmax}$ points, then the total complexity of this part is $O((\log_4(N)-\log_4(NP))*N/NP)$.

Combining the calculations from the two different parts of the parallel decomposition algorithm, we conclude that the transfer complexity is of order O(2*jp*(jp+1)) and the computational complexity is of order $O(nmax*2^{Xmax}*2^{Ymax}/4^{jp})$.

Thus, if $N=4^{nmax}$ is the number of points (Xmax=Ymax=nmax) and $NP=4^{jp}$ is the number of processes then the transfer complexity is $O(2*[\log_4(NP)]^2)$ and the computational complexity is $O(\log_4(N)*N/NP)$.

8.3. PARALLEL WAVELET PACKET BEST BASIS SELECTION

At the end of the decomposition, we have obtained the complete discrete analysis of the input data. We must now extract a subset of wavelet packets forming a basis. There are many methods to extract a basis : keeping only the wavelet packets at a certain scale, for example, gives one solution.

But if we consider a family of wavelet packets ${}^{f}C^{j}\ with\ j \in J\ and\ f \in F_J$, selecting a basis among all those packets means that if a packet ${}^{f_0}C^{j_0}$ belongs to the basis, all its descendants, the packets ${}^{4*f_0+d_0}C^{j_0+1}$, ${}^{4*(4*f_0+d_0)+d_1}C^{j_0+2}$, etc ... (with d_0, d_1 equals 0,1,2 or 3) cannot belong to the same basis.

We will maximize a concentration criterion to find a "best basis". For each packet ${}^{f}C^{j}$, we compute a value $M\left({}^{f}C^{j}\right)$ using a cost function M on the set.

For example, we can take a measure M which counts the non null wavelet packet coefficients of a wavelet packet. Another commonly used measure is Me, which evaluates the entropy of a wavelet packet expansion :

$$\left\{\begin{array}{rcl} Me(\varnothing) & = & 0 \\ Me({}^{f}C^{j}) = -\sum\limits_{k,l}({}^{f}C^{j}_{k,l})^2 * \ln\left[({}^{f}C^{j}_{k,l})^2\right] \end{array}\right\}.$$

Once the measure is chosen, we shall produce a family of wavelet packets that satisfies the condition for being a basis and that minimizes the value of the measure.

This is done by an iterative algorithm. We define ${}^{f}C^{j}_{value}$ to be the value of the measure of the best basis of wavelet packets in the subset below ${}^{f}C^{j}$. First, for all the frequencies f at scale j=nmax, we initialize ${}^{f}C^{j}_{value} = M\left({}^{f}C^{j}\right)$. Then we start the algorithm at scale j=nmax-1. For all the frequencies f at scale j, we compare $v_0 = M\left({}^{f}C^{j}\right)$ to $v_1 = \sum\limits_{d=0}^{3} {}^{4*f+d}C^{j+1}_{value}$. If v_0 is smaller than or equal to v_1 we keep the wavelet packet ${}^{f}C^{j}$, we set ${}^{f}C^{j}_{value} = v_0$, and we discard all its descendants. If v_0 is greater than v_1, we keep the four children packets ${}^{4*f+d}C^{j+1}$, and we set ${}^{f}C^{j}_{value}$ to v_1. We decrement j by one and iterate until j equals 0. At scale 0, all the kept packets form the "best basis" of the signal.

To select the best basis in parallel, we will use how the wavelet packets are stored on the different processes after the decomposition. Thus, if we have 4^{jp} processes, then from scale 0 to scale jp-1 the wavelet packets will be shared by more than one process, while from scale jp to nmax they will be part of a single process.

From the definition of the best basis selection, one can see that there will be two different algorithms, depending on whether j is greater or smaller than jp. In order to know if a packet is kept or not, we define a variable ${}^{f}C^{j}_{status}$ (${}^{f}lC^{j}_{status}$

for a local packet) equal to *KEPT* if the wavelet packet ${}^{f}C^{j}$ is part of the best basis, and *NOT_KEPT* if it is not.

Before selecting the best basis, we will suppose that the user has already chosen a measure M and has already computed the value $M\left({}^{f}C^{j}\right)$ for every packet of the decomposition. Hence it will not be counted in the computational complexity.

First, let us consider the case when j ≥ jp.
In that case, the best basis algorithm is the same as the sequential one except that it is done on each subset of wavelet packets which belongs to each process.

The computational complexity of this algorithm is of order O(4/3*($4^{nmax-jp}$ - 1)). Therefore, if we have NP=4^{jp} processes and N=4^{nmax} points, this complexity becomes O(4/3*(N/NP - 1)).

Second, let us consider the case when j < jp.
From the parallel decomposition, we have seen that at scale j, process p contains the local packet ${}^{\left[\frac{p}{4^{jp-j}}\right]}lC^{j}$. At scale j, the packet ${}^{f}C^{j}$ is then shared by the processes $p=4^{jp-j}\cdot f$ to $p=4^{jp-j}\cdot(f+1)-1$. With another small calculation, one can see that the packets ${}^{4*f+d}C^{j+1}$ $(d=0..3)$ are shared by this same list of processes.

From these three remarks, we deduce that if we add up $lv_0 = M\left({}^{\left[\frac{p}{4^{jp-j}}\right]}lC^{j}\right)$ for all the processes $p=4^{jp-j}\cdot f\_base$ to $p=4^{jp-j}\cdot(f\_base+1)-1$ where $f\_base=\left[\frac{p}{4^{jp-j}}\right]$, we will obtain $v_0 = M\left({}^{f\_base}C^{j}\right)$. Similarly, if we add up ${}^{\left[\frac{p}{4^{jp-j-1}}\right]}lC^{j+1}_{value}$ for the same processes, we will get $v_1 = \sum_{d=0}^{3} {}^{4*f\_base+d}C^{j+1}_{value}$.

One solution consists in adding up all those local values in order to get v_0 and v_1. However, this is redundant : we do the same calculation at scale j+1 and at scale j, once for calculating v_0 and the second time for v_1. But we can easily change the sequential algorithm to improve this if at each scale j, after having calculated v_0 and v_1, we set the local value ${}^{\left[\frac{p}{4^{jp-j}}\right]}lC^{j}_{value}$ to the winner.

Then, each process sharing ${}^{4 \cdot f\_base+d}C^{j+1}$ will have the same local value ${}^{4*f\_base+d}lC^{j+1}_{value}$, so the calculation of v_1 as we have already explained it, does not give v_1 but gives $4^{jp-j-1} \cdot v_1$. Although it seems worse than the first solution, this approach is in fact better, because the calculation of v_1 needs only the addition of four local values instead of 4^{jp-j}. Indeed, we can now calculate v_1 by the following formula : $v_1 = \sum_{i=0}^{i=3} {}^{\left[\frac{depl+\left(p \bmod 4^{jp-j-1}\right)+4^{jp-j-1}\cdot i}{4^{jp-j-1}}\right]}lC^{j+1}_{value}$ with $depl = 4^{jp-j} \cdot f\_base$.

An example might clarify this formula :

> Suppose that jp=2, j=0 and p=5. To calculate v_1, the previous formula gives $v_1 = {}^{1}lC^{j+1}_{value} + {}^{5}lC^{j+1}_{value} + {}^{9}lC^{j+1}_{value} + {}^{13}lC^{j+1}_{value}$ so process 5 only has to exchange its data with processes 1, 9 and 13.

If we store the information costs locally in this way, then the parallel best-basis search algorithm for process p at levels j < jp becomes the following :

- We start at j = jp-1.
- First, we calculate $f\_base = \left[\frac{p}{4^{jp-j}}\right]$ and $depl = 4^{jp-j} \cdot f\_base$.

  Second, we calculate $v_0 = \sum_{i=0}^{4^{jp-j}-1} M\left({}^{\left[\frac{depl+i}{4^{jp-j}}\right]}lC^{j}\right)$ by adding all the values of the local packets at scale j. Third, we calculate $v_1 = \sum_{i=0}^{i=3} {}^{\left[\frac{depl+\left(p \bmod 4^{jp-j-1}\right)+4^{jp-j-1}\cdot i}{4^{jp-j-1}}\right]}lC^{j+1}_{value}$ by adding all the values of the local packets at scale j+1. Finally, we compare v_0 to v_1. If v_0 is smaller than or equal to v_1 then we keep the local wavelet packet ${}^{\left[\frac{p}{4^{jp-j}}\right]}lC^{j}$, we set ${}^{\left[\frac{p}{4^{jp-j}}\right]}lC^{j}_{value}$ to v_0, and we discard all its descendants. If v_0 is greater than v_1, we keep the local packet ${}^{\left[\frac{p}{4^{jp-j-1}}\right]}lC^{j+1}$, and we set ${}^{\left[\frac{p}{4^{jp-j}}\right]}lC^{j}_{value}$ to v_1.
- We decrement j by one and iterate until j equals 0.

One way to exchange the local values is to take all the processes involved and perform pairwise exchanges until every process has the total sum. Figure 8.3.1 depicts this algorithm graphically.

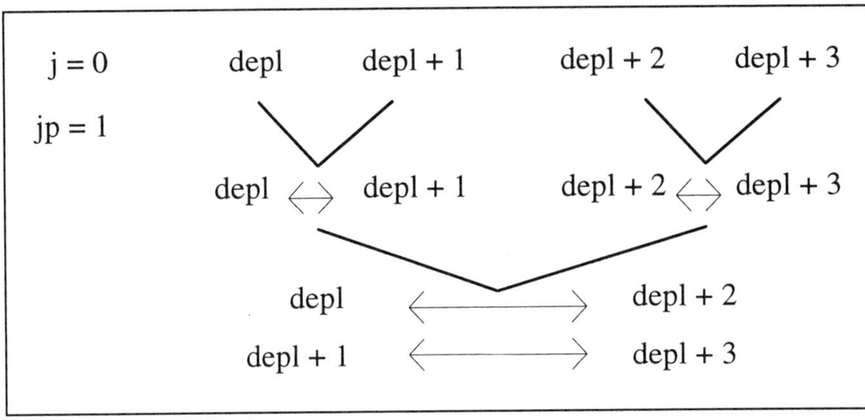

Figure 8.3.1 :
Transfer of the local values v_0 between processes depl to depl+3

To calculate v_1, we will use the same principle, except that the first transfer may not be anymore between process depl and depl+1, but might instead be between depl and depl+4^{jp-j-1}. The transfer complexity of this function is of order $O(2*\log_2(np/p\_shift))$ and the computation complexity is of order $O(\log_2(np/p\_shift))$.

The pseudo-code representation of the algorithms described above is given in the annex 2 "Best-basis".

Since we do the external loop jp times and we call the function *Local value transfer* twice, the first time with *np/p_shift* equaling 4^{jp-j} and the second time with *np/p_shift* equaling 4, we deduce that the transfer complexity is of order $O(2*jp*(jp+1))$ and the computational complexity is half of that, or $O(jp*(jp+1))$. If $NP=4^{jp}$ is the number of processes, then the transfer complexity is of order $O(2*[\log_4(NP)]^2)$ and the computational complexity is of order $O([\log_4(NP)]^2)$.

For the whole parallel best basis algorithm, the transfer complexity is of order $O(2*jp*(jp+1))$ and the computational complexity is of order $O(4/3*(4^{nmax-jp}-1) + jp*(jp+1))$.

In the case when we have $N=4^{nmax}$ points and $NP=4^{jp}$ processes, the transfer complexity is $O(2*[\log_4(NP)]^2)$ and the computational complexity is $O(4/3*N/NP + [\log_4(NP)]^2)$.

8.4. PARALLEL WAVELET PACKET RECONSTRUCTION

The reconstruction is the opposite of the decomposition, from a family of kept wavelet packets ${}^{f_j}C^j$ with $j \in J$ (where $J \subset \mathrm{N}_{nmax} = \{n \in \mathrm{N}; 0 \le n \le nmax\}$) and $f_j \in F_J$ (where for each $j \in J$, $F_j \subset \mathrm{N}_{4^j-1}$), we want to obtain $s_{k,l} = {}^0C^0_{k,l}$ (where $0 \le k < 2^{Xmax}$ and $0 \le l < 2^{Ymax}$) which is the corresponding signal representation (S). We obtain S by applying the following recursive formula until we get to ${}^0C^0$:

$$^{f_{j-1}=\left[\frac{f_j+d}{4}\right]}C^{j-1}_{k_{j-1},l_{j-1}} = \sum_{d=0}^{3}\sum_{n_j,m_j} F\left((f_j+d) \bmod 4, \left[\frac{f_j+d}{4}\right]\right)_{k_{j-1}-2n_j,\, l_{j-1}-2m_j} * {}^{f_j+d}C^j_{n_j,m_j}$$

where $F(d,f_j)_{n,m} = Filter(gray\_code2d(d,f_j))_{n,m}$ is exactly the same as for the decomposition.

Usually we do not program this formula directly, rather we reconstruct a whole branch of the tree, one level at a time. Instead of having its entire father $^{\left[\frac{f_j}{4}\right]}C^{j-1}$, we only have one part of it, but if we reconstruct from all the packets at scale j, we will obtain their entire fathers. The formula we program is then :

$$^{f_{j-1}=\left[\frac{f_j}{4}\right]}C^{j-1}_{k_{j-1},l_{j-1}} = {}^{f_{j-1}}C^{j-1}_{k_{j-1},l_{j-1}} + \sum_{n_j,m_j} F\left(f_j \bmod 4, \left[\frac{f_j}{4}\right]\right)_{k_{j-1}-2n_j,\, l_{j-1}-2m_j} * {}^{f_j}C^j_{n_j,m_j}$$

and the first time we encounter the packet ${}^{f_{j-1}}C^{j-1}_{k_{j-1},l_{j-1}}$ we must initialize it to the null packet (its coefficients being all zeros) if it is not a kept packet.

We suppose again that we have 4^{jp} processes, and that the data are dispatched to each of them as before. In the same way as the sequential algorithm, we will start at scale nmax and reconstruct the ancestors of each kept packet until we reach scale 0. There will again be two different algorithms, one when scale j is greater than jp, and the other when scale j is less than or equal to jp.

First, we consider the case when j > jp, the algorithm is the same as the sequential one except that it is done on the subset of wavelet packets that belong to a process.

In the worst case, the best basis is composed of all the packets at scale nmax. In that case, the computational complexity is of order $O((nmax-jp) * 2^{Xmax}*2^{Ymax}/4^{jp})$. In all other cases it is less than this.
If we have $N=4^{nmax}$ points and $NP=4^{jp}$ processes, the computational complexity is therefore smaller than or equal to $O((\log_4(N)-\log_4(NP))*N/NP)$.

Second, we consider the case when scale j ≤ jp, we suppose initially that each process has an entire data set. Each of them will do the reconstruction locally using the sequential algorithm, and when they reach scale 0, each of them will have a partial signal. We will then need to add all these partial signals in order to get the whole signal. Unfortunately, even if we do not have a lot of transfers, the calculation complexity is exactly the same as for the sequential algorithm, so the only advantage is that we need less memory. Instead we will propose another solution which has a lower computational complexity but might require a larger number of transfers.

We proceed in the following way. At scale j (1 ≤ j ≤ jp), we reconstruct the local packet ${}^{f=\left[\frac{p}{4^{jp-j}}\right]}lC^{\,j}$ in a full size temporary father packet ${}^{f/4}tC^{\,j-1}$ using the formula : ${}^{f/4}tC^{\,j-1}_{k,l} = {}^{f/4}tC^{\,j-1}_{k,l} + \sum_{n,m} FF_{n-2(base\_x+k),m-2*(base\_y+l)} * {}^{f}lC^{\,j}_{n,m}$ where FF is the filter F(f mod 4,f / 4), and base_x and base_y are the x- and y-offsets of the local packet ${}^{f}lC^{\,j}$. At this point we must add the local packet ${}^{f/4}lC^{\,j-1}$ to this temporary packet if it is a kept packet. Then, each process sharing the packet ${}^{f/4}C^{\,j-1}$ exchanges and adds all the temporary packets in order to have the whole one. After the transfer, they keep their own part of this father ${}^{f/4}lC^{\,j-1}$. Then we decrement j by one and iterate until j equals 0.

One problem occurs when all concerned processes have to exchange the temporary packets : what happens when not all the processes are waiting for the transfer ? This case is not impossible, it happens when there is no kept packet when scale j is greater than or equal to j0 (with 0 ≤ j0 ≤ jp) on at least one process. We handle this problem in a very simple manner : if the packet ${}^{p}lC^{\,jp}$ does not exist, we pretend it is the null packet. The propagation of this packet to all scales j smaller than jp will handle the problem once and for all.

We first discuss the function which transfers packets between the concerned processes, then we explain the computational part of the reconstruction algorithm.

The function doing these transfers is similar to the local value transfer function we have presented in the best basis selection algorithm. Here, we consider the transfer and the addition of local arrays instead of local values.

The pseudo-code representation of the algorithms described above is given in the annex 3 "Reconstruction".

The transfer complexity of this function is of order O(2*log$_2$(np/p_shift)) and the computational complexity is of order O(size_array*log$_2$(np/p_shift)).

As the function *Local array transfer* is done jp times, we deduce that the transfer complexity of the algorithm is of order O(2*jp*(jp+1)) and the computational complexity is of order O(jp*(jp+1)*2^{Xmax}*2^{Ymax}/4^{jp}).
If N=4^{nmax} points and NP=4^{jp} processes, the transfer complexity is O(2*[$\log_4$(NP)]^2) and the computational complexity is O([$\log_4$(NP)]^2*N/NP).

In the worst case, if we consider that the best basis is composed of all the wavelet packets at scale nmax, the transfer complexity is O(2*jp*(jp+1)) and the computational complexity is O((nmax+jp^2)*2^{Xmax}*2^{Ymax}/4^{jp}).

If we now suppose that N=4^{nmax} points and NP=4^{jp} processes, the transfer complexity is smaller than or equal to O(2*[$\log_4$(NP)]^2) and the computational complexity is smaller than or equal to O(($\log_4$(N)+[$\log_4$(NP)]^2)*N/NP).

8.5. PARALLEL MATRIX-VECTOR MULTIPLICATION IN THE WAVELET PACKET BASIS EXPANSION

We wish to compute the product of a matrix C = ($c_{i,j}$) by a vector D = (d_i) in the wavelet packet expansion. We suppose that $0 \le i < 2^{Xmax}$ and $0 \le j < 2^{Ymax}$ so the result is a vector E = (e_j). We perform the multiplication in 3 steps.

During the first step we decompose the matrix C in a wavelet packet best basis, we obtain a family of kept wavelet packets ${}^{cf_j}C^j$ with $j \in J$ (where $J \subset \mathrm{N}_{nmax} = \{n \in \mathrm{N}; 0 \le n \le nmax\}$) and $cf_j \in F_J$ (where for each $j \in J$, $F_j \subset \mathrm{N}_{4^j-1}$). In the second step, we decompose the vector D in all the possible one dimensional wavelet packets ${}^{df_j}D^j$ with $0 \le j \le nmax$ and $0 \le df_j < 2^j$.

Each frequency cf_j is the combinaison of a frequency cf_x_j along the x-axis and a frequency cf_y_j among the y-axis. We write that $cf_j = (cf\_x_j, cf\_y_j)$ and we calculate cf_x_j and cf_y_j by :

$$cf\_x_j = \sum_{jj=0}^{j-1} 2^{j-jj-1} \cdot \left(\left\lceil \frac{cf_j}{4^{j-jj-1}} \right\rceil \bmod 2 \right)$$

$$cf\_y_j = \sum_{jj=0}^{j-1} 2^{j-jj-1} \cdot \left(\left\lceil \frac{\left\lfloor \frac{cf_j}{4^{j-jj-1}} \right\rfloor}{2} \right\rceil \bmod 2 \right)$$

The multiplication in the wavelet packet expansion consists in obtaining all the packets ${}^{cf\_y_j}E^j$ which are the solution of the multiplication of ${}^{cf_j=(cf\_x_j,cf\_y_j)}C^j$ by ${}^{cf\_x_j}D^j$.

The third and final step is to reconstruct all the packets ${}^{cf\_y_j}E^j$ to obtain the vector E.

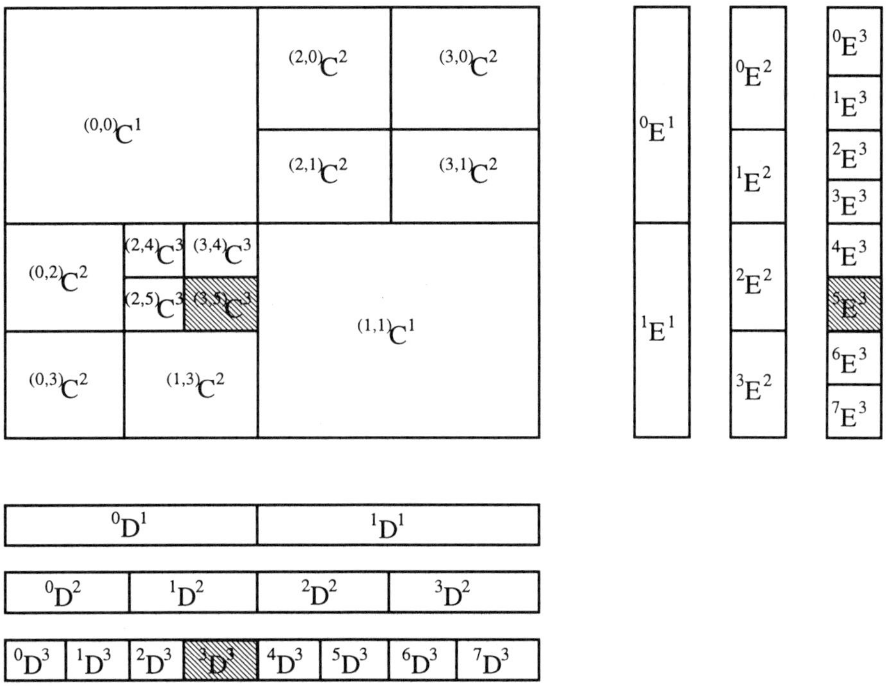

Figure 8.5.1 :
Example of the multiplication in the wavelet packet expansion,
we consider ${}^5E^3 = {}^{(3,5)}C^3 \times {}^3D^3$

We can use the parallel algorithms previously described in order to perform the first step and the multiplication.

At the beginning, we use 4^{jp} processes to decompose thc matrix C in all the wavelet packets and we select its best basis. On each process, we decompose the vector D in all the one dimensional wavelet packets.

The multiplication between the wavelet packets can now be done as follows :

- for each kept packet ${}^fC^j$ at scale j, we first find the frequencies fx and fy corresponding to f, then

- if j is smaller than jp we do a multiplication between ${}^{f}lC^{j}$ and the corresponding local packet of ${}^{fx}D^{j}$ (named ${}^{fx}lD^{j}$) to obtain the local packet ${}^{fy}lE^{j}$. We insert it in the null packet ${}^{fy}E^{j}$.
- if j is greater than or equal to jp, we do the multiplication between ${}^{f}C^{j}$ and ${}^{fx}D^{j}$ to have ${}^{fy}E^{j}$.

At the end of the multiplication, each process has a partial vector E in a wavelet packet expansion (it is not usually a basis). They will then reconstruct all the kept packets to get their partial vector E.

In a last part, each process exchanges and adds all the partial vectors E in order to get the complete result of the multiplication of C by D.

The pseudo-code representation of the algorithm described above are given in the annex 4 "Matrix-vector multiplication".

To calculate the transfer and computational complexities, we have to add the complexities of one parallel 2d decomposition, one parallel 2d best basis selection, one sequential 1d decomposition, one matrix vector multiplication, one sequential 1d recomposition and one array transfer.

If we consider the case when all the coefficients in the best basis are non negligible, this is of course much worse than the direct computation. But, if instead of all the coefficients N, we consider that only R are non negligible, the matrix-vector multiplication becomes of order O(R), and if R << N then this method is more efficient than the ordinary multiplication.

8.6. APPLICATIONS OF THE WAVELET PACKET TRANSFORM TO THE COMPRESSION AND ANALYSIS OF A TWO-DIMENSIONAL TURBULENT FLOW

8.6.1. Methodology

We will now apply the wavelet packet transform algorithm to compress a two-dimensional turbulent field which has been computed using a pseudo-spectral Galerkin integration of the Navier-Stokes equations. We will represent this field as a sequence $S_{k,l}$ where $0 \le k < 2^{Xmax}$ and $0 \le l < 2^{Ymax}$.

Our first goal will be to decompose this turbulent field into two distinctive parts : its inhomogenous and organized components $S^{>}$, called coherent structures, and its homogenous and random components $S^{<}$ called background noise. Both components contain energy at all scales, so finding an optimal representation to perform this segmentation is not straightforward. An example is the Fourier representation; although the spatial information is kept in the phases, it is quite complicated to use it in that form. The grid-point representation is not a good one either, because we want to be able to compute the power spectrum of both components. The decay of energy at small scales, or the scaling law, provides an information very relevant for the stastistical theory of turbulence. To segment the field, we will look for the fewest number of wavelet packet coefficients necessary to adequately describe the coherent structures. We will select the fewest wavelet packet coefficients such that the discarded is homogeneous and its Fourier spectrum is a broad-band noise.

Our second goal will be to study an individual coherent structure. From the non compressed wavelet packet representation of the field, we will extract one coherent structure and compute its local spectrum to analyze its scaling law. The number of wavelet packet coefficients defining this structure will then give us an estimate of the number of degrees of freedom attached that it has. This is a kind of "theoretical dimension" of the coherent structure.

8.6.2. The compression technique

From the wavelet packet best-basis representation of the field, namely a family of kept wavelet packets ${}^{f_j}C^j$ with $j \in J$ (where $J \subset \mathrm{N}_{nmax} = \{n \in \mathrm{N}\,; 0 \leq n \leq nmax\}$) and $f_j \in F_J$ (where for each $j \in J$, $F_j \subset \mathrm{N}_{4^j-1}$), we will only keep the most energetic coefficients containing a given percentage of the total energy.

In sequential, the compression algorithm is implemented as follows : first, we sort the squares of all the wavelet packet coefficients in a decreasing order. Then, we keep as many coefficients as needed in order that the remaining field has a certain percentage of the initial energy (sum of the squares of the field's coefficients). This gives us a deterministic way of selecting a threshold ε underneath which we set all the wavelet packet coefficients to zero.

The parallel algorithm is not very different : in a first step, each process will sort the squares of the wavelet packet coefficients that belong to them. Then, two by two they will exchange their sorted list, and will perform a merge sort on those two lists to obtain the sorted list corresponding to both processes. They will repeat this step until each process has the whole sorted list of coefficients. This exchange of data is done as presented in paragraph 8.4 in the function *Local_array_transfer*, the difference is that here we merge the two arrays instead of adding them. In a last part, each process selects the coefficients as in the sequential algorithm and only keeps those that belong to itself.

The pseudo-code algorithm is given in the annex 5 "Compression".

8.6.3. The local extraction and local spectrum

The purpose of local extraction is to keep the wavelet packet coefficients ${}^{f}C_{k,l}^{j}$ that are positioned at a certain point (a_0, b_0) where $.(a_0, b_0) \in \left[0, 2^{Xmax}\right[\times\left[0, 2^{Ymax}\right[$, or in a certain interval $[a_1, a_2]\times[b_1, b_2]$ where $[a_1, a_2]\times[b_1, b_2]\subset\left[0, 2^{Xmax}\right[\times\left[0, 2^{Ymax}\right[$.

Each wavelet packet coefficient ${}^{f}C_{k,l}^{j}$ is localized in the space domain in the interval $\left[k_0-2^{j-1}, k_0+2^{j-1}\right]\times\left[l_0-2^{j-1}, l_0+2^{j-1}\right]$ where (k_0, l_0) is the center of mass of the wavelet packet coefficient and is defined as follows :

$$\begin{cases} k_0 = 2^j \cdot k + {}^{f}ShiftX^{j} \\ l_0 = 2^j \cdot l + {}^{f}ShiftY^{j} \end{cases} \quad \text{and}$$

$$\begin{cases} {}^{f}ShiftX^{j} = 2 * {}^{f/4}ShiftX^{j-1} + F(f \bmod 4, f/4)_{xshift} \\ {}^{f}ShiftY^{j} = 2 * {}^{f/4}ShiftY^{j-1} + F(f \bmod 4, f/4)_{yshift} \end{cases} \quad \text{and}$$

$$\begin{cases} F(d,f)_{xshift} = \dfrac{\sum\limits_{i,j} i * F(d,f)_{i,j}^{2}}{\sum\limits_{i,j} F(d,f)_{i,j}^{2}} \\ F(d,f)_{yshift} = \dfrac{\sum\limits_{i,j} j * F(d,f)_{i,j}^{2}}{\sum\limits_{i,j} F(d,f)_{i,j}^{2}} \end{cases}$$

We will keep this coefficient if the intersection between this interval and the interval $\left[a_0-2^{j-1}, a_0+2^{j-1}\right]\times\left[b_0-2^{j-1}, b_0+2^{j-1}\right]$ (respectively the interval $\left[a_1-2^{j-1}, a_2+2^{j-1}\right]\times\left[b_1-2^{j-1}, b_2+2^{j-1}\right]$) is not empty.

Extracting the local coefficients consists in keeping, for each wavelet packet belonging to the best-basis, the coefficients which verify this property.

To perform the local extraction, the algorithm will go over the wavelet packet tree in a depth-first order. This allows us to calculate ${}^{f}ShiftX^{j}$ and ${}^{f}ShiftY^{j}$ directly using the above formula. Then, for each kept wavelet packet of the tree, we will retain its local coefficients, namely those which are positionned in (a_0, b_0) (respectively $[a_1, a_2] \times [b_1, b_2]$).

The pseudo-code algorithm is given in the annex 6 "Local extraction".

After having locally extracted some wavelet packet coefficients, we can compute their corresponding local spectrum. Once we have chosen what area of the field we want to represent in the Fourier domain, either at a point (a_0, b_0) or in an interval $[a_1, a_2] \times [b_1, b_2]$, we perform the local extraction of that particular area and we reconstruct the filtered field by using only the coefficients kept during the extraction. Once the reconstruction is done, we compute the Fourier transform of this filtered field. We name it the local spectrum of the original field for the area (a_0, b_0) (respectively $[a_1, a_2] \times [b_1, b_2]$) as it is the spectrum of the coefficients corresponding to that area.

8.6.4. Compression and analysis of a two-dimensional turbulent flow

The Navier-Stokes equations describe the dynamics of a two-dimensional incompressible flow.
Considering the periodic plane $S = (0, 2\pi) \times (0, 2\pi) \in R^2$, and in the absence of external forcing, the Navier-Stokes equations are :

$$\begin{cases} \dfrac{\partial u}{\partial t} + (u \cdot \nabla) u + \nabla P - \nu \nabla^2 u = 0 & in\, S \times R^+ \\ \nabla \cdot u = 0 & in\, S \times R^+ \\ u(x,0) = u_0(x) & in\ S \end{cases}$$

where $\boldsymbol{u}$ is the velocity field, $\boldsymbol{P}$ the pressure field, and ν the kinematic viscosity.
We can rewrite these equations in terms of the vorticity $\omega = \dfrac{\partial u}{\partial x_1} - \dfrac{\partial u}{\partial x_2}$ and of

the stream function ψ defined such that $u = \begin{pmatrix} u_1 \\ u_2 \end{pmatrix} = \begin{pmatrix} -\dfrac{\partial \psi}{\partial x_2} \\ \dfrac{\partial \psi}{\partial x_1} \end{pmatrix}$.

The Navier-Stokes equations then become :

$$\left\{\begin{array}{ll}\frac{\partial \omega}{\partial t}+J(\psi,\omega)-\nu\nabla^2\omega=0 & in\, S\times R^+\\ \omega=\nabla^2\psi & in\, S\times R^+\\ \omega(x,0)=\omega_0(x) & in\, S\end{array}\right\} \text{ with } J(\psi,\omega)=\frac{\partial\psi}{\partial x_1}\cdot\frac{\partial\omega}{\partial x_2}-\frac{\partial\psi}{\partial x_2}\cdot\frac{\partial\omega}{\partial x_1}$$

The vorticity field we will consider was computed by Thierry Philipovitch[4] on a 512x512 grid, with a hyperdissipative operator $-(-\nabla^2)^4$ and no external forcing, using the numerical code developped by Claude Basdevant[4]. The time integration was done with an Adams-Bashforth scheme and the space integration with a pseudo-spectral Galerkin method. The code ran for 38000 time steps $\Delta t=10^{-3}\left(T^1\right)$, which corresponds to 635 initial eddy-turnover times, starting from a random distribution of vorticity with energy $E=0.5\left(M\,L^2\,T^{-2}\right)$ and an initial enstrophy $Z=279\left(T^2\right)$.

We first compressed the vorticity field and observed as shown in figure 8.6.4.2 that :

- the 3000 strongest wavelet packet coefficients, corresponding to a compression factor of 87, retain 99.9 % of the energy[5],
- the 739 strongest wavelet packet coefficients, corresponding to a compression factor of 355, retain 99 % of the energy,
- the 73 strongest wavelet packet coefficients, corresponding to a compression factor of 3591, retain 90 % of the energy,
- the 14 strongest wavelet packet coefficients, corresponding to a compression factor of 18724, retain 50 % of the energy.

---

[4] LMD, Ecole Normale Supérieure, Paris, France

[5] In turbulence the L^2-norm of the vorticity field is usually called enstrophy. In this paragraph, we have called it energy to conform with the notation of the first part of this chapter.

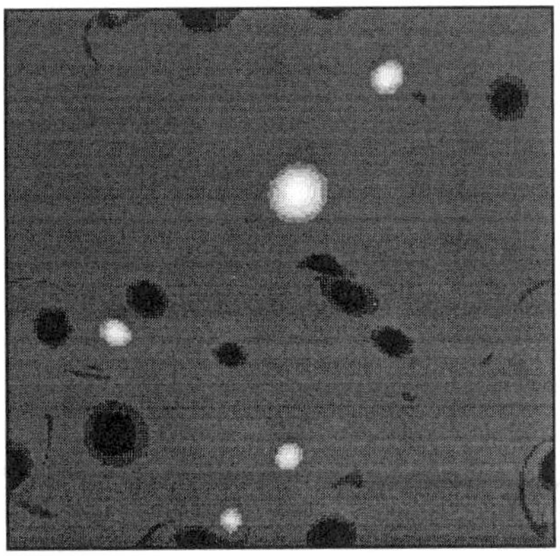

Figure 8.6.4.1 : Initial Vorticity field, 512x512 (262144) coefficients

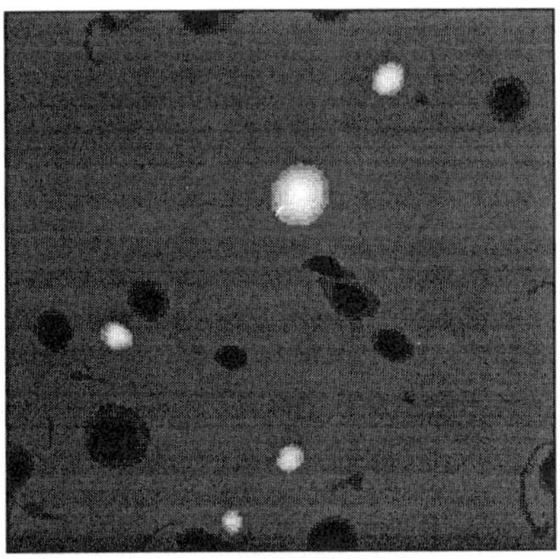

Part a : 99.9 % >

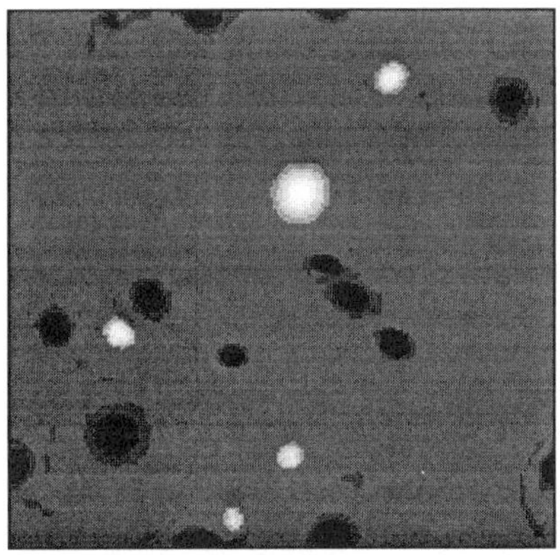

Part b : 99 % >

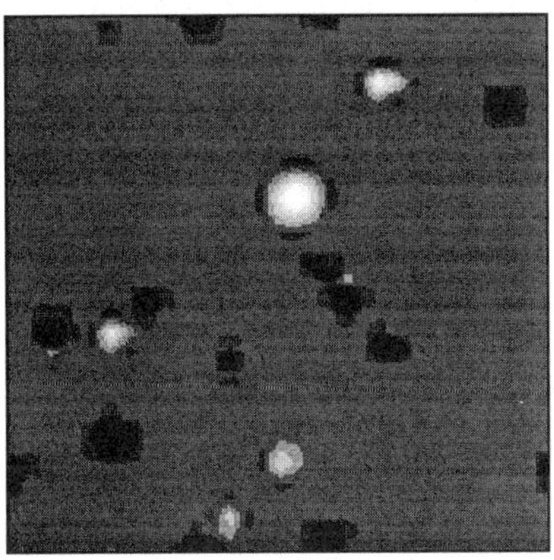

Part c : 90 % >

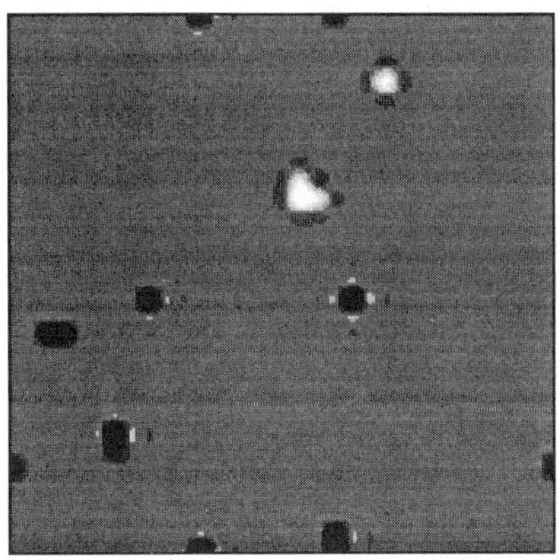

Part d : 50 % >

Figure 8.6.4.2 : The compressed fields $S^{>}$
Pallette : maximum 36.3 in white, minimum -38.1 in black

Part a : 0.1 % <

Part b : 1 % <

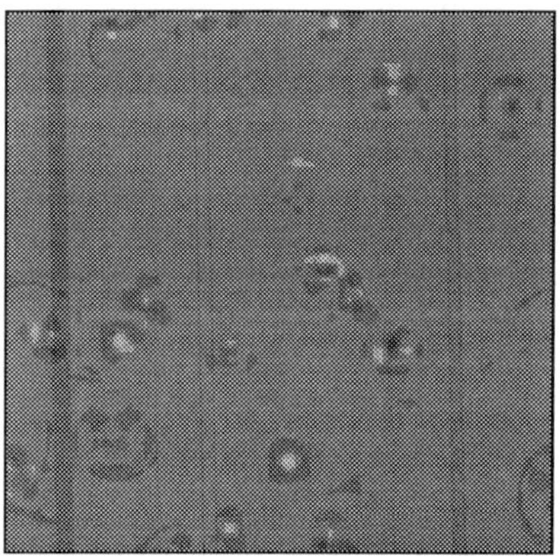

Part c : 10 % <

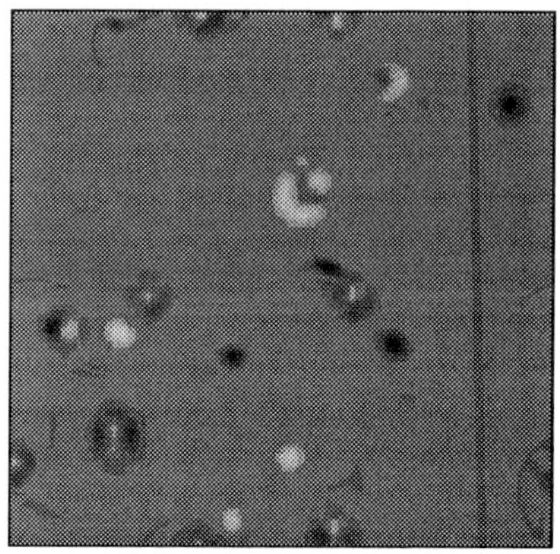

Part d : 50 % <

Figure 8.6.4.3 :
The discarded fields $S^{<}$
Pallette : maximum 36.3 in white, minimum -38.1 in black

In looking at the corresponding discarded fields and their power spectrum (Figure 8.6.4.3 and 8.6.4.4), the field containing 99% of the energy (compression factor of 355) is selected as the optimal one, because its discarded field remains quite a homogeneous broad-band noise. Therefore, the fewest number of wavelet packet coefficients necessary to represent the set of coherent structures of this vorticity field is 739, which gives us an estimate of the number of degrees of freedom we will have to consider to compute the flow evolution.

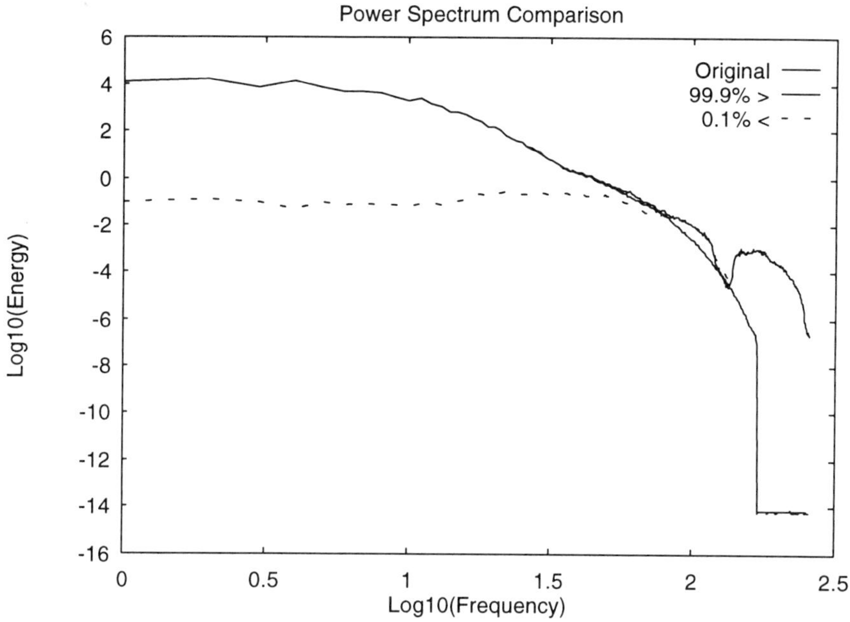
Power Spectrum Comparison
Original
99.9% >
0.1% <
Log10(Energy)
Log10(Frequency)
6
4
2
0
-2
-4
-6
-8
-10
-12
-14
-16
0
0.5
1
1.5
2
2.5

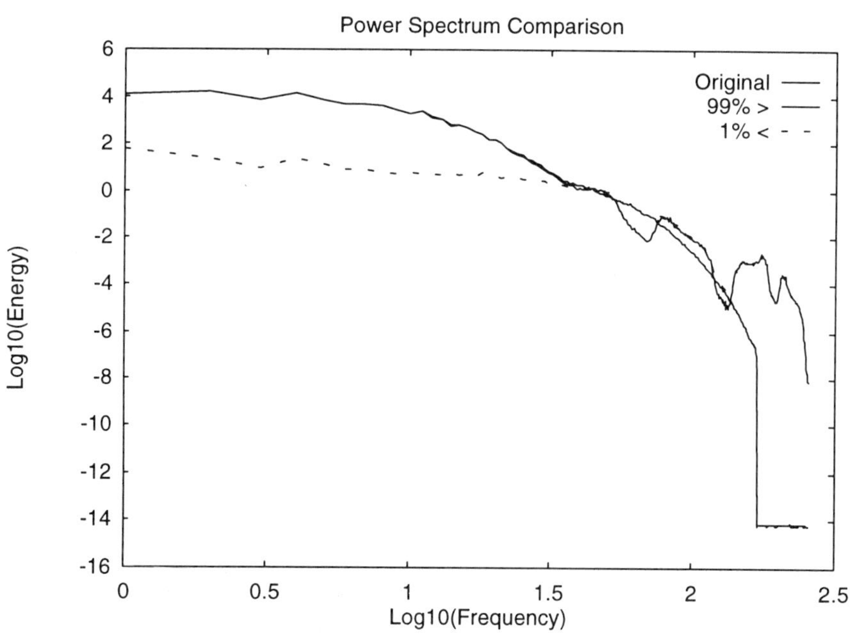
Power Spectrum Comparison
Original
99% >
1% <
Log10(Energy)
Log10(Frequency)
6
4
2
0
-2
-4
-6
-8
-10
-12
-14
-16
0
0.5
1
1.5
2
2.5

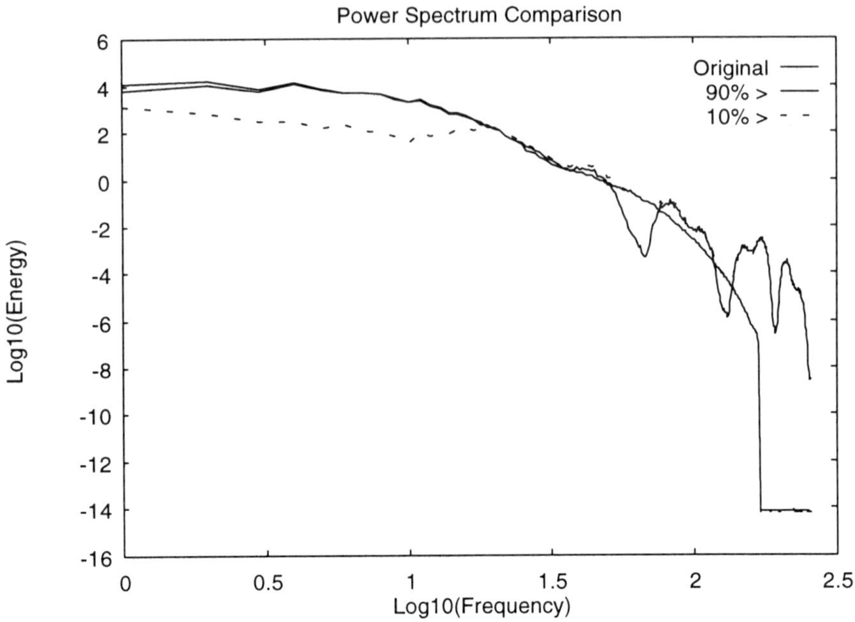

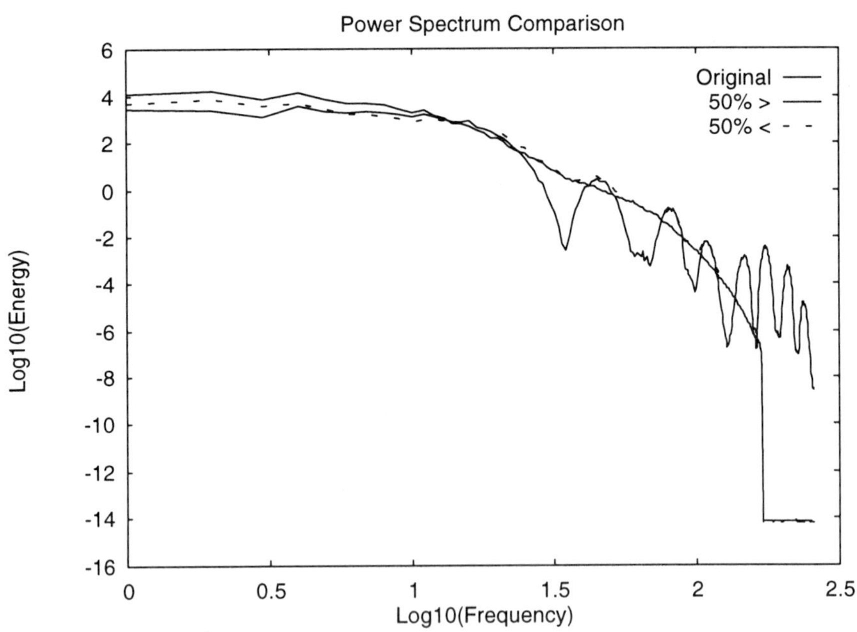

Figure 8.6.4.4 : $S^{>}$ and $S^{<}$ power spectrum comparisons

Second, We extracted the maximum positive coherent structure from the whole set of wavelet packet coefficients (Figure 8.6.4.5). There are 9000 wavelet packet coefficients defining the structure.

Figure 8.6.4.5 : Local extraction of the maximum positive vortex
Pallette : maximum 36.3 in white, minimum -38.1 in black

We then performed the fourier transform to this extracted field and we computed its power spectrum (Figure 8.6.4.6).

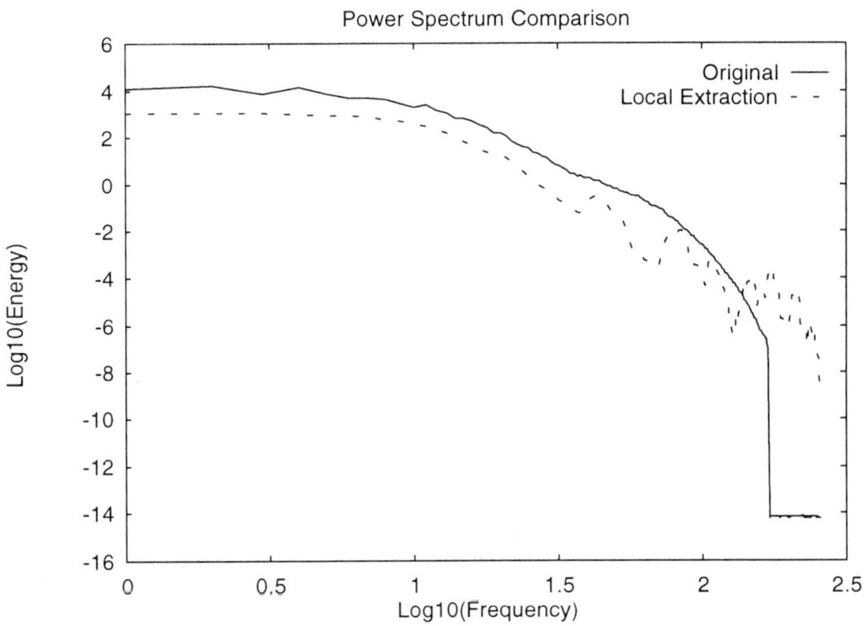

Figure 8.6.4.6 : Local extraction power spectrum comparison

8.7. ACKNOWLEDGEMENT

We are very grateful to Thierry Philipovitch for having provided us with the fully developed turbulent field we analyzed in this chapter.

8.8. REFERENCES

[1] Mladen Victor Wickerhauser, "Computation with Adapted Time-Frequency Atoms", pp. 175--184 in "Progress in Wavelet Analysis and Applications : Proceedings of the International Conference in Toulouse, 8--13 June 1992", edited by Yves Meyer and Sylvie Roques, Editions Frontières, ISBN 2-86332-130-7.

[2] Gregory Belkin, Ronald R. Coifman and Vladimir Rokhlin, "Fast Wavelet Transform and Numerical Algorithms I, "Communications on Pure and Applied Math, XLIV (1991), 141--183.

[3] Stephane G. Mallat, "Multiresolution Approximation and Wavelet Orthonormal Bases of $L^2(R)$", Transactions of the American Mathematical society, 315 (1989), 69--87.

[4] Ronald R. Coifman, Yves Meyer, Stephen R. Quake and Mladen Victor Wickerhauser, "Signal Processing and Compression with Wavelet Packets", pp. 77--93 in "Progress in Wavelet Analysis and Applications : Proceedings of the International Conference in Toulouse, 8--13 June 1992", edited by Yves Meyer and Sylvie Roques, Editions Frontières, ISBN 2-86332-130-7.

[5] Ronald R. Coifman and Mladen Victor Wickerhauser, "Entropy Based Algorithms for Best Basis Selection", IEEE Transactions on Information Theory, 32 (March 1992), 712--718.

8.9. ANNEXES : Algorithms written in pseudo-code

8.9.1. Annex 1 : Decomposition

The function doing the data transfers has the following representation in pseudo-code :

The parameters of function *Data transfers* are :

p is the process_name
jp is such that 4^{jp} is the number of processes
j is the current scale
f is the current frequency
width is the width of the signal at scale 0
height is the height of the signal at scale 0
Local_packet is the local packet to transfer

First we initialize the parameters :

f = p / 4^{jp-j}
factor = 1
f_child = p
llx = width / 2^{jp}
lly = height / 2^{jp}
base_x = cx(p,jp,width)-cx(f,j,width)
base_y = cy(p,jp,height)-cy(f,j,height)

$${}^{f}C^{j}_{base\_x+0..wh-1,base\_y+0..ht-1} = Local\_packet_{0..wh-1,0..ht-1}$$

Then we transfer data :

FOR jj = jp-1 TO j BY STEP -1
  f_father = f_child / 4
  depl = 4 * f_father
  IF (((p/factor) equals depl) OR ((p/factor) equals (depl+2)))
    bx=cx(f_child*factor,jp,width)-cx(f,j,width)
    by=cy(f_child*factor,jp,height)-cy(f,j,height)
    Send ${}^{f}C^{j}_{bx+0..llx-1,by+0..lly-1}$ to process *p+factor*

    bx=cx((f_child+1)*factor,jp,width)-cx(f,j,width)
    by=cy((f_child+1)*factor,jp,height)-cy(f,j,height)
    Receive ${}^{f}C^{j}_{bx+0..llx-1,by+0..lly-1}$ from process *p+factor*
  ELSE
    bx=cx((f_child-1)*factor,jp,width)-cx(f,j,width)

by=cy((f_child-1)*factor,jp,height)-cy(f,j,height)
Receive ${}^{f}C^{j}_{bx+0..llx-1,by+0..lly-1}$ from process *p-factor*

bx=cx(f_child*factor,jp,width)-cx(f,j,width)
by=cy(f_child*factor,jp,height)-cy(f,j,height)
Send ${}^{f}C^{j}_{bx+0..llx-1,by+0..lly-1}$ to process *p-factor*
ENDIF

llx = llx *2

IF (((p/factor) equals depl) OR ((p/factor) equals (depl+1)))
bx=cx(depl*factor,jp,width)-cx(f,j,width)
by=cy(depl*factor,jp,height)-cy(f,j,height)
Send ${}^{f}C^{j}_{bx+0..llx-1,by+0..lly-1}$ to process *p+2*factor*

bx=cx((depl+2)*factor,jp,width)-cx(f,j,width)
by=cy((depl+2)*factor,jp,height)-cy(f,j,height)
Receive ${}^{f}C^{j}_{bx+0..llx-1,by+0..lly-1}$ from process *p+2*factor*
ELSE
bx=cx((depl-2)*factor,jp,width)-cx(f,j,width)
by=cy((depl-2)*factor,jp,height)-cy(f,j,height)
Receive ${}^{f}C^{j}_{bx+0..llx-1,by+0..lly-1}$ from process *p-2*factor*

bx=cx(depl*factor,jp,width)-cx(f,j,width)
by=cy(depl*factor,jp,height)-cy(f,j,height)
Send ${}^{f}C^{j}_{bx+0..llx-1,by+0..lly-1}$ to process *p-2*factor*
ENDIF

lly = lly*2
f_child = f_father
factor = 4*factor
ENDFOR j

Data transfers returns the packet ${}^{f}C^{j}$.

The algorithm of the decomposition when j < jp has the following pseudo-code implemetation :

p is the process name
4^{jp} is the number of processes
wh = width / 2^{jp}
ht = height / 2^{jp}

```
FOR j = 0 TO jp-1
    f = p / 4^(jp-j)
    fc = p / 4^(jp-j-1)
    d = fc modulo 4

    //First we exchange data to have ^f C^j
```

$$ {}^{f}C^{j} = Data\ transfers\left(p, jp, j, f, 2^{Xmax}, 2^{Ymax}, {}^{f}lC^{j}\right) $$

```
    //Second, we define the filter to use :
        FF = F(d,f)

    //Third, we calculate the x- and y- offsets :
        base_x = cx(p,jp,width)-cx(f,j,width)
        base_y = cy(p,jp,height)-cy(f,j,height)

    // Fourth we calculate ^fc lC^(j+1)
```

$$ {}^{fc}lC^{j+1}_{k,l} = \sum_{kk,ll} FF_{kk,ll}\ {}^{f}C^{j}_{kk+2*(base\_x+k),\, ll+2*(base\_y+l)} $$

```
ENDFOR j
```

The algorithm of the decomposition when j ≥ jp has the following pseudo-code implemetation :

```
p is the process name
FOR j = jp TO nmax
    FOR f = 4^(j-jp) * p TO 4^(j-jp) * (p+1) - 1
        FOR d = 0 TO 3
            First we set the filter to use : FF = F(d,f)
            The we calculate ^(4f+d) C^(j+1) :
```

$$ {}^{4f+d}C^{j+1}_{k,l} = \sum_{kk,ll} FF_{kk,ll}\ {}^{f}C^{j}_{kk+2*k,\, ll+2*l} $$

```
        ENDFOR d
    ENDFOR f
ENDFOR j
```

8.9.2. Annex 2 : Best Basis Selection

First, we consider the case when j ≥ jp :

```
FOR f = 4^(nmax-jp) * p TO 4^(nmax-jp) * (p+1) - 1
```

$$ {}^{f}C^{j}_{status} = KEPT $$

$$ {}^{f}C^{j}_{value} = M\left({}^{f}C^{j}\right) $$

```
ENDFOR f

FOR j = nmax-1 TO jp BY STEP -1
      FOR f = 4^(j-jp) * p TO 4^(j-jp) * (p+1) - 1
```

$$v_1 = \sum_{d=0}^{3} {}^{4*f+d}C_{value}^{j+1}$$

$$v_0 = M\left({}^{f}C^{j}\right)$$

```
            IF (v0 ≤ v1)
```

$${}^{f}C_{status}^{j} = KEPT$$

$${}^{f}C_{value}^{j} = v_0$$

```
                  factor=1
                  FOR jj=j+1 TO nmax
                        factor = factor*4
                        f_base = factor*f
                        FOR ff = 0 TO factor
```

$${}^{f\_base+ff}C_{status}^{jj} = NOT\_KEPT$$

```
                        ENDFOR ff
                  ENDFOR jj
            ELSE
```

$${}^{f}C_{status}^{j} = NOT\_KEPT$$

$${}^{f}C_{value}^{j} = v_1$$

```
            ENDIF
      ENDFOR f
ENDFOR j
```

Second, we consider the case when j < jp :
To implement this algorithm,

First set the parameters of function *Local value transfer* :

p is the name of the current process
np is the number of processes doing the transfer
p_base is the pivot process
p_shift is the shift of processes to do the transfer
$local\_value_p$ is the value of the process p to transfer

We must also store some side data :

test=np
stepsize=p_shift
pvalue=$local\_value_p$

Then, we do the transfers :

```
WHILE (test > p_shift)
      IF (((p - p_base) mod (2*stepsize)) < stepsize)
```

```
            p2=p+stepsize
            start=1
        ELSE
            p2=p-stepsize
            start=0
        ENDIF

        IF (start = 1)
            Send pvalue to process p2
            Receive p2value from process p2
        ELSE
            Receive p2value from process p
            Send pvalue to process p
        ENDIF

        pvalue = pvalue + p2value

        test = test/2
        stepsize = stepsize * 2
    ENDWHILE
    total_value=pvalue
```

Local value transfer returns total_value

The best basis selection algorithm when j is smaller than jp is then :

p is the name of the current process
4^{jp} is the maximum number of processes

FOR j = jp-1 TO 0 BY STEP -1
    f = p / 4^{jp-j}
    f_child = p / 4^{jp-j-1}

    v_0 = Local value transfer(p,4^{jp-j},f,1, $M\left({}^{f}lC^{j}\right)$)
    v_1 = Local value transfer(p,4^{jp-j},f,4^{jp-j-1}, ${}^{f\_child}lC^{j+1}_{value}$)

    IF ($v_0 \le v_1$)
        ${}^{f}lC^{j}_{status} = KEPT$
        ${}^{f}lC^{j}_{value} = v_0$
        factor=1
        FOR jj=j+1 TO nmax
            factor = factor*4

```
                f_base = factor*f
                FOR ff = 0 TO factor
                        f_base+ff lC^jj_status = NOT_KEPT
                ENDFOR ff
            ENDFOR jj
        ELSE
            f lC^j_status = NOT_KEPT
            f lC^j_value = v_1
        ENDIF
    ENDFOR j
```

8.9.3. Annex 3 : Reconstruction

First, we consider the case when j > jp.
In this case, the algorithm is the same as the sequential one :

```
p is the process name
FOR j = nmax TO jp+1 BY STEP -1
    FOR f = 4^(j-jp-1) * p TO 4^(j-jp-1) * (p+1) - 1
        IF (f C^(j-1) exists)
            IF (f C^(j-1)_status = NOT_KEPT)
                We set f C^(j-1)_(k,l) = 0, ∀ k,l
            ENDIF
        ENDIF
    ENDFOR f

    FOR f = 4^(j-jp) * p TO 4^(j-jp) * (p+1) -1
        IF (f C^j exists)
            IF (f C^j_status = KEPT)
                1/ We set the filter to use for reconstructing :
                    FF = F(f mod 4,f / 4)
                2/ We create f/4 C^(j-1) is it doesn't exist
                    by setting all its coefficients to zero.
                3/ We set f/4 C^(j-1)_status = KEPT
                4/ We compute the partial reconstruction :
                   f/4 C^(j-1)_(k,l) = f/4 C^(j-1)_(k,l) + Σ_(n,m) FF_(k-2n, l-2m) * f C^j_(n,m)
                5/ We don't need any more f C^j
            ENDIF
        ENDIF
```

```
        ENDFOR f
    ENDFOR j
```

Second, we consider the case when scale $j \leq jp$.

Function *Local array transfer* has the following parameters :

p is the name of the current process
np is the number of processes doing the transfer
p_base is the pivot process
p_shift is the shift of processes to do the transfer
$local\_array_p$ is the array of the process p to transfer
size_array is the size of the local array

It contains the following local variables :

test=np
stepsize=p_shift
pvalue[i]=$local\_value_p$[i] for $0 \leq i < size\_array$

It then does these transfers :

```
WHILE (test > p_shift)
    IF (((p - p_base) mod (2*stepsize)) < stepsize)
        p2=p+stepsize
        start=1
    ELSE
        p2=p-stepsize
        start=0
    ENDIF

    IF (start = 1)
        Send array pvalue to process p2
        Receive array p2value from process p2
    ELSE
        Receive array p2value from process p
        Send array pvalue to process p
    ENDIF

    pvalue[i] = pvalue[i] + p2value[i] for 0≤i<size_array

    test = test/2
    stepsize = stepsize * 2
ENDWHILE

total_value[i] =pvalue[i]
```

Local array transfer returns the array total_value

The reconstruction algorithm can be implemented as follows :

p is the name of the process
$2^{Xmax}*2^{Ymax}$ is the size of the field at scale 0
wh = $2^{Xmax-jp}$
ht = $2^{Ymax-jp}$

FOR j = jp TO 1 BY STEP -1
np = 4^{jp-j}
f = p / 4^{jp-j}
f2 = f / 4
IF ($^{f}lC^{j}$ exists)
IF ($^{f}lC^{j}_{status} = NOT\_KEPT$)
1/ We set $^{f}lC^{j}_{k,l} = 0,\ \forall 0 \le k < wh,\ 0 \le l < ht$
2/ We set $^{f}lC^{j}_{status} = KEPT$
ENDIF
ELSE
1/ We create $^{f}lC^{j}$ by setting all its coefficients to zero
2/ We set $^{f}lC^{j}_{status} = KEPT$
ENDIF

IF ($^{f}lC^{j}$ exists)
IF ($^{f}lC^{j}_{status} = KEPT$)
1/ We calculate local shifts
$bx = cx(p, jp, 2^{Xmax}) - cx(f2, j-1, 2^{Xmax})$
$by = cy(p, jp, 2^{Ymax}) - cy(f2, j-1, 2^{Ymax})$
$base\_x = cx(p, jp, 2^{Xmax}) - cx(f, j, 2^{Xmax})$
$base\_y = cy(p, jp, 2^{Ymax}) - cy(f, j, 2^{Ymax})$

2/ We create a real size temporary father $^{f2}tC^{j-1}$ with null coefficients

3/ We update it if needed
IF ($^{f2}lC^{j-1}$ exists)
IF ($^{f2}lC^{j-1}_{status} = KEPT$)
$^{f2}tC^{j-1}_{bx+0..wh-1,by+0..ht-1} = {}^{f2}lC^{j-1}_{0..wh-1,0..ht-1}$
ENDIF
ENDIF

4/ We calculate the partial reconstruction

$$^{f2}tC^{j-1}_{k,l} = {}^{f2}tC^{j-1}_{k,l} + \sum_{n,m} FF_{k-2*(n+base\_x),\, l-2*(m+base\_y)} * {}^{f}lC^{j}_{n,m}$$

5/ We calculate the sum of all $^{f2}tC^{j-1}$

len = $2^{Xmax-j+1} * 2^{Xmax-j+1}$

$$^{f2}tC^{j-1} = Local\ array transfer\left(p, 4^{jp-j+1}, f2*4^{jp-j+1}, 1, {}^{f2}tC^{j-1}, len\right)$$

6/ If $^{f2}lC^{j-1}$ doesn't exist we set it to the null vector

7/ We set $^{f2}lC^{j-1}_{status} = KEPT$

8/ We transfer the data from $^{f2}tC^{j-1}$ to $^{f2}lC^{j-1}$

$$^{f2}lC^{j-1}_{0..wh-1,0..ht-1} = {}^{f2}tC^{j-1}_{bx+0..wh-1,by+0..ht-1}$$

9/ We don't need any more $^{f2}tC^{j-1}$ and $^{f}lC^{j}$

ENDIF

ENDIF

ENDFOR

8.9.4. Annex 4 : Matrix-Vector Multiplication

The multiplication algorithm may be implemented as follows :

p is the process name
2^{Xmax} by 2^{Ymax} is the size of the matrix C
2^{Xmax} is the size of the vector D
so 2^{Ymax} is the size of the vector E

wh=$2^{Xmax-jp}$ and ht=$2^{Ymax-jp}$

A/ In parallel, using all the processes
We decompose C in all the 2D-wavelet packets,
We select the best basis for C

B/ On each process,
We decompose D in all the 1D-wavelet packets

C/ We do the multiplication when a packet is
shared by more than one process
FOR j = 0 TO jp-1
f = p / 4^{jp-j}

IF ($^{f}lC^{j}$ exists) THEN

IF ($^{f}lC^{j}_{status} = KEPT$) THEN

1/ We set the signal shifts

$$base\_x = cx\left(p, jp, 2^{Xmax}\right) - cx(f, j, 2^{Xmax})$$

$$base\_y = cy\left(p, jp, 2^{Ymax}\right) - cy(f, j, 2^{Ymax})$$

2/ We set the frequencies among the x- and y- axis

$$fx = \sum_{jj=0}^{j-1} 2^{j-jj-1} \cdot \left(\left\lfloor \frac{f}{4^{j-jj-1}} \right\rfloor \bmod 2 \right)$$

$$fy = \sum_{jj=0}^{j-1} 2^{j-jj-1} \cdot \left(\left\lfloor \frac{\left\lfloor \frac{f}{4^{j-jj-1}} \right\rfloor}{2} \right\rfloor \bmod 2 \right)$$

3/ We keep the local vector

$$^{fx}lD^{j}_{0..wh-1} = {}^{fx}D^{j}_{base\_x+0..wh-1}$$

4/ We perform the smaller matrix-vector multiplication

$$^{fy}lE^{j} = {}^{f=(fx,fy)}lC^{j} \times {}^{fx}lD^{j}$$

5/ We insert $^{fy}lE^{j}$ into $^{fy}E^{j}$

$$^{fy}E^{j}_{base\_y+0..ht-1} = {}^{fy}lE^{j}_{0..ht-1}$$

6/ We set $^{fy}E^{j}_{status} = KEPT$

ENDIF

ENDIF

ENDFOR j

D/ We do the multiplication when a packet belongs to a process

FOR j = jp TO nmax

FOR f = 4^{j-jp} * p TO 4^{j-jp} * (p+1) -1

IF ($^{f}C^{j}$ exists) THEN

IF ($^{f}C^{j}_{status} = KEPT$) THEN

1/ We perform the smaller matrix-vector multiplication

$$^{fy}E^{j} = {}^{f=(fx,fy)}C^{j} \times {}^{fx}D^{j}$$

2/ We set ${}^{fy}E^{j}_{status} = KEPT$

ENDIF

ENDIF

ENDFOR f

ENDFOR j

E/ On each process,

We Reconstruct from all the kept wavelet packets ${}^{fy}E^{j}$

Each process has its own part of the partial vector E

F/ We exchange and combine the partial vectors E to have the whole vector E

$$E = Local\,array\,transfer\left(p, 4^{jp}, 0, 1, E, 2^{Ymax}\right)$$

G/ The multiplication is finished,

Each process has the whole vector E.

8.9.5. Annex 5 : Compression

Here is first the function doing the merge of the local sorted list of wavelet packet coefficients :

Function *Merge local list* has the following parameters :

p is the name of the current process
np is the number of processes doing the transfer
p_base is the pivot process
p_shift is the shift of processes to do the transfer
$local\_list_p$ is the list of the process p to transfer
size_list is the size of the local list

It contains the following local variables :

test=np
stepsize=p_shift
plist[i] = $local\_list_p[i]$ for $0 \le i < size\_list$

It then does these transfers :

```
WHILE (test > p_shift)
     IF (((p - p_base) mod (2*stepsize)) < stepsize)
          p2=p+stepsize
          start=1
     ELSE
          p2=p-stepsize
```

```
            start=0
        ENDIF

        IF (start = 1)
            Send size_list to process p2
            Send list plist to process p2
            Receive size_list2 to process p2
            Receive list  p2list from process p2
        ELSE
            Receive size_list2 to process p2
            Receive list  p2list from process p2
            Send size_list to process p2
            Send list plist to process p2
        ENDIF

        // We perform a merge sort in order that pvalue be the sorted
         list corresponding to plist and p2list
        plist = merge(plist, p2list)
        size_list = size_list + size_list2

        test = test/2
        stepsize = stepsize * 2
    ENDWHILE

    total_list[i] =plist[i] for 0≤i<size_list
```

Merge local list returns total_list.

Then here is the pseudo-code for the compression algorithm :

p is the process name
4^{jp} is the number of processes
2^{Xmax} x 2^{Ymax} is the size of the field at scale 0
max_energy if the sum of the square of the coefficients that we would like to keep. (For example, a certain percentage of the initial field energy).

1/ First, the process groups all its wavelet packet coefficients and sort them such as their squares be in a decreasing order. The process has now a list $L1 = \left({}^{f_0}C^{j_0}_{k_0,l_0}, {}^{f_1}C^{j_1}_{k_1,l_1}, \ldots, {}^{f_{n1-1}}C^{j_{n1-1}}_{k_{n1-1},l_{n1-1}} \right)$ such that all the coefficients verify $\left({}^{f_0}C^{j_0}_{k_0,l_0} \right)^2 > \left({}^{f_1}C^{j_1}_{k_1,l_1} \right)^2 > \ldots > \left({}^{f_{n1-1}}C^{j_{n1-1}}_{k_{n1-1},l_{n1-1}} \right)^2$.

2/ It exchanges its list with other processes until it has the entire sorted list.

$$TL = Merge\ local\ list\left(p, 4^{jp}, 0, 1, L1, n1\right)$$

$$TL = \left({}^{f_0}TC^{j_0}_{k_0,\, l_0},\ {}^{f_1}TC^{j_1}_{k_1,\, l_1}, \ldots,\ {}^{f_{tn-1}}TC^{j_{tn-1}}_{k_{tn-1},\, l_{tn-1}} \right)$$

3/ It selects only the coefficients needed to have the specified max_energy.

sum = 0
nc = 0
WHILE ((sum < max_energy) AND (nc < tn))

$$sum = sum + \left({}^{f_{nc}}C^{j_{nc}}_{k_{nc},\, l_{nc}} \right)^2$$

nc = nc + 1
ENDWHILE

4/ It keeps only the coefficients that belong to itself

FOR i = 0 TO nc-1
IF (${}^{f_i}C^{j_i}$ exists)
IF (${}^{f_i}C^{j_i}_{status} = KEPT$)
1/ We create ${}^{f_i}CC^{j_i}$ to the null vector if it does not exist

2/ We set the coefficient to keep

$${}^{f_i}CC^{j_i}_{k_i,\, l_i} = {}^{f_i}TC^{j_i}_{k_i,\, l_i}$$

ENDIF
ENDIF
ENDFOR

5/ The new set of wavelet packets ${}^{f}CC^{j}$ represent the compressed version of the initial field ${}^{f}C^{j}$.

8.9.6. Annex 6 : Local extraction

First, here is the pseudo-code for the function doing the sequential extraction on the interval $[a_1, a_2] \times [b_1, b_2]$ (We consider the extraction at a point by setting a_2 to a_1 and b_2 to b_1) :

Function *Extract_local_Coefficients* has the following parameters :
j is the scale and
f is the frequency defining the wavelet packet to consider
ShiftX and ShiftY are the x- and y- shifts at scale j-1 and frequency

f/4.

First, we compute the x- and y- shifts for the considered packet

$$ShiftX = 2 * ShiftX + F(f \bmod 4, f/4)_{xshift}$$

$$ShiftY = 2 * ShiftY + F(f \bmod 4, f/4)_{yshift}$$

Then, if the wavelet packet is part of the best basis, we extract its local coefficients

IF (${}^{f}C^{j}$ exists)

IF (${}^{f}C^{j}_{status} = KEPT$)

1/ We initialize ${}^{f}CL^{j}$ to the null vector

2/ We set its status to KEPT

$${}^{f}CL^{j}_{status} = KEPT$$

3/ We set the boundaries for the local extraction

// *Round(x)* returns the nearest integer to *x*

$$ipx1 = Round\left(\frac{a_1 - 2^{j-1} - ShiftX}{2^j}\right)$$

$$ipx2 = Round\left(\frac{a_2 + 2^{j-1} - ShiftX}{2^j}\right)$$

$$ipy1 = Round\left(\frac{b_1 - 2^{j-1} - ShiftY}{2^j}\right)$$

$$ipy2 = Round\left(\frac{b_2 + 2^{j-1} - ShiftY}{2^j}\right)$$

4/ We keep only the local coefficients

FOR ipx = ipx1 TO ipx2

FOR ipy = ipy1 TO ipy2

$${}^{f}CL^{j}_{ipx \bmod 2^{Xmax-j},\, ipy \bmod 2^{Ymax-j}} = {}^{f}C^{j}_{ipx \bmod 2^{Xmax-j},\, ipy \bmod 2^{Ymax-j}}$$

ENDFOR

ENDFOR

ENDIF

ENDIF

In a last part, we recursively extract the local coefficients in the child packets

IF ($j \neq nmax$)

FOR d = 0 TO 3

$$Extract\_Local\_Coefficients(j+1, 4*f+d, ShiftX, ShiftY)$$

ENDFOR

ENDIF

The result is the new set of wavelet packets ${}^{f}CL^{j}$

Then, here is the pseudo-code algorithm for extracting the interval $[a_1, a_2] \times [b_1, b_2]$ ($a_1 \le a_2$ and $b_1 \le b_2$) on each processor (We consider the extraction at a point by setting a_2 to a_1 and b_2 to b_1) :

p is the process name
4^{jp} is the number of processes
2^{Xmax} x 2^{Ymax} is the size of the field at scale 0

First, we consider the case when j < jp
IF (${}^{0}lC^{0}$ exists)
IF (${}^{0}lC^{0}_{status} = KEPT$)
1/ We initialize ${}^{0}lCL^{0}$ to the null vector
2/ ${}^{0}lCL^{0}_{status} = KEPT$
3/ We compute the base coordinates
$bx = cx\left(p, jp, 2^{Xmax}\right)$
$by = cy\left(p, jp, 2^{Ymax}\right)$
4/ We set the boundaries of the local extraction
$OK = 1$
$ipx1 = \max(a_1 - bx, 0)$
IF ($ipx1 \ge 2^{Xmax-jp}$) $OK = 0$
$ipy1 = \max(b_1 - by, 0)$
IF ($ipy1 \ge 2^{Ymax-jp}$) $OK = 0$
$ipx2 = \min(a_2 - bx, 2^{Xmax-jp} - 1)$
IF ($ipx2 < 0$) $OK = 0$
$ipy2 = \min(b_2 - by, 2^{Ymax-jp} - 1)$
IF ($ipy2 < 0$) $OK = 0$
5/ We extract the local coefficients
IF ($OK = 1$)
FOR ipx = ipx1 TO ipx2
FOR ipy = ipy1 TO ipy2
${}^{f}lCL^{j}_{ipx, ipy} = {}^{f}lC^{j}_{ipx, ipy}$
ENDFOR
ENDFOR
ENDIF
ENDIF
ENDIF

$ShiftX = 0$
$ShiftY = 0$
IF ($^{0}lCL^{0}$ does not exist)
FOR j = 1 TO jp-1
f = p / 4jp
$ShiftX = 2 * ShiftX + F(f \bmod 4, f/4)_{xshift}$
$ShiftY = 2 * ShiftY + F(f \bmod 4, f/4)_{yshift}$
IF ($^{f}lC^{j}$ exists)
IF ($^{f}lC^{j}_{status} = KEPT$)
1/ We initialize $^{f}lCL^{j}$ to the null vector
2/ $^{f}lCL^{j}_{status} = KEPT$
3/ We compute the base coordinates
$bx = cx(p, jp, 2^{Xmax}) - cx(f, j, 2^{Xmax})$
$by = cy(p, jp, 2^{Ymax}) - cy(f, j, 2^{Ymax})$
4/ We set the boundaries of the local extraction
$OK = 1$

$$ipx1 = Round\left(\frac{a_1 - 2^{j-1} - ShiftX}{2^j}\right)$$

$$ipx1 = \max\left(\left(ipx1 \bmod 2^{Xmax-j}\right) - bx, 0\right)$$

IF ($ipx1 \geq 2^{Xmax-jp}$) $OK = 0$

$$ipy1 = Round\left(\frac{b_1 - 2^{j-1} - ShiftY}{2^j}\right)$$

$$ipy1 = \max\left(\left(ipy1 \bmod 2^{Ymax-j}\right) - by, 0\right)$$

IF ($ipy1 \geq 2^{Ymax-jp}$) $OK = 0$

$$ipx2 = Round\left(\frac{a_2 + 2^{j-1} - ShiftX}{2^j}\right)$$

$$ipx2 = \min\left(\left(ipx2 \bmod 2^{Xmax-j}\right) - bx, 2^{Xmax-jp}\right)$$

IF ($ipx2 < 0$) $OK = 0$

$$ipy2 = Round\left(\frac{b_2 + 2^{j-1} - ShiftY}{2^j}\right)$$

$$ipy2 = \min\left(\left(ipy2 \bmod 2^{Ymax-j}\right) - by, 2^{Ymax-jp}\right)$$

IF ($ipy2 < 0$) $OK = 0$
5/ We extract the local coefficients

```
                IF (OK=1)
                    FOR ipx = ipx1 TO ipx2
                        FOR ipy = ipy1 TO ipy2
                            f lCL^j_{ipx,ipy} = f lC^j_{ipx,ipy}
                        ENDFOR
                    ENDFOR
                ENDIF
            ENDIF
        ENDIF
    ENDFOR
ENDIF
```

Then, the case when j ≥ jp

$Extract\_Local\_Coefficients(jp, p, ShiftX, ShiftY)$

The result is the new set of wavelet packets ${}^{f}CL^{j}$ and ${}^{f}lCL^{j}$

Index